Engineering Geology

Engineering Geology

FRED G. BELL

Department of Geology and Applied Geology
University of Natal, Durban, South Africa

OXFORD

Blackwell Scientific Publications

LONDON EDINBURGH BOSTON
MELBOURNE PARIS BERLIN VIENNA

© 1993 by
Blackwell Scientific Publications
Editorial Offices:
Osney Mead, Oxford OX2 0EL
25 John Street, London WC1N 2BL
23 Ainslie Place, Edinburgh EH3 6AJ
238 Main Street, Cambridge
    Massachusetts 02142, USA
54 University Street, Carlton
    Victoria 3053, Australia

Other Editorial Offices:
Librairie Arnette SA
2, rue Casimir Delavigne
75006 Paris
France

Blackwell Wissenschafts-Verlag GmbH
Meinkestrasse 4
D-1000 Berlin 15
Germany

Blackwell MZV
Feldgasse 13
A-1238 Wien
Austria

First published 1993

Set by Setrite Typesetters, Hong Kong
Printed and bound in Great Britain at
the University Press, Cambridge

DISTRIBUTORS

Marston Book Services Ltd
PO Box 87
Oxford OX2 0DT
(*Orders*: Tel: 0865 791155
            Fax: 0865 791927
            Telex: 837515)

USA
Blackwell Scientific Publications, Inc.
238 Main Street
Cambridge, MA 02142
(*Orders*: Tel: 800 759−6102
            617 876−7000)

Canada
Oxford University Press
70 Wynford Drive
Don Mills
Ontario M3C 1J9
(*Orders*: Tel: 416 441−2941)

Australia
Blackwell Scientific Publications Pty Ltd
54 University Street
Carlton, Victoria 3053
(*Orders*: Tel: 03 347−5552)

A catalogue record for this title
is available from the British Library

ISBN 0−632−03223−5

Library of Congress
Cataloging in Publication Data

Bell, Fred G.
    Engineering geology/Fred G. Bell.
        p.   cm.
    Includes bibliographical references
    and index.
    ISBN 0−632−03223−5
    1. Engineering geology. I. Title.
TA705.B328   1993
624.1'51 − dc20

Contents

Preface

Engineering geology has been simply defined as the application of geology to engineering practice, in other words it is concerned with those geological factors which influence the location, design, construction and maintenance of engineering works. As such it draws upon several geological disciplines such as petrology, sedimentology, structural geology, geomorphology and, to a lesser extent, stratigraphy. Accordingly these subjects are considered, from an engineering point of view, in the first three chapters of the text. Engineering geology is intimately associated with hydrogeology, soil mechanics and rock mechanics. Hence chapters four and five deal with the engineering geological aspects of these topics. The applied aspects of engineering geology are dealt with in the remaining chapters, namely, construction materials, site investigation, planning and construction.

The text is written for those who come in contact with engineering geology from undergraduates and post graduates to those engaged in the professions. The latter involves engineering geology itself but also civil engineering, structural engineering, mining engineering, water engineering, quarrying, architecture, surveying and building to a greater or lesser extent, that is, all those who are involved with the ground. It is not simply a student textbook to be discarded when one leaves university. It contains more detail than most courses in engineering geology require, except those which are highly specialist. It therefore should form part of an academic's and a professional person's library.

Obviously the information contained within any book is limited by its size. With this in mind, a list of more advanced texts is provided along with numerous references. All the major journals which an engineering geologist needs to consult are mentioned in these references. So the reader is set on the right track and knows where to go to follow up a topic and to keep up to date.

Although the text, no doubt, has a bias which reflects the author's background and viewpoint, within the constraints of its size, it is as well balanced as any other. It therefore should be of value to anyone engaged in engineering geology wherever he or she is.

Lastly, a great debt is owed to Mrs Roma Brackley, for without her stout-hearted efforts at the keyboard the manuscript would never have been produced in the time it was.

F.G. Bell
1992

1   Rock types and stratigraphy

Rocks are divided according to their origin, into three groups; the igneous, metamorphic, and sedimentary rocks.

1.1 IGNEOUS ROCKS

Igneous rocks are formed when hot, molten rock material called magma solidifies. Magmas are developed either within or beneath the Earth's crust, that is, in the mantle. They comprise hot solutions of several liquid phases, the most conspicuous of which is invariably a complex silicate phase. Thus, the igneous rocks are principally composed of silicate minerals. Furthermore, of the silicate minerals, six families, the olivines, the pyroxenes (e.g. augite), the amphiboles (e.g. hornblende), the micas (e.g. biotite and muscovite), the feldspars (e.g. orthoclase, albite, and anorthite), and the silica minerals (e.g. quartz), are quantitatively by far the most important constituents. Figure 1.1 shows the approximate distribution of these minerals in the commonest igneous rocks.

The magmas which are generated when melting occurs in the mantle or crust are named primary magmas. They tend to be basaltic in composition and represent the parent material from which secondary or derived magmas may arise due to differentiation or contamination. Differentiation is brought about due to the fact that different minerals crystallize at different temperatures so that an order of crystallization can be distinguished. When those minerals which crystallize at high temperatures have formed, the composition of the remaining magma is changed. This process, known as fractional crystallization, can produce different types of rock from the original magma. A magma becomes contaminated when country rock is incorporated into it. This can alter its composition. Evidence of contamination is exhibited, for example, by the presence of fragments of country rock, termed xenoliths, which have not been completely assimilated by the host magma.

It would appear, however, that most granitic rocks are developed by other processes, that is, granitization and anatexis. Granitization is a process by

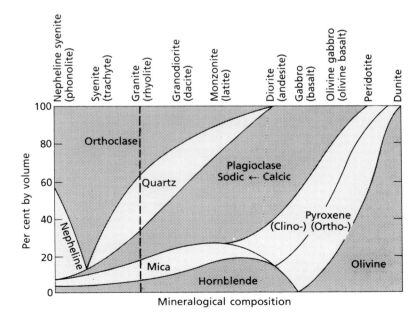

Fig. 1.1 Approximate mineralogical compositions of the more common types of igneous rocks, e.g. granite approximately 40% orthoclase, 33% quartz, 13% plagioclase, 9% mica, and 5% hornblende (plutonic types without brackets; volcanic equivalents in brackets).

which solid rocks are converted to rocks of granitic character without passing through a magmatic stage. Anatectic processes, on the other hand, lead to the remelting of rocks. Such rocks frequently have a mixed or hybrid appearance. They have been termed migmatites.

Igneous rocks may be divided into intrusive and extrusive types according to their mode of occurrence. In the former type, the magma crystallizes within the Earth's crust, whereas in the latter it solidifies at the surface, having been erupted as lavas and/or pyroclasts from a volcano. The intrusions may be further subdivided on a basis of their size, into major and minor categories: the former are developed in a plutonic (deep-seated) environment; the latter in a hypabyssal environment. About 95% of the plutonic intrusions have a granite−granodiorite composition, and basaltic rocks account for approximately 98% of the extrusives.

1.1.1 Igneous intrusions

The form which intrusions adopt may be influenced by the structure of the country rocks. This applies particularly to minor intrusions such as dykes and sills.

Dykes are discordant igneous intrusions, that is, they traverse their host rocks at a high angle and are steeply dipping (Fig. 1.2). As a consequence their surface outcrop is little affected by topography and they usually strike in a straight course. Dykes range in width up to several tens of metres, but on average, they are a few metres wide. The length of their surface outcrop also varies, for example, the Cleveland dyke in the north of England can be traced over some 200 km. Dykelets may extend from and run parallel to large dykes; irregular offshoots may also branch away from large dykes. Dykes may act as feeders for lava flows and sills, and often occur along faults, which provide a natural path of escape for the injected magma. Most dykes are of basaltic composition. Dykes may be multiple or composite. Multiple dykes are formed by two or more injections of the same material which occur at different times. A composite dyke involves two or more injections of magma of different compositions.

Sills, like dykes, are generally thin, parallel-sided igneous intrusions which occur over relatively extensive areas. Their thickness, however, can vary up to several hundred metres. However, unlike dykes, they are injected in a roughly horizontal direction, although their attitude may subsequently be altered by folding. When sills form in a series of sedimentary rocks, the magma is injected along bedding planes (Fig. 1.3). Nevertheless, an individual sill may transgress upwards from one horizon to another. Because sills are intruded along bedding planes they are said to be concordant and their outcrop is similar to that of the country rocks. Sills may be fed from dykes and small dykes may arise from sills. Most sills are composed of basic igneous material. Nonetheless, some sills may be multiple or composite in character.

The major intrusions include batholiths, stocks, and bosses. Batholiths are very large, and are generally composed of granite or granodiorite. Many batholiths have immense surface exposures. For

Fig. 1.2 Dykes on the shore, south-west coast, Isle of Arran, Scotland.

Fig. 1.3 The Whin Sill near Barden Mill, Northumberland, England.

instance, the Coast Range batholith of Alaska and British Columbia is exposed over 1000 km in length and approximately 130 to 190 km in width. Batholiths are associated with orogenic regions. They appear to have no visible base and their contacts are well defined and dip steeply outwards. However, some granitic batholiths appear to be made up of composite irregular sheets. They are more or less stratified and are not bottomless, and have been termed stratiform granitic massifs. Bosses are distinguished from stocks in that they have a circular outcrop. Both their surface exposures are of limited size,

frequently less than 100 km². They may represent upward extensions from deep-seated batholiths.

Certain structures are associated with granite massifs, tending to be best developed at their margins, for example, particles of elongate habit are aligned with their long axes parallel to each other, giving rise to linear flow structures (Fig. 1.4). Platy flow structure occurs where schlieren (layers which possess the same minerals as the rock itself but in different proportions), are orientated parallel to one another. Most joints and minor faults in batholiths possess a relationship with the shape of the intrusion.

Fig. 1.4 Block diagram showing the types of structures in a batholith. Q, cross-joints; S, longitudinal joints; L, flat-lying joints; STR, planes of stretching; F, linear-flow structures; A, aplite dykes. (After Balk, 1938.)

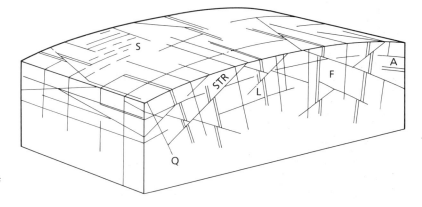

Fractures are first developed in the solidified margins of plutonic masses and may be filled with material from the still liquid interior. Cross-joints, or Q joints, lie at right angles to the flow lines. They tend to radiate from the centre of the massif. Joints which strike parallel to the flow lines and are steeply dipping are known as longitudinal, or S joints. Pegmatites or aplites (section 1.1.4) may be injected along both types of joints mentioned. Diagonal joints are orientated at 45° to the direction of the flow lines. Flat-lying joints may be developed during or after formation of the intrusion and may be distinguished as primary and secondary, respectively. Normal faults and thrusts occur in the marginal zones of large intrusions and the adjacent country rocks. Numerous thrusts are arranged *en echelon* and the displacement along them is usually measured in a few centimetres. The angle at which a marginal thrust dips is dependent upon its position in relation to the intrusion, for example, against steep contacts the dip is low whilst it is higher in the roof zone. Minor sets of pinnate shear joints may be found along marginal thrusts. Flat-lying normal faults form as a result of tension developed parallel to the flow lines. They are generally restricted to the upper parts of a massif.

1.1.2 Volcanic activity and extrusive rocks

Volcanic zones are associated with the boundaries of the crustal plates (Fig. 1.5). Plates can be largely continental, oceanic, or both. Oceanic crust is composed of basaltic material, whereas continental crust varies from granitic to basaltic in composition. At destructive plate margins oceanic plates are overridden by continental plates. The descent of the oceanic plate, together with any associated sediments, into zones of higher temperature leads to melting and the formation of magmas. Such magmas vary in composition, and some, which are richer in silica, are often responsible for violent eruptions. By contrast, at constructive plate margins where plates are diverging, the associated volcanic activity is a consequence of magma formation in the upper mantle. The magma is basaltic and is less viscous than andesitic or rhyolitic magmas, hence, there is relatively little explosive activity and associated lava flows are more mobile. However, certain volcanoes, for example, those of the Hawaiian Islands, are located in the centres of plates. Obviously these volcanoes are totally unrelated to plate boundaries. They owe their origins to hot spots in the Earth's

internal structure which have 'burned' holes through the overlying plates.

Volcanic activity is a surface manifestation of a disordered state within the Earth's interior which has led to the melting of rock material and the consequent formation of magma. This magma travels to the surface where it is extravasated either from a fissure or a central vent. In some cases, instead of flowing from the volcano as a lava the magma is exploded into the air by the rapid escape of the gases from within it. The fragments produced by explosive activity are collectively known as pyroclasts.

Eruptions from volcanoes are spasmodic rather than continuous. Between eruptions, activity may still be witnessed in the form of steam and vapours issuing from small vents named fumaroles or solfataras. However, in certain volcanoes even this form of surface manifestation ceases and such a dormant state may continue for centuries — to all intents and purposes these volcanoes appear extinct. In old age the activity of a volcano becomes limited to emissions of gases from fumaroles and hot water from geysers and hot springs.

When a magma is erupted it separates at low pressures into incandescent lava and a gaseous phase. Steam may account for over 90% of the gases emitted during a volcanic eruption. Other gases present include carbon dioxide, carbon monoxide, sulphur dioxide, sulphur trioxide, hydrogen sulphide, hydrogen chloride, and hydrogen fluoride. Small quantities of methane, ammonia, nitrogen, hydrogen thiocyanate, carbonyl sulphide, silicon tetrafluoride, ferric chloride, aluminium chloride, ammonium chloride, and argon have also been noted in volcanic gases. It has often been noted that hydrogen chloride is, next to steam, the major gas evolved during an eruption, but that in the later stages the sulphurous gases take over this role. The rate at which hydrogen chloride escapes determines the explosiveness of the eruption. An explosive eruption occurs when magma, because of its high viscosity (the viscosity is to a large extent governed by the silica content), cannot readily allow the escape of gas. The amount of gas a magma holds is only a secondary factor.

Pyroclasts may consist of fragments of lava exploded on eruption, of fragments of pre-existing solidified lava or pyroclasts, or of fragments of country rock which, in both latter instances, have been blown from the neck of the volcano.

The size of pyroclasts varies enormously and depends upon the viscosity of the magma, the violence

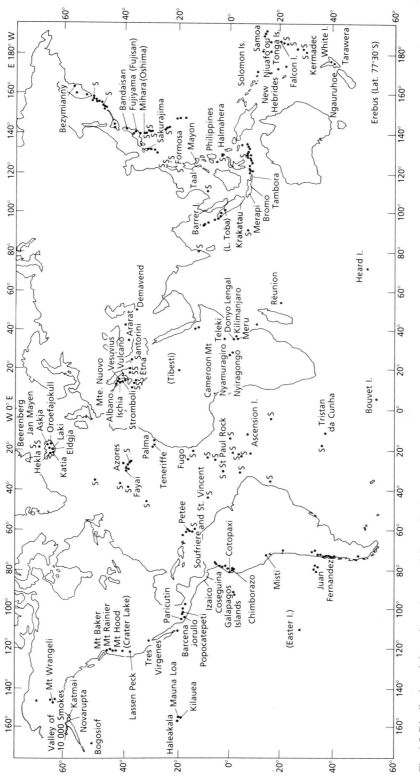

Fig. 1.5 Distribution of the active volcanoes of the world. S, submarine eruptions.

of the explosive activity, the amount of gas coming out of solution during the flight of the pyroclast, and the height to which it is thrown. The largest blocks thrown into the air may weigh over 100 tonnes, whereas the finest pyroclastic material consists of very fine ash which may take years to fall back to the Earth's surface. The largest pyroclasts are referred to as volcanic bombs. These consist of clots of lava or of fragments of wall rock. The term lapilli is applied to pyroclastic material which has a diameter varying from approximately 10 to 50 mm. Cinders or scoria are irregular-shaped materials of lapilli size. They are usually glassy and fairly to highly vesicular. Much more ash is produced on the eruption of acidic magma rather than basaltic. This is because acidic material is more viscous and so gas cannot escape as readily from it as it can from basaltic lava. Beds of ash commonly show lateral variation as well as vertical. In other words, with increasing distance from the volcanic vent the ash becomes finer — *lateral variation*; because the heavier material falls first, ashes frequently exhibit graded bedding, with coarser material occurring at the base of a bed — it becoming finer towards the top — *vertical variation*. Reverse grading may occur as a consequence of an increase in the violence of eruption or changes in wind velocity. The spatial distribution of ash is very much influenced by wind direction, and deposits on the leeside of a volcano may be much more extensive than on the windward; indeed they may be virtually absent from the latter side.

After pyroclastic material has fallen back to the surface it eventually becomes indurated. It is then described as tuff. According to the material of which tuff is composed, distinction can be drawn between ash tuff, pumiceous tuff, and tuff breccia. Tuffs are usually well bedded and the deposits of individual eruptions may be separated by thin bands of fossil soil or old erosion surfaces. Pyroclast deposits which accumulate beneath the sea are often mixed with a varying amount of sediment and are referred to as tuffites. They are generally well sorted and well bedded. Rocks which consist of fragments of volcanic ejectamenta set in a fine-grained groundmass are referred to as agglomerate or volcanic breccia, depending on whether the fragments are rounded or angular, respectively.

When clouds or showers of intensely heated, incandescent lava spray fall to the ground, they weld together. Because the particles become intimately fused with each other they attain a largely pseudo-viscous state, especially in the deeper parts of the deposit. The term ignimbrite is used to describe these rocks. If ignimbrites are deposited on a steep slope, then they begin to flow, hence they resemble lava flows. Ignimbrites are associated with *nuées ardentes* (avalanches of intensely hot fragments of lava and gas which explode from the side of a volcano).

Lavas are emitted from volcanoes at temperatures only slightly above their freezing points. During the course of their flow the temperature falls from within outwards, until solidification takes place somewhere between 600 and 900°C, depending upon their chemical composition and gas content. Basic lavas solidify at a higher temperature than do acidic ones. Generally, flow within a lava stream is laminar. The rate of lava flow is determined by the gradient of the slope down which it moves and by its viscosity which, in turn, is governed by its composition, temperature, and volatile content. As already mentioned, the greater the silica content of a lava, the greater is its viscosity. Thus, basic lavas flow much faster and further than do acidic lavas — the former type have been known to travel at speeds of up to 80 km/h.

The upper surface of a recently solidified lava flow develops either: a hummocky, ropy (pahoehoe); a rough, fragmental, clinkery, spiny (aa); or a blocky structure. The reasons for the formation of these different structures are not fully understood but the physical properties of the lava and the amount of disturbance it has to undergo play an important part. The pahoehoe is the most fundamental type, but some way downslope from the vent it may give way to aa or block lava. In other cases aa or block lava may be traceable into the vent. The surface of lava solidifies before the main body of the flow beneath. If this surface crust cracks before the lava has completely solidified, then the fluid lava below may ooze up through the crack to form a squeeze-up. Pressure ridges are built on the surface of lava flows where the solidified crustal zone is pushed into a linear fold. Tumuli are upheavals of domelike shape whose formation may be aided by a localized increase in hydrostatic pressure in the fluid lava beneath the crust. Pipes, vesicle trains, or spiracles may be developed in the lava depending on: (i) the amount of gas given off; (ii) the resistance offered by the lava; and (iii) the speed at which the flow is travelling. Pipes are tubes which project upwards from the base and are usually several centimetres in length and a centimetre or less in diameter. Vesicle trains form when gas action has not been strong

Fig. 1.6 Columnar jointing in basalt. Giants Causeway, Antrim, Northern Ireland. Note the 'organ' in the background showing colonnade overlain by entablature.

enough to produce pipes. Spiracles are openings formed by explosive disruption of the still-fluid lava by gas generated beneath it.

Thin lava flows are broken by joints which may run either at right angles or parallel to the direction of flow. Joints do occur with other orientations but are much less common. Those joints which are normal to the surface usually display a polygonal arrangement but only rarely do they give rise to columnar jointing. The joints develop as the lava cools. Primary joints form first, from which secondary joints arise, and so it continues.

Typical columnar jointing is developed in thick flows of basalt (Fig. 1.6). Columnar-jointed flows may exhibit a two- or three-tiered arrangement. The columns in columnar jointing are interrupted by cross-joints which may be either flat or saucer-shaped. The latter may be convex, up or down. These are not to be confused with platy joints which are developed in lavas as they become more viscous on cooling, so that slight shearing occurs along flow planes.

1.1.3 Texture of igneous rocks

The degree of crystallinity is one of the most important items of texture. An igneous rock may be composed of an aggregate of crystals, of natural glass, or of crystals and glass in varying proportions. This depends on the rate of cooling and composition of the magma and the environment under which the rock developed. If a rock is completely composed of crystalline mineral material, then it is described as holocrystalline. Most igneous rocks are holocrystalline. Conversely, rocks which consist entirely of glassy material are referred to as holohyaline. The terms hypocrystalline, hemicrystalline, or merocrystalline are given to rocks which are made up of intermediate proportions of crystalline and glassy material.

When referring to the size of individual crystals they are described as cryptocrystalline if they can just be seen under the highest resolution of a microscope or as microcrystalline if they can be seen at a lower magnification. These two types, together with glassy rocks, are collectively described as aphanitic, which means that the individual minerals cannot be distinguished with the naked eye. When the minerals of which a rock is composed are megascopic or macroscopic, that is, they can be recognized with the unaided eye, it is described as phanerocrystalline. Three grades of macroscopic texture are usually distinguished (Anon, 1977a): fine-grained (under 0.06 mm diameter), medium-grained (between 0.06 and 2.0 mm diameter), and coarse-grained (over 2.0 mm diameter).

A granular texture is one in which there is no glassy material and the individual crystals have a grainlike appearance. If the minerals are approximately the same size, the texture is described as equigranular. If they are of uneven size, the texture is referred to as inequigranular. Many volcanic and hypabyssal rocks display inequigranular textures, the two most important types being porphyritic and poikilitic. In the former case, large crystals or phenocrysts are set in a fine-grained groundmass. A porphyritic texture may be distinguished as macro-porphyritic or microporphyritic according to whether or not it may be observed with the unaided eye. The poikilitic texture is characterized by the presence of small crystals enclosed within larger ones.

The most important rock forming minerals are often referred to as felsic (light-coloured) and mafic (dark-coloured). Felsic minerals include quartz, muscovite mica, feldspars, and feldspathoids; whilst olivines, pyroxenes, amphiboles, and biotite mica are mafic minerals. The colour index of a rock is an expression of the percentage of mafic minerals which it contains. Four categories have been distinguished:
1 leucocratic rocks, containing less than 30% dark minerals;
2 mesocratic rocks, containing between 30 and 60% dark minerals,
3 melanocratic rocks, containing between 60 and 90% dark minerals; and
4 hypermelanic rocks, containing over 90% dark minerals.
Usually acidic rocks are leucocratic whilst basic and ultrabasic rocks are melanocratic and hypermelanic, respectively.

1.1.4 Igneous rock types

Granites and granodiorites are by far the commonest rocks of the plutonic association. They are characterized by a coarse-grained, holocrystalline granular texture. Although the term granite lacks precision, a normal granite has been defined as a rock in which quartz forms more than 5% and less than 50% of the quarfeloids (quartz, feldspar, feldspathoid content). Potash feldspar constitutes 50 to 95% of the total feldspar content, the plagioclase is sodi-calcic, and the mafites form more than 5% and less than 50% of the total constituents (Fig. 1.7).

In granodiorite the plagioclase is oligoclase or andesine and is at least double the amount of potash feldspar present, the latter forming 8 to 20% of the rock. The plagioclases are nearly always euhedral (i.e. have well-formed crystal faces), as may be biotite and hornblende. These minerals are set in a quartz–potash feldspar matrix.

The term pegmatite refers to coarse or very coarse grained rocks which are formed during the last stages of differentiation of a magma. Pegmatites, although commonly associated with granitic rocks, are found in association with all types of plutonites.

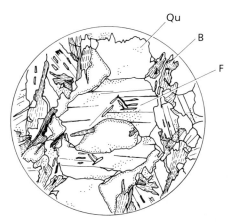

Fig. 1.7 Biotite granite containing quartz, feldspar, biotite, and some hornblende. Qu, quartz; F, feldspar; B, biotite (×14).

They occur as dykes, sills, veins, lenses, or irregular pockets in the host rocks, with which they rarely have sharp contacts (Fig. 1.8).

Aplites occur as veins (usually several tens of millimetres thick) in granites, although like pegmatites they are found in association with other plutonites. They possess a fine-grained, equigranular texture. There is no important chemical difference between aplite and pegmatite and it is assumed that they both have crystallized from residual magmatic solutions. Indeed, they may be associated with each other in composite intrusions.

Rhyolites are acidic extrusive rocks which are commonly associated with andesites. They are generally regarded as representing the volcanic equivalents of granite. They are usually leucocratic and sometimes exhibit flow banding. They may be holocrystalline but very often they contain an appreciable amount of glass. Rhyolites are frequently porphyritic, the phenocrysts varying in size and abundance. They occur in a glassy, cryptocrystalline, or microcrystalline groundmass.

Acidic rocks of hypabyssal occurrence are often porphyritic. Quartz porphyry is the commonest example and is similar in composition to rhyolite but it occurs in sills and dykes.

Syenites are plutonic rocks which have a granular texture and consist of potash feldspar, a subordinate amount of sodic plagioclase, and some mafic minerals (usually biotite or hornblende). Diorite has been defined as an intermediate, plutonic granular rock composed of plagioclase and hornblende, although

Fig. 1.8 Pegmatite vein traversing Shap granite (just left of centre), Shap quarry, Shap, Cumbria, England.

at times the latter may be partially or completely replaced by biotite and/or pyroxene. Plagioclase, in the form of oligoclase and andesine, is the dominant feldspar. If orthoclase is present, it acts only as an accessory mineral.

Trachytes and andesites are the fine-grained equivalents of syenites and diorites, respectively. Andesite is the commoner of the two types. Trachytes are extrusive rocks, which are often porphyritic, in which alkali feldspars are dominant. Most phenocrysts are composed of alkali feldspar, and to a lesser extent of alkali–lime feldspar. More rarely biotite, hornblende, and/or augite may form phenocrysts. The groundmass is usually a holocrystalline aggregate of sanidine (feldspar) laths.

Andesites are commonly porphyritic with a holocrystalline groundmass. Plagioclase (oligoclase–andesine), which is the dominant feldspar, forms most of the phenocrysts. The plagioclases of the groundmass are more sodic than those of the phenocrysts. Sanidine and anorthoclase rarely form phenocrysts but the former mineral does occur in the

groundmass and may encircle some of the plagioclase phenocrysts. Hornblende is the commonest of the ferromagnesian minerals and may occur as phenocrysts or in the groundmass, as may biotite and pyroxene.

Gabbros are dark-coloured, plutonic igneous rocks with granular textures. Plagioclase, commonly labradorite (but bytownite also occurs), is usually the dominant mineral. The pyroxenes found in gabbros are typically augite, diopsidic augite, and diallage.

Basalts are the extrusive equivalents of gabbros and are principally composed of basic plagioclase and pyroxene in roughly equal amounts, or there may be an excess of plagioclase. Those basalts which contain olivine are distinguished as olivine basalts (Fig. 1.9). Basalts are by far the most important type of extrusive rock but they also occur in dykes and sills. They exhibit a great variety of textures and may be holocrystalline or merocrystalline, equigranular, or macro- or microporphyritic.

Dolerites are commonly found in minor intrusions. They consist primarily of plagioclase (usually labradorite) and pyroxene (usually augite). The plagioclase occurs both as phenocrysts and in the groundmass. Dolerites are fine- to medium-grained. Usually they are equigranular but as they grade towards basalts they tend to become porphyritic, especially in the case of plagioclase. Nevertheless, the phenocrysts generally constitute less than 10% of the rock.

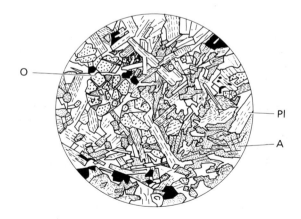

Fig. 1.9 Olivine basalt containing entire calcic plagioclase, olivine, and augite. Pl, calcic plagioclase; O, olivine; A, augite (×24).

1.2 METAMORPHISM AND METAMORPHIC ROCKS

Metamorphic rocks are derived from pre-existing rock types and have undergone mineralogical, textural, and structural changes. The latter have been brought about by changes which have taken place in the physical and chemical environments in which the rocks existed. The processes responsible for change give rise to progressive transformation which takes place in the solid state. The changing conditions of temperature and/or pressure are the primary agents causing metamorphic reactions in rocks. Individual minerals are stable over limited temperature–pressure conditions which means that when these limits are exceeded mineralogical adjustment has to be made to establish equilibrium with the new environment. Grade refers to the range of temperature under which metamorphism occurred.

When metamorphism occurs there is usually little alteration in the bulk chemical composition of the rocks involved with the exception of water and volatile constituents such as carbon dioxide. Little material is lost or gained and this type of alteration is described as an isochemical change. By contrast, allochemical changes are brought about by metasomatic processes which introduce or remove material from the rocks they affect. Metasomatic changes are brought about by hot gases or solutions permeating through rocks.

Two major types of metamorphism may be distinguished on the basis of geological setting. One type is of local extent whereas the other extends over a large region. The first type includes thermal or contact metamorphism and the latter refers to regional metamorphism.

1.2.1 Metamorphic textures and structures

Most deformed metamorphic rocks possess some kind of preferred orientation, commonly exhibited as mesoscopic linear or planar structures which allow the rocks to split more easily in one direction than another. One of the most familiar examples is cleavage in a slate; a similar type of structure in metamorphic rocks of higher grade is schistosity. Foliation comprises a segregation of particular minerals into inconstant bands or contiguous lenticles which exhibit a common parallel orientation.

Slaty cleavage (frequently equated with flow cleavage) is probably the most familiar type of preferred orientation and occurs in rocks of low metamorphic grade (Fig. 1.10). It is characteristic of slates and phyllites, is independent of bedding (which it commonly intersects at high angles), and reflects a highly developed preferred orientation of mineral boundaries, particularly of those belonging to the mica family.

Fracture cleavage is a parting defined by closely spaced parallel fractures which are usually independent of any planar preferred orientation of mineral boundaries that may be present in a rock mass (Chapter 2). Unlike slaty cleavage, fracture cleavage is not restricted to one type of rock.

Harker (1939) maintained that schistosity was developed in a rock when it was subjected to in-

Fig. 1.10 Old workings making use of the near vertical cleavage in slate, near Llanberis, North Wales.

creased temperatures and stress which involved its reconstitution, which was brought about by localized solution of mineral material and recrystallization. In all types of metamorphism the growth of new crystals takes place in an attempt to minimize stress. When recrystallization occurs under conditions which include shearing stress, then a directional element is imparted to the newly formed rock. Minerals are arranged in parallel layers along the direction normal to the plane of shearing stress giving the rock its schistose character (Fig. 1.11). The most important minerals responsible for the development of schistosity are those which possess an acicular, flaky, or tabular habit, the micas being the principal family involved. The more abundant flaky and tabular minerals are in such rocks, the more pronounced is the schistosity.

Foliation in a metamorphic rock (e.g. gneiss) is a most conspicuous feature consisting of parallel bands or tabular lenticles formed of contrasting mineral assemblages such as quartz–feldspar and biotite–hornblende (Fig. 1.12). This parallel orientation agrees with the direction of schistosity, if any is present in nearby rocks. Foliation would therefore seem to be related to the same system of stress and strain responsible for the development of schistosity. However, at higher temperatures the influence of stress becomes less and so schistosity tends to disappear in rocks of high grade metamorphism. By contrast, foliation becomes a more significant feature.

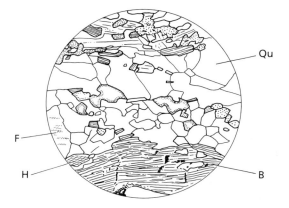

Fig. 1.12 Gneiss in which bands of quartz and feldspar are separated more or less from biotite and hornblende. Qu, quartz, F, feldspar; B, biotite; H, hornblende (×24).

1.2.2 Thermal or contact metamorphism

Thermal metamorphism occurs around igneous intrusions so that the principal factor controlling these reactions is temperature. The encircling zone of metamorphic rocks around an igneous intrusion is referred to as the contact aureole (Fig. 1.13), the size of which depends upon the temperature and size of the intrusion, the quantity of hot gases and hydrothermal solutions which emanated from it, and the type of country rocks involved. Aureoles developed in argillaceous sediments are more impressive than those found in arenaceous or calcareous rocks. This is because clay minerals, which account for a large proportion of the composition of argillaceous rocks, are more susceptible to temperature changes than quartz or calcite. Aureoles formed in igneous or previously metamorphosed terrains are also less significant than those developed in argillaceous sediments. Nevertheless the capricious nature of thermal metamorphism must be emphasized for even within one formation of the same rock type the width of the aureole may vary.

Within a contact aureole there is usually a sequence of mineralogical changes from the country rocks to the intrusion, which have been brought about by the effects of a decreasing thermal gradient whose source was in the hot magma. Indeed aureoles in argillaceous rocks may be concentrically zoned with respect to the intrusion. A frequently developed sequence varies inward from spotted slates to schists then hornfelses. Hornfelses are characteristic products of thermal metamorphism. They are dark-

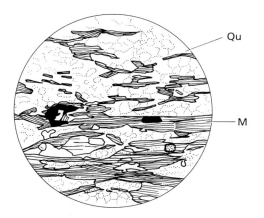

Fig. 1.11 Mica schist in which quartz and muscovite are segregated. Qu, quartz; M, muscovite (×24).

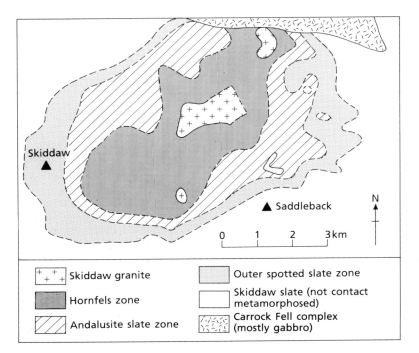

Fig. 1.13 Geological sketch map of the Skiddaw granite and its contact aureole.

Legend:

+ + Skiddaw granite

Hornfels zone

Andalusite slate zone

Outer spotted slate zone

Skiddaw slate (not contact metamorphosed)

Carrock Fell complex (mostly gabbro)

coloured rocks with a fine-grained decussate (i.e. interlocking) texture.

Aureoles formed in calcareous rocks frequently exhibit less regularity. The width of the aureole appears to be related to the chemical composition and permeability of the parent calcareous beds. Marbles may be found in such aureoles forming when limestones undergo metamorphism.

The reactions which occur when arenaceous sediments are subjected to thermal metamorphism are usually less complicated than those which take place in their argillaceous or calcareous counterparts. For example, the metamorphism of a quartz arenite leads to the recrystallization of quartz to form a quartzite with a mosaic texture and the higher the grade, the coarser the fabric. At high grades foliation tends to develop and a gneissose rock is produced.

The acid and intermediate igneous rocks are resistant to thermal metamorphism and are usually only affected at very high grades. For example, when granites are intruded by basic igneous masses, total recrystallization may be brought about in the immediate neighbourhood of the contact to produce a gneissose rock.

Basic igneous rocks undergo a number of changes when subjected to thermal metamorphism. In the outermost region of an aureole the pyroxenes are changed but the plagioclases remain unaffected thereby leaving the parental igneous texture intact. As the intrusion is approached, the rocks become completely recrystallized. At medium grade metamorphism hornblende hornfelses are common. Nearest the contact the high grade rocks are typically represented by pyroxene hornfelses.

1.2.3 Dynamic metamorphism*

Dynamic metamorphism is produced on a comparatively small scale and is usually highly localized, for example, its effects may be found in association with large faults or thrusts. On a larger scale it is associated with folding. However, in the latter case it is difficult to distinguish between the processes and effects of dynamic metamorphism and those of low grade regional metamorphism. What can be said is that at low temperatures recrystallization is at a minimum and the texture of a rock is largely governed by the mechanical processes which have been operative. The processes of dynamic metamorphism include brecciation, cataclasis, granu-

* It has been argued that dynamic metamorphism is not true metamorphism in that the processes involved give rise to deformation rather than transformation.

lation, mylonitization, pressure solution, partial melting, and slight recrystallization.

Stress is the most important factor in dynamic metamorphism. When a body is subjected to stresses which exceed its limit of elasticity, it is permanently strained or deformed. If the stresses are equal in all directions, then the body simply undergoes a change in volume, whereas if they are directional its shape is changed.

Brecciation is the process by which a rock is fractured, the angular fragments produced being of varying size. It is commonly associated with faulting and thrusting. The fragments of a crush breccia may themselves be fractured and the mineral components may exhibit permanent strain phenomena. If during the process of fragmentation pieces are rotated, then they are eventually rounded and embedded in the worn-down powdered material. The resultant rock is referred to as a crush conglomerate.

Mylonites are produced by the pulverization of rocks, which not only involves extreme shearing stress but also considerable confining pressure. Mylonitization is therefore associated with major faults. Mylonites are composed of strained porphyroblasts† set in an abundant matrix of fine-grained or cryptocrystalline material. Quartzes in the groundmass are frequently elongated. Those mylonites which have suffered great stress lack porphyroblasts, having a laminated structure with a fine, granular texture. The individual laminae are generally distinguishable because of their different colour. Protomylonite is transitional between microcrush breccia and mylonite, whilst ultramylonite is a banded or structureless rock in which the material has been reduced to powder.

In the most extreme cases of dynamic metamorphism the resultant crushed material may be fused to produce a vitrified rock referred to as a pseudotachylite. It usually occurs as very small discontinuous lenticular bodies or branching veins in granite, quartzite, amphibolite, and gneiss. Quartz and feldspar fragments are usually found in a dark-coloured glassy base.

1.2.4 Regional metamorphism

Metamorphic rocks extending over hundreds or even thousands of square kilometres are found exposed in

† The suffix -blast (or -blastic) indicates a metamorphic texture, hence porphyroblastic is the metamorphic equivalent of porphyritic.

the Pre-Cambrian shields and the eroded roots of fold mountains. As a consequence the term regional has been applied to this type of metamorphism. Regional metamorphism involves both the processes of changing temperature and stress. The principal factor is temperature of which the maximum figure concerned in regional metamorphism is probably in the neighbourhood of 800°C. Igneous intrusions are found within areas of regional metamorphism, but their influence is restricted. Regional metamorphism may be regarded as taking place when the confining pressures are in excess of 3 kbar, whilst below that figure, certainly below 2 kbar, falls within the field of contact metamorphism. What is more, temperatures and pressures conducive to regional metamorphism must have been maintained over millions of years. That temperatures rose and fell is indicated by the evidence of repeated cycles of metamorphism. These are demonstrated not only by mineralogical evidence but also by that of structures. For instance, cleavage and schistosity are the result of deformation which is approximately synchronous with metamorphism, but many rocks show evidence of more than one cleavage or schistosity which implies repeated deformation and metamorphism.

Regional metamorphism is a progressive process, that is, in any given terrain formed initially of rocks of the same composition, zones of increasing grade may be defined by different mineral assemblages. Each zone is defined by a significant mineral and their mineralogical variation can be correlated with changing temperature–pressure conditions. The boundaries of each zone therefore can be regarded as isograds — lines of equal metamorphic conditions (Fig. 1.14).

Slates are the products of low grade regional metamorphism of argillaceous sediments. As the metamorphic grade increases, slate gives way to phyllite in which somewhat larger crystals of chlorite and mica occur. This, in turn, gives way to mica schists. A variety of minerals such as garnet, kyanite, and staurolite may be present in these schists indicating formation at increasing temperatures.

When sandstones are subjected to regional metamorphism, a quartzite develops which has a granoblastic texture. A micaceous sandstone or one in which there is an appreciable amount of argillaceous material, on metamorphism yields a quartz–mica schist. Metamorphism of arkoses and feldspathic sandstones leads to the recrystallization of feldspar and quartz so that granulites, with a granoblastic texture, are produced.

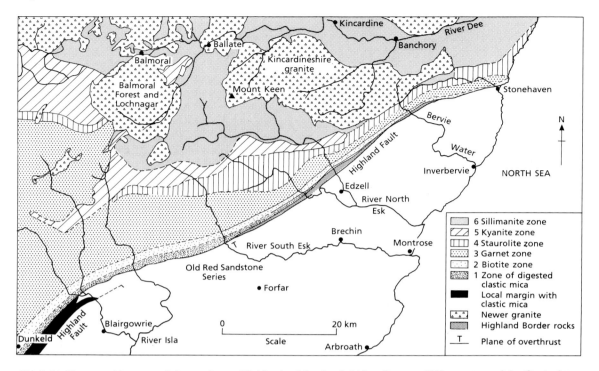

Fig. 1.14 Metamorphic zones of the south-east Highlands of Scotland. (After Barrow, 1912, courtesy of the Geologists Association.)

Pure carbonate rocks when subjected to regional metamorphism simply recrystallize to form either a calcite or dolomite marble with a granoblastic texture. Any silica present in a limestone tends to reform as quartz. The presence of micas in these rocks tends to give them a schistose appearance, schistose marbles or calc-schists being developed. Where mica is abundant it forms lenses or continuous layers giving the rock a foliated structure.

In regionally metamorphosed rocks derived from acid igneous parents quartz and white mica are important components, muscovite–quartz schist being a typical product of the lower grades. Conversely, at high grades white mica is converted to potash feldspar. In the medium and high grades, quartzofeldspathic gneisses and granulites are common. Some of the gneisses are strongly foliated.

Basic rocks are converted into greenschists by low grade regional metamorphism, to amphibolites at medium grade and to pyroxene granulites and eclogites at high grades.

1.2.5 Metasomatism

Metasomatic activity involves the introduction of material into and removal from, a rock mass by a gaseous or aqueous medium. The resultant chemical reactions lead to mineral replacement. Thus, two types of metasomatic action can be distinguished:
1 pneumatolytic — that of hot gas; and
2 hydrothermal — that of hot solutions.
Replacement is brought about by atomic or molecular substitution so that there is usually little change in rock texture. The composition of the transporting medium is continuously changing because of material being dissolved out of, and emplaced into, the rocks which are affected.

The hot gases and solutions involved usually emanate from an igneous source and the effects of metasomatism are often particularly notable about a granitic intrusion. Acid magmas have a greater concentration of volatiles than basic magmas. Both gases and solutions make use of any structural weaknesses such as faults or joint planes, in the rocks they invade. Because these provide easier paths of escape, metasomatic activity is concentrated along them. They also travel through the pore spaces in rocks, the rate of infiltration being affected by the porosity, the shape of the pores, and the temperature−pressure gradients.

Metasomatic action, especially when it is concentrated along fissures and veins, may bring about severe alteration of certain minerals. For example feldspars in granite or gneiss may be highly kaolinized as a result of metasomatism; limestone may be reduced to a weakly bonded granular aggregate.

1.3 SEDIMENTARY ROCKS

The sedimentary rocks form an outer skin on the Earth's crust, covering three-quarters of the continental areas and most of the sea floor. They vary in thickness up to 10 km. Nevertheless they only comprise about 5% of the crust.

Most sedimentary rocks are of secondary origin in that they consist of detrital material derived by the breakdown of pre-existing rocks. Indeed it has been variously estimated that shales and sandstones, both of mechanical derivation, account for between 80 and 95% of all sedimentary rocks. Certain sedimentary rocks are the products of chemical or biochemical precipitation whilst others are of organic origin. Thus, the sedimentary rocks can be divided into two principal groups:
1 clastic or exogenetic types; and
2 non-clastic or endogenetic types.
However, one factor which all sedimentary rocks have in common is that they are deposited and this gives rise to their most noteworthy characteristic, i.e. they are bedded or stratified.

Most sedimentary rocks are formed from the breakdown products of pre-existing rocks. Accordingly, the rate at which denudation takes place acts as a control on the rate of sedimentation, which in turn affects the character of a sediment. However, the rate of denudation is determined not only by the agents at work, that is, by weathering, or by river, marine, wind, or ice action, but also by the nature of the surface — upland areas are more rapidly worn away than are lowlands. Denudation may be regarded as a cyclic process, in that it begins with or is furthered by the elevation of a land surface and as this is gradually worn down the rate of denudation slackens. Each cycle of erosion is accompanied by a cycle of sedimentation. Also, the harder the rock, the more able it is to resist denudation. Geological structure also influences the rate of breakdown.

Furthermore the amount of sedimentation is affected by the amount of subsidence which occurs in a basin of deposition.

The particles of which most sedimentary rocks are composed have undergone varying amounts of transportation. This, together with the agent responsible, be it water, wind, or ice, plays an important role in determining the character of a sediment: transport over short distances usually means that the sediment is unsorted (the exception being beach sands), as does transportation by ice. With lengthier transport by water or wind, not only does the material become sorted but it is further reduced in size. The character of a sedimentary rock also is influenced by the type of environment in which it has been deposited, witness the presence of ripple marks and cross-bedding in sediments which accumulate in shallow water.

The composition of a sedimentary rock depends: (i) on the composition of the parent material and the stability of its component minerals; (ii) on the type of action to which the parent rock was subjected; and (iii) on the length of time it had to suffer such action. The least stable minerals tend to be those which are developed in environments very different from those experienced at the Earth's surface. In fact quartz, and to a much lesser extent, mica, are the only common constituents of igneous and metamorphic rocks which are found in abundance in sediments. Most of the others ultimately give rise to clay minerals. The more mature a sedimentary rock is, the more it approaches a stable end product and very mature sediments are likely to have experienced more than one cycle of sedimentation.

The type of climatic regime in which a deposit accumulates and the rate at which this takes place also affect the stability and maturity of the resultant sedimentary product. For example, chemical decay is inhibited in arid regions so that less stable minerals are more likely to survive than in humid regions. However, even in humid regions immature sediments may form when basins are rapidly filled with detritus derived from neighbouring mountains, the rapid burial affording protection against the attack of subaerial agencies.

In order to turn an unconsolidated sediment into a solid rock it must be lithified. Lithification involves two processes, namely, consolidation and cementation. The amount of consolidation which takes place within a sediment depends upon its composition and texture and upon the pressures acting on it, notably that due to the weight of overburden.

Consolidation of sediments deposited in water also involves dewatering, that is, the expulsion of connate water from the sediments. The porosity of a sediment is reduced as consolidation takes place and as the individual particles become closely packed they may even be deformed. Pressures developed during consolidation may lead to the differential solution of minerals and the authigenic growth of new ones.

Fine-grained sediments possess a higher porosity than do coarser types and therefore undergo a greater amount of consolidation. For example, muds and clays may have original porosities ranging up to 80% compared with 45 to 50% in sands and silts. Hence, if muds and clays could be completely consolidated (they never are) they would occupy only 20 to 45% of their original volume. The amount of consolidation which takes place in sands and silts varies from 15 to 25%.

Cementation involves the bonding together of sedimentary particles by the precipitation of material in the pore spaces. This reduces the porosity. The cementing material may be derived by partial intrastratal solution of grains or may be introduced into the pore spaces from an extraneous source by circulating waters. Conversely, cement may be removed from a sedimentary rock by leaching. The type of cement and, more importantly, the amount, affect the strength of a sedimentary rock. The type also influences its colour, sandstones with a siliceous or calcium carbonate cement are usually whitish grey; those with a sideritic cement are buff-coloured; a red colour is indicative of a haematitic cement and a brown colour indicates limonite. However, sedimentary rocks are frequently cemented by more than one material.

The matrix of a sedimentary rock refers to the fine material trapped within the pore spaces between the particles. It helps to bind the latter together.

The texture of a sedimentary rock refers to the size, shape, and arrangement of its constituent particles. Size is a property which is not easy to assess accurately, for the grains and pebbles of which clastic sediments are composed are irregular shaped, three-dimensional objects. Direct measurement can only be applied to large individual fragments where the length of the three principal axes can be recorded. But even this rarely affords a true picture of size. Estimation of volume by displacement may provide a better measure. Because of their smallness, the size of grains of sands and silts has to be measured indirectly by sieving and sedimentation techniques, respectively. The size of gravels is also assessed by sieving. Individual particles of clay have to be measured with the aid of an electron microscope. If a rock is strongly indurated, then its disaggregation is impossible without fracturing many of the grains. In such a case a thin section of the rock is made and size analysis is carried out with the aid of a petrological microscope and micrometer. The results of a size analysis may be represented graphically by a frequency curve or histogram. More frequently, however, they are used to draw a cumulative curve. The latter may be drawn on semilogarithmic paper (Fig. 1.15). Various statistical parameters such as median and mean size, deviation, skewness, and kurtosis can be calculated from data derived from cumulative curves. The median or mean size permits the determination of the grade of gravel, sand, or silt, or their lithified equivalents. Deviation affords a measure of sorting. However, the latter can be quickly and simply estimated by visual examination of the curve in that the steeper it is, the better the sorting of the sediment. The size of the particles of a clastic sedimentary rock allows it to be placed in one of three groups:

1 rudaceous or psephitic (over 2.0 mm);
2 arenaceous or psammitic (0.06 to 2.0 mm);
3 argillaceous or pelitic (less than 0.06 mm).

Reference to the American and British scales is made in Chapter 5 where a description of mixed aggregates is also provided.

Shape is probably the most fundamental property of any particle but unfortunately it is one of the most difficult to quantify. Frequently, shape is assessed in terms of roundness and sphericity, which may be estimated visually by comparison with standard images (Fig. 1.16).

A sedimentary rock is an aggregate of particles and some of its characteristics depend upon the position of these particles in space. The degree of grain orientation within a rock varies between perfect preferred orientation, in which all the long axes run in the same direction, and perfect random orientation, where the long axes point in all directions. The latter is only infrequently found as most aggregates possess some degree of grain orientation.

The arrangements of particles in a sedimentary rock involves the concept of packing, which refers to the spatial density of the particles in the aggregate, and has been defined as the mutual spatial relationship among the grains. It includes grain-to-grain contacts and the shape of the contact. The packing density involves the closeness or spread of particles, that is, how much space in a given area is occupied

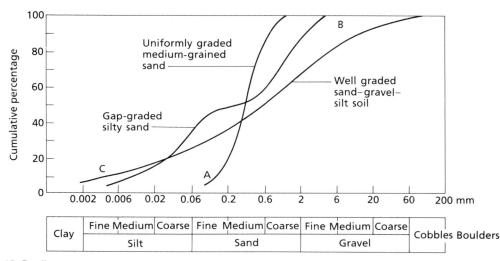

Fig. 1.15 Grading curves.

by grains. Packing is an important property of sedimentary rocks for it is related to their degree of consolidation, density, porosity, and strength.

1.3.1 Bedding and sedimentary structures

Sedimentary rocks are characterized by their stratification, and bedding planes are frequently the dominant discontinuity in sedimentary rock masses (Fig. 1.17). As such their spacing and character (are they irregular, waved or straight, tight or open, rough or smooth?) are of particular importance to the engineer. Several spacing classifications have been advanced, that given in Table 2.3 being one of the most commonly accepted and systematic. An individual bed may be regarded as a thickness of

sediment of the same composition which was deposited under the same conditions.

Lamination results from minor fluctuations in the velocity of the transporting medium or the supply of material, both of which produce alternating thin layers of slightly different grain size. Generally, lamination is associated with the presence of thin layers of platy minerals, notably micas. These have a marked preferred orientation, usually parallel to the bedding planes, and are responsible for the fissility of the rock. The surfaces of these laminae are usually smooth and straight. Although lamination is most characteristic of shales it may also be present in siltstones and sandstones, and occasionally in some limestones.

Cross- or current bedding is a depositional feature

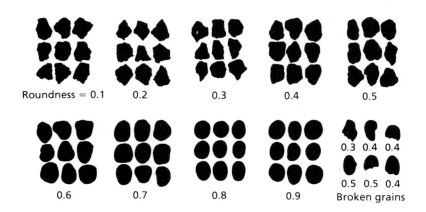

Fig. 1.16 Images for estimating roundness values. (After Krumbein, 1941.)

Fig. 1.17 Bedding in the Wick Flagstones, coast of Caithness, Scotland.

which occurs in sediments of fluvial, littoral, marine, and aeolian origin and is most notably found in sandstones (Fig. 1.18). In wind-blown sediments it is generally referred to as dune bedding. Cross-bedding is confined within an individual sedimentation unit and consists of cross-laminae inclined to the true bedding planes. The original dip of these cross-laminae is frequently between 20 and 30°. The thick-

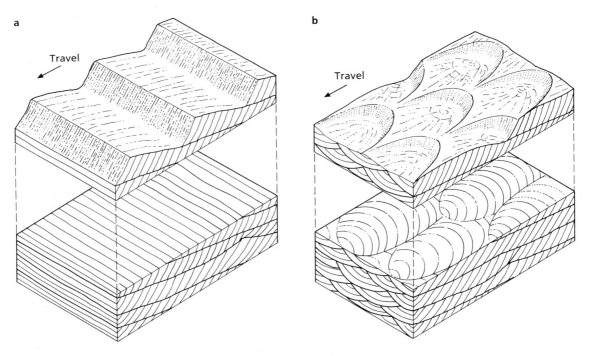

Fig. 1.18 Relationship of cross-bedding to surface shape of ripples: (a) planar cross-bedding formed by bed forms with long, substantially straight crests; (b) trough cross-bedding due to bed forms with short, curved crests.

ness of the sedimentation unit in which they occur varies enormously, for example, in micro-cross-bedding it measures only a few millimetres, whilst in dune bedding the unit may exceed 100 m.

Although graded bedding occurs in several different types of sedimentary rock, it is characteristic of greywackes. As the name suggests the sedimentation unit exhibits a grading from coarser grain size at the bottom to finer at the top. Individual graded beds range in thickness from a few millimetres to several metres. Usually the thicker the bed, the coarser it is overall.

1.3.2 Sedimentary rock types

A gravel is an unconsolidated accumulation of rounded fragments, the lower size limit of which is 2 mm. The term rubble has been used to describe those deposits with angular fragments. The composition of a gravel deposit reflects not only the source rocks of the area from which it was derived but is also influenced by the agent(s) responsible for its formation and the climatic regime in which it was, or is, being deposited. The latter two factors have a varying tendency to reduce the proportion of unstable material. Relief also influences the nature of a gravel deposit, for example, gravel production under low relief is small and the pebbles tend to be inert residues such as vein quartz, quartzite, chert, and flint. Conversely, high relief and accompanying rapid erosion yield coarse, immature gravels.

When a gravel becomes indurated it forms a conglomerate, when a rubble is indurated it is termed a breccia. Those conglomerates in which the fragments are in contact and so make up a framework are referred to as orthoconglomerates. By contrast those deposits in which the larger fragments are separated by matrix are referred to as paraconglomerates (they are in effect conglomeratic sandstones or mudstones).

Sands consist of a loose mixture of mineral grains and rock fragments. Generally, they tend to be dominated by a few minerals, the chief of which is quartz. Usually the grains show some degree of orientation, presumably related to the direction of flow of the transporting medium. The process by which a sand is turned into a sandstone is partly mechanical, involving grain fracturing, bending, and deformation. However, chemical activity is much more important and includes decomposition and solution of grains, precipitation of material from pore fluids, and intergranular reactions.

Silica is the most common cementing agent in sandstones, particularly in older ones. Various carbonate cements, especially calcite, are also common cementing materials and ferruginous and gypsiferous cements are also found. Cement, notably the carbonate and gypsiferous types, may be removed in solution by percolating pore fluids. This brings about varying degrees of decementation.

Quartz, feldspar, and rock fragments are the principal detrital components of which sandstones are composed, and consequently they have been used to define the major classes of sandstone. Pettijohn et al. (1972) also used the type of matrix in their classification. In other words those sandstones with more than 15% matrix were termed wackes. The chief type of wacke is greywacke which can be subdivided into lithic and feldspathic varieties (Fig. 1.19). Those sandstones with less than 15% matrix were divided into three families: the orthoquartzites or quartz arenites contain 95% or more of quartz (Fig. 1.20); 25% or more of the detrital material in arkoses consists of feldspar; and in lithic sandstones 25% or more of the detrital material consists of rock fragments.

Silts are clastic sediments derived from pre-existing rocks, chiefly by mechanical breakdown processes. They are mainly composed of fine quartz material. Silts may occur in residual soils but in such instances they are not important. However, silts are commonly found in alluvial, lacustrine, and marine deposits. These silts tend to interdigitate with deposits of sand and clay. Silts are also present with sands and clays in estuarine and deltaic sediments. Lacustrine silts are often banded. Marine silts also may be banded. Wind-blown silts are generally uniformly sorted.

Siltstones may be massive or laminated, the individual laminae being picked out by mica and/or carbonaceous material. Micro-cross-bedding is frequently developed and in some siltstones the laminations may be convoluted. Siltstones have a high quartz content with a predominantly siliceous cement and are often interbedded with shales or fine-grained sandstones, the siltstones occurring as thin ribs.

Deposits of clay are principally composed of fine quartz and clay minerals. The latter represent the commonest breakdown products of most of the chief rock-forming silicate minerals. The three major types of clay mineral are kaolinite, illite and montmorillonite. They are all hydrated aluminium silicates.

Kaolinite is formed by the alteration of feldspars, feldspathoids, and other aluminium silicates due to

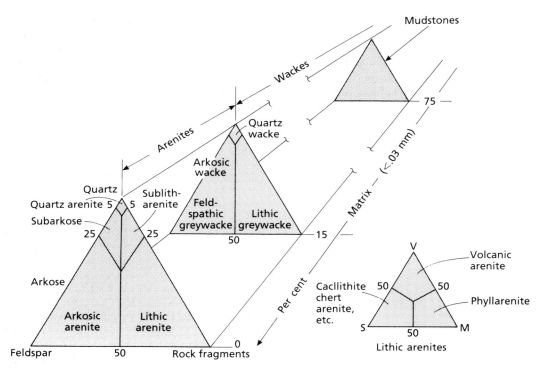

Fig. 1.19 Classification of sandstones. M, metamorphic; S, sedimentary; V, volcanic. (After Pettijohn *et al.*, 1972.)

hydrothermal action. Weathering under acidic conditions is also responsible for the kaolinization. Kaolinite is the chief clay mineral in most residual and transported clays, is important in shales, and is

Fig. 1.20 Quartz arenite from the Fell Sandstone Group, Carboniferous, Rothbury, Northumberland, England.

found in variable amounts in fireclays, laterites, and soils. It is the most important clay mineral in china clays and ball clays. Deposits of kaolin are associated with acid igneous rocks such as granites, granodiorites, and tonalites, and with gneisses and granulites.

Illite is of common occurrence in clays and shales, and is found in variable amounts in tills and loess, but is less common in soils. It develops as an alteration product of feldspars, micas, or ferromagnesian silicates upon weathering or may form from other clay minerals during diagenesis. Like kaolinite, illite also may be of hydrothermal origin. The development of illites, both under weathering and by hydrothermal processes, is favoured by an alkaline environment.

Montmorillonite develops when basic igneous rocks, in badly drained areas, are subjected to weathering. Magnesium is necessary for this mineral to form, but if the rocks were well drained, then it would be carried away and kaolinite would develop. An alkaline environment favours the formation of montmorillonite. Montmorillonite occurs in soils and argillaceous sediments such as shales formed from basic igneous rocks. It is the principal constituent of bentonitic clays, which are formed by the weathering of basic volcanic ash, and of Fuller's earth, which is also formed when basic igneous rocks are weathered. Hydrothermal action may also lead to the development of montmorillonite.

Residual clays develop in place and are the products of weathering. In humid regions residual clays tend to become enriched in hydroxides of ferric iron and aluminium, and impoverished in lime, magnesia, and alkalis. Even silica is removed in hot humid regions, resulting in the formation of hydrated alumina or iron oxide, as in laterite (Chapter 5).

The composition of transported clays varies because these materials consist mainly of abrasion products (usually silty particles) and transported residual clay material.

Shale is the commonest sedimentary rock and is characterized by its lamination. Sedimentary rock of similar size range and composition, but which is not laminated, is usually referred to as mudstone. In fact there is no sharp distinction between shale and mudstone, one grading into the other. An increasing content of siliceous or calcareous material decreases the fissility of a shale whereas shales which have a high organic content are finely laminated. Laminae range from 0.05 to 1.0 mm in thickness, with most in the range of 0.1 to 0.4 mm.

Clay minerals and quartz are the principal constituents of mudstones and shales. Feldspars often occur in the siltier shales. Shales may also contain appreciable quantities of carbonate, particularly calcite and gypsum, and calcareous shales frequently grade into shaly limestones. Carbonaceous black shales are usually rich in organic matter, contain a varying amount of pyrite, and are finely laminated.

The term limestone is applied to those rocks in which the carbonate fraction exceeds 50%, over half of which is calcite or aragonite. If the carbonate material is made up chiefly of dolomite, then the rock is named dolostone (this rock is generally referred to as dolomite but this term can be confused with that of the mineral of the same name). Limestones and dolostones constitute about 20 to 25% of the sedimentary rocks according to Pettijohn (1975). This figure is much higher than some of the estimates provided by previous authors.

Limestones are polygenetic. Some are of mechanical origin representing carbonate detritus which has been transported and deposited or that has accumulated in situ. Others represent chemical or biochemical precipitates which have formed in place. Allochthonous or transported limestone has a fabric similar to that of sandstone and may also display current structures such as cross-bedding and ripple marks. By contrast carbonate rocks which have formed in situ (i.e. autochthonous types) show no evidence of sorting or current action and at best possess a poorly developed stratification.

Lithification of carbonate sediments is often initiated as cementation at points of intergranular contact rather than as consolidation. In fact, carbonate muds consolidate very little because of this early cementation. The rigidity of the weakest carbonate rocks, such as chalk, may be attributed to mechanical interlocking of grains with little or no cement. Nevertheless, cementation may take place more or less at the same time as deposition, but cemented and uncemented assemblages may be found within short, horizontal distances. Indeed, a recently cemented carbonate layer may overlie uncemented material. Because cementation occurs concurrently with, or soon after deposition, carbonate sediments can support high overburden pressures before consolidation takes place. Hence high values of porosity may be retained to considerable depths of burial. Eventually, however, the porosity is greatly reduced by post-depositional changes which bring about recrystallization. Thus a crystalline limestone is formed.

Folk (1973) distinguished two types of dolostone. First, he recognized an extremely fine-grained crystalline dolomicrite (less than 20 μm grain diameter), and second, a more coarsely grained or saccharoidal dolostone in which there was plentiful evidence of replacement. He regarded the first type as of primary origin, and the second as being formed as a result of diagenetic replacement of calcite by dolomite in limestone. Primary dolostones tend to be thinly laminated and are generally unfossiliferous. They are commonly associated with evaporates and may contain either nodules or scattered crystals of gypsum or anhydrite. In those dolostones formed by dolomitization the original textures and structures may be obscured or even have disappeared.

Carbonate rocks may contain various amounts of impurities, notably quartz and clay minerals. A classification of impure limestones and dolostones can be made with reference to a ternary diagram (Fig. 1.21a), as can a classification of carbonate, sand (quartz), and clay mixtures (Fig. 1.21b). Fine-grained argillaceous limestones are frequently wavy-bedded or nodular-bedded, the beds being separated by thin shaly partings.

Evaporitic deposits are quantitatively unimportant as sediments. They are formed by precipitation from

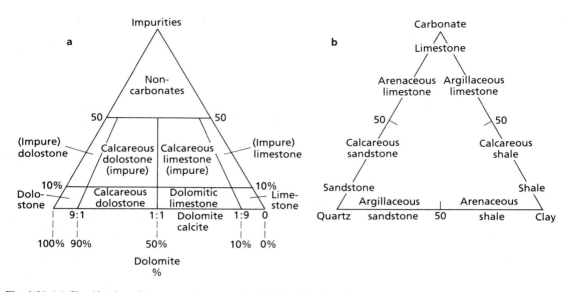

Fig. 1.21 (a) Classification of impure carbonate rocks; (b) classification of carbonate−sand−clay mixtures.

saline waters, the high salt content being brought about by evaporation from inland seas or lakes in arid areas. Salts can also be deposited from subsurface brines, brought to the surface of a playa or sabkha flat (Chapter 3) by capillary action.

Sea water contains approximately 3.5%, by weight, of dissolved salts, about 80% of which is sodium chloride. Experimental work has shown that when the original volume of sea water is reduced by evaporation to about half, then a little iron oxide and some calcium carbonate are precipitated. Gypsum begins to form when the volume is reduced to about one-fifth of the original; rock salt begins to precipitate when about one-tenth of the volume remains; and when only 1.5% of the sea water is left, potash and magnesium salts start to crystallize. This order agrees in a general way with the sequence found in some evaporitic deposits. However, many exceptions are known. Many complex replacement sequences occur amongst evaporitic rocks, for example, carbonate rocks may be replaced by anhydrite and sulphate rocks by halite.

Organic residues which accumulate as sediments are of two major types:
1 peaty material (which when buried gives rise to coal);
2 sapropelic residues.
Sapropel is rich in silt, or composed wholly of organic compounds which collect at the bottom of still bodies of water. Such deposits may give rise to cannel or boghead coals. Sapropelic coals usually have a significant amount of inorganic matter as opposed to humic coals in which the inorganic content is low. The former are not generally extensive and are not underlain by seatearths.

Peat deposits accumulate in poorly drained environments where the formation of humic acid gives rise to deoxygenated conditions. These inhibit the bacterial decay of organic matter. Peat accumulates wherever the deposition of plant debris exceeds the rate of its decomposition. A massive deposit of peat is required to produce a thick seam of coal, for example, a seam 1 m thick probably represents 15 m of peat.

Chert and flint are the two most common siliceous sediments of chemical origin. Chert is a dense rock composed of one or more forms of silica such as opal, chalcedony or microcrystalline quartz. It may occur as thin beds or nodules in carbonate host rocks. Carbonate material may be scattered throughout impure varieties. Gradations occur from chert to sandstone with chert cement, although sandy cherts are not common. Cherts may suffer varying degrees of devitrification.

Some sediments may have a high iron content. The iron carbonate, siderite, often occurs interbedded with chert or mixed in varying proportions with clay, as in clay ironstones. Some iron-bearing formations are formed mainly of iron oxide, haematite being the most common mineral. Haematite-rich beds are generally oolitic. Limonite occurs in oolitic form in some ironstones. Bog iron ore is chiefly an earthy mixture of ferric hydroxides. Siliceous iron ores include chamositic ironstones, which are also typically oolitic. Glauconitic sandstones and limestones may contain 20% or more FeO and Fe_2O_3. On rare occasions bedded pyrite is found in black shale.

1.4 STRATIGRAPHY AND STRATIFICATION

Stratigraphy is that branch of geology which deals with the study and interpretation of stratified rocks and with the identification, description, sequence (both vertical and horizontal), mapping, and correlation of stratigraphic rock units. As such it begins with the discrimination and description of stratigraphical units such as formations. This is necessary so that the complexities present in every stratigraphical section may be simplified and organized.

Deposition involves the build-up of material on a given surface, either as a consequence of chemical or biological growth or, far more commonly, due to mechanically broken particles being laid down on such a surface. Hence, this surface exerts an important influence on the attitude of the beds which are formed. Gravity may also exert an influence on the attitude of the bedding planes in mechanically derived sediments. At the start of deposition the layers of sediment more or less conform to the surface on which accumulation is occurring, provided this is not too irregular. With continued deposition any irregularities in the original surface are filled and the strata which then form tend to lie in a horizontal plane. However, it should be borne in mind that once a layer is formed, and before lithification occurs, it may be disturbed by subsequent deposition. Furthermore, differential consolidation of different materials (e.g. sand and mud) or differential consolidation over buried hills may give rise to inclined bedding.

The changes which occur during deposition are responsible for stratification, that is, the layering which characterizes sedimentary rocks. The simple cessation of deposition ordinarily does not produce stratification. The most obvious change which gives rise to stratification is that in the composition of the material being deposited. Even minor changes in the type of material may lead to distinct stratification, especially if they affect the colour of the rocks concerned. Changes in grain size also may cause notable layering and changes in other textural characteristics may help distinguish one bed from another, as may variations in the degree of consolidation or cementation.

The extent and regularity of beds of sedimentary rocks vary within wide limits. This is because lateral persistence and regularity of stratification reflect the persistence and regularity of the agent responsible for deposition, for example, sands may have been deposited in one area and muds in a neighbouring area. Hence, lateral changes in lithology reflect differences in the environments in which deposition took place. On the other hand, a formation with a particular lithology, which is mappable as a stratigraphic unit, may not have been laid down at the same time wherever it occurs. The base of such a formation is described as diachronous (Fig. 1.22). Diachronism is brought about when a basin of deposition is advancing or retreating, as in a marine transgression or regression. In an expanding basin the lowest sediments to accumulate are not as extensive as those succeeding. The latter are said to overlap the lower-most deposits. Conversely, if the basin of deposition is shrinking the opposite situation arises in that succeeding beds are less extensive. This phenomenon is termed offlap. Examples of overlap and offlap generally are found in association with cyclic sedimentation due to oscillating crustal movements.

Agents which are confined to channels or deposited over small areas produce irregular strata that are not persistent. By contrast, strata that are very persistent are produced by agents which operate over wide areas. In addition, folding and faulting of strata, along with subsequent erosion, give rise to discontinuous outcrops.

Since sediments are deposited it follows that the topmost layer in any succession of strata is the youngest. Also, any particular stratum in a sequence can be dated by its position in the sequence relative to other strata. This is the Law of Superposition. This principle applies to all sedimentary rocks except, of course, those which have been inverted by folding or where strata have been thrust over younger rocks.

1.4.1 Unconformities

An unconformity represents a break in the stratigraphical record and occurs when changes in the palaeogeographical conditions led to a cessation of deposition. Such a break may correspond to a rela-

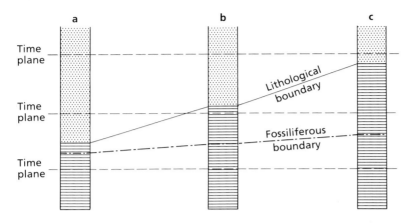

Fig. 1.22 Diachronism of a lithological boundary and the migration time of a fossil assemblage. The fossiliferous horizon may be regarded as a time plane if the localities (a), (b), and (c) are not far distant. As a rule, time planes cannot be identified.

tively short interval of geological time or a very long one. An unconformity normally means that uplift and erosion have taken place, resulting in some previously formed strata being removed. The beds above and below the surface of unconformity are described as unconformable.

The structural relationship between unconformable units allows four types of unconformity to be distinguished. In Fig. 1.23a stratified rocks rest upon igneous or metamorphic rocks. This type of feature has frequently been referred to as a nonconformity (it has also been called a heterolithic unconformity). An angular unconformity is shown in Fig. 1.23b, where an angular discordance separates the two units of stratified rocks. In an angular unconformity the lowest bed in the upper sequence of strata usually rests on beds of differing ages (see Fig. 3.16, p. 69). This is referred to as overstep. In a disconformity, as illustrated in Fig. 1.23c, the beds lie parallel both above and below the unconformable surface but the contact between the two units concerned is an uneven surface of erosion. When deposition is interrupted for a significant period but there is no apparent erosion of sediments or tilting or folding, then subsequently formed beds are deposited parallel to those already existing. In such a case the interruption in sedimentation may be demonstrable only by the incompleteness of the

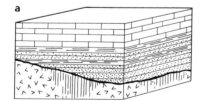
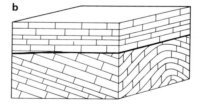
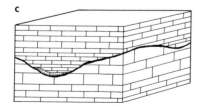
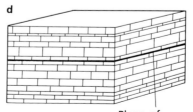

Plane of unconformity

Fig. 1.23 Types of unconformity: (a) nonconformity or heterolithic unconformity; (b) angular unconformity; (c) disconformity; (d) paraconformity.

fossil sequence. This type of unconformity has been termed a paraconformity (Fig. 1.23d).

One of the most satisfactory criteria for the recognition of unconformities is evidence of an erosion surface between two formations. Such evidence may take the form of pronounced irregularities in the surface of the unconformity. Evidence may also take the form of weathered strata beneath the unconformity, weathering having occurred prior to the deposition of the strata above. Fossil soils provide a good example. The abrupt truncation of bedding planes, folds, faults, dykes, joints, etc., in the beds below the unconformity is characteristic of an unconformity, although large-scale thrusts will give rise to a similar structural arrangement. Post-unconformity sediments often commence with a conglomeratic deposit. The pebbles in the conglomerate may be derived from the older rocks below the unconformity.

1.4.2 Rock and time units

Stratigraphy distinguishes rock units and time units. A rock unit, such as a stratum or a formation, possesses a variety of physical characteristics which enable it to be recognized as such, and so measured, described, mapped, and analysed. A rock unit is sometimes termed a lithostratigraphical unit.

A particular rock unit required a certain interval of time for it to form. Hence stratigraphy not only deals with strata but also deals with age, and the relationship between strata and age. Accordingly time units and time−rock units have been recognized. Time units are simply intervals of time, the largest of which are eons, although this term tends to be used infrequently. There are two eons, representing Pre-Cambrian time and Phanerozoic time. Eons are divided into eras, and eras into periods (Table 1.1). Periods are in turn divided into epochs and epochs into ages. Time units and time−rock units are directly comparable, that is, for each time unit there is a corresponding time−rock unit. For example, the time−rock unit corresponding to a period is a system and systems are divided into series. Indeed, the time allotted to a time unit is determined from the rocks of the corresponding time−rock unit. Most time units in local exposures, except possibly some of the smallest, are incomplete.

A time−rock unit has been defined as a succession of strata bounded by theoretically uniform time planes, regardless of the local lithology of the unit.

Fossil evidence usually provides the basis for the establishment of time planes. Ideal time−rock units would be bounded by completely independent time planes. However, practical time−rock units depend on whatever evidence is available.

The geological systems are time−rock units which are based on stratigraphical successions present in certain historically important areas. In other words, in their type localities the major time−rock units are also rock units. The boundaries of major time−rock units generally are important structural or faunal breaks or are placed at highly conspicuous changes in lithology. Major unconformities frequently are chosen as boundaries. Away from their type areas major time−rock units may not be so distinctive or easily separated. In fact although systems are regarded as of global application there are large regions where the recognition of some of the systems has not proved satisfactory.

1.4.3 Correlation

The process by which the time relationship between strata in different areas is established is referred to as correlation. Correlation is therefore the demonstration of equivalency of stratigraphical units. Fossil and lithological evidence are the two principal criteria used in correlation.

The principle of physical continuity may be of some use in local correlation, that is, it can be assumed that a given bed, or bedding plane, is roughly contemporaneous throughout an outcrop of bedded rocks. Tracing of bedding planes laterally, however, is severely limited since individual beds or bedding planes die out, are interrupted by faults, are missing in places due to removal by erosion, are concealed by overburden, or merge with others laterally. Consequently, outcrops are rarely good enough to permit an individual bed to be traced laterally over an appreciable distance. A more practicable procedure is to trace a member of a formation. However, this can also prove misleading if beds are diachronous.

Where outcrops are discontinuous, physical correlation depends on lithological similarity, that is, matching rock types across the breaks in the hope of identifying the beds involved. The lithological characters used to make comparison in such situations include gross lithology, subtle distinctions within one rock type such as a distinctive heavy mineral suite, notable microscopic features or dis-

Table 1.1 The geological time-scale

Eras	Periods and systems	Derivation of names	Duration of period (Ma)	Total from beginning (Ma)
CAINOZOIC	Quaternary			
	Recent or Holocene*	Holos = complete whole		
	Glacial or Pleistocene*	Pleiston = most	2 or 3	2 or 3
	Tertiary			
	Pliocene*	Pleion = more	9 or 10	12
	Miocene*	Meion = less (i.e. less than in Pliocene) — 'cene' from Kainos = recent	13	25
	Oligocene*	Oligos = few	15	40
	Eocene*	Eos = dawn	20	60
	Paleocene*	Palaios = old	10	70
		The above comparative terms refer to the proportions of modern marine shells occurring as fossils		
MESOZOIC	Cretaceous	Creta = chalk	65	135
	Jurassic	Jura Mountains	45	180
	Triassic	Three-fold division in Germany	45	225
	(New Red Sandstone = desert sandstones of the Triassic Period and part of the Permian)			
PALAEOZOIC	Permian	Permia, ancient kingdom between the Urals and the Volga	45	270
	Carboniferous	Coal (carbon)-bearing	80	350
	Devonian	Devon (marine sediments)	50	400
	(Old Red Sandstone = land sediments of the Devonian period)			
	Silurian	Silures, Celtic tribe of Welsh Borders	40	440
	Ordovician	Ordovices, Celtic tribe of North Wales	60	500
	Cambrian	Cambria, Roman name for Wales	100	600
PRECAMBRIAN ERA				
ORIGIN OF EARTH				5000

* Frequently regarded as epochs or stages.

tinctive key beds. The greater the number of different, and especially unusual, characters that can be linked, the better are the chances of reliable correlation. Even so, such factors must be applied with caution and wherever possible such correlation should be verified by the use of fossils.

If correlation can be made from one bed in a particular outcrop to one in another, it can be assumed that the beds immediately above and below are also correlative, provided, of course, that there was no significant break in deposition at either exposure. Better still, if two beds can be correlated between two local exposures, then the intervening beds are presumably correlative, even if the character of the intervening rocks is different in the two outcrops. This again depends on there being no important break in deposition at either of the locations.

The Law of Faunal Succession was formulated at the end of the eighteenth century, and states that strata of different ages are characterized by different fossils or suites of fossils. That is, each formation can be identified by its distinctive suite of fossils without the need of lateral tracing.

As far as correlation is concerned good fossils should have a wide geographical distribution and a limited stratigraphical range. In general groups of organisms which possessed complicated structures provide better guides for correlation than those that were simple. The usefulness of fossils is enhanced if the group concerned evolved rapidly, for where morphological changes take place rapidly, individual species are of short duration. These fossils provide a more accurate means of subdividing the geological column and therefore provide more precise correlation. Groups which were able to swim or float prove especially useful since they ranged widely and were little restricted in distribution by the conditions on the sea floor.

The principal way in which fossils are used in correlation is based on the recognition of characteristic species in strata of a particular age. This method can be applied in two ways:
1 index fossils can be established, which in turn allows a particular bed to be identified; and
2 fossils may be used to distinguish zones.
A zone may be defined as the strata laid down during a particular interval of time when a given fauna or flora existed. In some cases zones have been based on the complete fauna present whilst in other instances they have been based on the members of a particular phylum or class. None the less, a zone is a division of time given in terms of rocks deposited. Although a faunal or floral zone is defined by reference to an assemblage of fossils, it is usually named after some characteristic species and this fossil is known as the zone fossil. Normally a faunal or floral zone is identifiable because certain species existed together for some time. It is assumed that these species have time ranges which are overlapping and that their time ranges are similar in different areas.

2 Geological structures

The two most important features which are produced when strata are deformed by earth movements are folds and faults, that is, the rocks are buckled or fractured, respectively. A fold is produced when a more or less planar surface is deformed to give a waved surface. A fault represents a surface of discontinuity along which the strata on either side have been displaced relative to each other. Such deformation principally takes place due to movements along shearing planes. When these are small and numerous, flexuring and folding result, whilst if they are few and large, they cause faulting.

2.1 FOLDS

2.1.1 Anatomy of folds

There are two important directions associated with folding, namely, dip and strike. True dip gives the maximum angle at which a bed of rock is inclined and should always be distinguished from apparent dip (Fig. 2.1). The latter is a dip of lesser magnitude whose direction can run anywhere between that of true dip and strike. Strike is the trend of a fold and is orientated at right angles to the true dip; it has no inclination (Fig. 2.1).

Folds are wavelike in shape and vary enormously in size. Simple folds are divided into two types — anticlines and synclines (Fig. 2.2). In the former the beds are convex upwards, whereas in the latter they are concave upwards. The crestal line of an anticline is the line which joins the highest parts of the fold whilst the trough line runs through the lowest parts of a syncline (Fig. 2.2). The amplitude of a fold is defined as the vertical difference between the crest and the trough, whilst the length of a fold is the horizontal distance from crest to crest or trough to trough. The hinge of a fold is the line along which

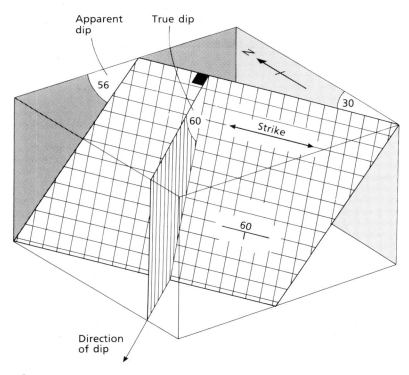

Fig. 2.1 Dip and stike: orientation of cross-hatched plane can be expressed as follows: strike 330°, dip 60°W, or dip 60° towards 240°.

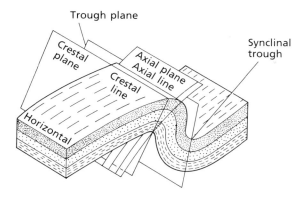

Fig. 2.2 Block diagram of a non-plunging overturned anticline and syncline, showing various fold elements.

the greatest curvature exists and it can be either straight or curved. However, the axial line is another term which has been used to describe the hinge line. The limbs of folds occur between the hinges, all folds having two limbs. The axial plane of a fold is commonly regarded as the plane which bisects the fold and passes through the axial or hinge line.

The interlimb angle, which is the angle measured between the two projected planes from the limbs of the fold, can be used to assess the degree of closure of a fold. Fleuty (1964) recognized five degrees of closure based on the interlimb angle: gentle folds are those with an interlimb angle greater than 120°; open folds, 120 to 70°; close folds, 70 to 30°; tight folds, less than 30°; and finally in isoclinal folds the limbs are parallel and so the interlimb angle is zero. He went on to express the attitude of a fold in terms of the dip of the axial plane (Table 2.1).

Folds are of limited extent and when one fades out the attitude of its axial line changes, that is, it dips away from the horizontal. This is referred to as the plunge or pitch of the fold (Fig. 2.3). The

Table 2.1 Attitude of folds

Dip of axial plane	Terms	
0	Recumbent fold	Horizontal
1–10°		Subhorizontal
10–30°		Gentle
30–60°	Inclined fold	Moderate
60–80°		Steep
80–89°		Subvertical
90°	Upright fold	Vertical

Table 2.2 Plunge of folds

Plunge	Terms	
0°		Horizontal
1–10°		Subhorizontal
10–30°		Gentle
30–60°	Plunging fold	Moderate
60–80°		Steep
80–89°		Subvertical
90°	Vertical fold	Vertical

amount of plunge can change along the strike of a fold and a reversal of plunge direction can occur. The axial line is then waved, concave-upwards areas being termed depressions whilst convex-upwards areas are known as culminations. Fleuty (1964) also classified plunge in a similar manner to that in which he classified the attitude of folding (Table 2.2).

2.1.2 Types of folding

Anticlines and synclines are symmetrical if both limbs are equally arranged about the axial plane so that the dips on opposing flanks are the same,

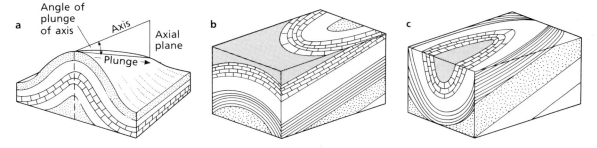

Fig. 2.3 (a) Block diagram of an anticlinal fold illustrating plunge; (b) plunging anticline; (c) plunging syncline.

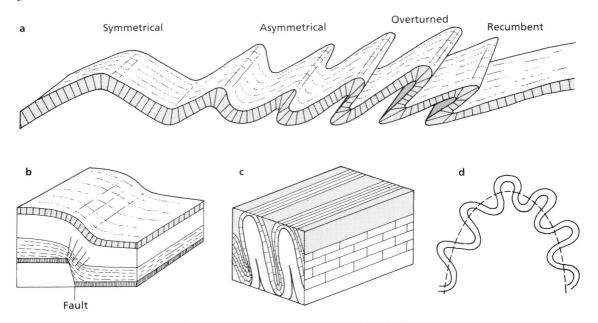

Fig. 2.4 (a) Types of folding; (b) monoclinal fold; (c) isoclinal folding; (d) fan folding.

otherwise they are asymmetrical (Fig. 2.4a). In symmetrical folds the axis is vertical whilst in asymmetrical folds it is inclined. If beds which are horizontal, or nearly so, suddenly dip at a high angle, then the feature they form is termed a monocline (Fig. 2.4b). When traced along its strike, a monocline may eventually flatten out or pass into a normal fault indeed monoclines are often formed as a result of faulting at depth. Isoclinal folds are those in which both the limbs and the axial plane are parallel (Fig. 2.4c). A fan fold is one in which both limbs are themselves folded (Fig. 2.4d).

As folding movements become intensified, overfolds are formed in which both limbs are inclined, together with the axis, in the same direction but at different angles (Fig. 2.4a). In a recumbent fold the beds have been completely overturned so that one limb is inverted, and the limbs, together with the axial plane, dip at a low angle (Fig. 2.4a).

2.1.3 Relationships of strata in folds

Parallel or concentric folds are those where the strata have been bent into more or less parallel curves in which the thickness of the individual beds remains the same. Figure 2.5(a) shows that because the thickness of the beds remains the same on folding, the shape of the folds changes with depth and in fact they fade out. Parallel folding occurs in competent (relatively strong) beds which may be interbedded with incompetent (relatively weak, plastic) strata.

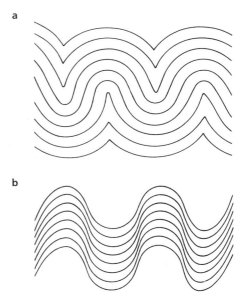

Fig. 2.5 (a) Parallel folding; (b) similar folding.

Similar folds retain their shape with depth as a result of flowage of material from the limbs into the crestal and trough regions (Fig. 2.5b). Similar folds develop in incompetent strata, but true similar folds are rare, for most change their shape to some extent along the axial plane. Most folds exhibit both the characteristics of parallel and similar folding.

Most folding is disharmonic in that the shape of the individual folds within the structure is not uniform, the fold geometry varying from bed to bed. Disharmonic folding occurs in interbedded competent and incompetent strata. Its essential feature is that incompetent horizons display more numerous and smaller folds than the more competent beds enclosing them. It is developed because competent and incompetent beds yield differently to stress.

Zig-zag or chevron folds have straight or nearly straight limbs with sharply curved or even pointed hinges (Fig. 2.6). Such folds possess features which are characteristic of both parallel and similar folds in that the strata in their limbs remain parallel, beds may be thinned but are never thickened, and the pattern of the folding persists with depth. Some bedding slip occurs and gives rise to a small amount of distortion in the hinge regions. The planes about which the beds are sharply bent are called kink planes and their attitude governs the geometry of the fold. Zig-zag folds are characteristically found in thin-bedded rocks, especially where there is a rapid alternation of more rigid beds such as sandstones, with interbedded shales.

2.1.4 Minor structures associated with folding

Cleavage is one of the most notable structures associated with folding and imparts to rocks the ability to split into thin slabs along parallel or slightly sub-parallel planes of secondary origin. The distance between cleavage planes varies according to the lithology of the host rock, that is, the coarser the texture, the further the cleavage planes are apart. Two principal types of cleavage have been recognized, namely, fracture cleavage and flow cleavage.

Flow cleavage occurs as a result of plastic deformation in which internal readjustments involving gliding, granulation, and the parallel reorientation of minerals of flaky habit — such as micas, chlorite, graphite, and haematite — together with the elongation of quartz and calcite, take place. The cleavage planes are commonly only a fraction of

Fig. 2.6 Chevron fold in limestone of Miocene age, Kaikuora, South Island, New Zealand.

a millimetre apart and when the cleavage is well developed, the original bedding planes may have partially or totally disappeared. Flow cleavage may develop in deeply buried rocks which are subjected to simple compressive stress, in which case the cleavage planes are orientated normal to the direction in which the stress was acting. As a result the cleavage planes run parallel to the axial planes of the folds. Many authors equate flow cleavage with true slaty cleavage.

Fracture cleavage can be regarded as closely spaced jointing, the distance between the planes being measured in millimetres or even in centimetres (Fig. 2.7). Unlike flow cleavage there is no parallel alignment of minerals, fracture cleavage having been caused by shearing forces. It therefore follows the

laws of shearing and develops at an angle of approximately 30° to the axis of maximum principal stress. However, fracture cleavage often runs almost normal to the bedding planes and in such instances it has been assumed that it is related to a shear couple. The external stress creates two potential shear fractures but since one of them trends almost parallel to the bedding, it is unnecessary for fractures to develop in that direction. The other direction of potential shearing is that in which fracture cleavage ultimately develops and this is facilitated as soon as the conjugate shear angle exceeds 90°. Fracture cleavage is frequently found in folded incompetent strata which lie between competent beds. For example, where sandstone and shale are highly folded, fracture cleavage occurs in the shale in order to fill the spaces left between the folds of the sandstone. However, fracture cleavage need not be confined to the in-beds. In competent rocks it forms a larger angle with the bedding planes than it does in the incompetent strata.

When brittle rocks are distorted, tension gashes may develop as a result of stretching over the crest of a fold or by local extension caused by drag exerted when beds slip over each other. Tension gashes resulting from bending of competent rocks usually appear as radial fractures concentrated at the crests of sharply folded anticlines. They represent failure following plastic deformation. Tension gashes formed by differential slip appear on the limbs of folds and are aligned approximately perpendicular to the local direction of extension. Tension gashes are distinguished from fracture cleavage and other types of fractures because their sides tend to gape. As a result they often contain lenticular bodies of vein quartz or calcite.

Tectonic shear zones lie parallel to the bedding and appear to be due to displacements caused by concentric folding. Such shear zones generally occur in clay beds with high clay-mineral contents. The shear zones range up to approximately 0.5 m in thickness and extend over hundreds of metres. Each shear zone exhibits a conspicuous principal slip which forms a gently undulating smooth surface. The interior of a shear zone is dominated by displacement shears and slip surfaces lying *en echelon* inclined at 10 to 30° to the *ab* plane (*a* is the direction of movement, *b* lies in the plane of shear, and *c* is at right angles to this plane). These give rise to a complex pattern of shear lenses, the surfaces of which are slickensided. Relative movement between the lenses is complicated, with many local variations.

Fig. 2.7 Fracture cleavage developed in highly folded Horton Flags, Silurian, near Stainforth, North Yorkshire, England. The inclination of the fracture cleavage is indicated by the near-vertical hammer. The other hammer indicates the dip of the bedding.

Thrust shears, and possibly fracture cleavage, have also been noted in these shear zones.

2.2 FAULTS

Faults are fractures in crustal strata along which the adjacent rock has been displaced (Fig. 2.8). The amount of displacement may vary from only a few tens of millimetres to several hundred kilometres. In many faults the fracture is a clean break but in others the displacement is not restricted to a simple fracture but is developed throughout a fault zone.

The dip and strike of a fault plane can be described in the same way as are those of a bedding plane. The angle of hade is the angle enclosed between the fault plane and the vertical. The hanging wall of a fault refers to the upper rock surface along which displacement has occurred, whilst the footwall is the term given to that below. The vertical shift along a fault plane is called the throw, whilst the term heave refers to the horizontal displacement. Where the displacement along a fault has been vertical, then the terms downthrow and upthrow refer to the relative movement of strata on opposite sides of the fault plane.

2.2.1 Classification of faults

A classification of faults can be made on a geometrical or a genetic basis, and as such can be based on the direction in which movement has taken place along the fault plane, on the relative movement of the hanging and footwalls, on the attitude of the fault in relation to the strata involved, and on the fault pattern. If the direction of slippage along the fault plane is used to distinguish between faults, then three types may be recognized:
1 dip-slip faults;
2 strike-slip faults;
3 oblique-slip faults.
In a dip-slip fault the slippage occurred along the dip, in a strike-slip fault it occurred along the strike, and in an oblique-slip fault movement occurred diagonally across the fault plane (Fig. 2.9). When the relative movement of the hanging and footwalls is used as a basis of classification, then normal, reverse, and wrench or tear faults can be recognized. The normal fault is characterized by the occurrence of the hanging wall on the downthrown side (Fig. 2.9a) whilst in the reverse fault the footwall occupies the down-thrown side. Reverse faulting involves a vertical duplication of strata (Fig. 2.9b), unlike

normal faults where the displacement gives rise to a region of barren ground. In the wrench fault neither the footwall nor the hanging wall have moved up or down in relation to one another (Fig. 2.9c). Considering the attitude of the fault to the strata involved, strike faults, dip (or cross-) faults, and oblique faults can be recognized. A strike fault is one which trends parallel to the beds it displaces, a dip or cross-fault is one which follows the inclination of the strata, and an oblique fault runs at an angle with the strike of the rocks it intersects. A classification based on the pattern produced by a number of faults does not take into account the effects on the rocks involved. Parallel faults, radial faults, peripheral faults, and *en echelon* faults are among the patterns which have been recognized.

In areas which have not undergone intense tectonic deformation reverse and normal faults generally dip at angles in excess of 45°. Their low-angled equivalents, termed thrusts and lags, are inclined at less than that figure. Splay faults occur at the extremities of strike-slip faults and strike-slip faults are commonly accompanied by numerous smaller parallel faults. Sinistral and dextral strike-slip faults can be distinguished in the following manner, if, when looking across a fault plane, the displacement on the far side has been to the left, it is sinistral; whereas if movement has been to the right, it is dextral.

Normal faults range in linear extension up to, occasionally, a few hundred kilometres in length. Generally, the longer faults do not form single fractures throughout their entirety but consist of a series of fault zones. The net slip on such faults may total over a thousand metres. Normal faults are commonly quite straight in outline but sometimes they may be sinuous or irregular with abrupt changes in strike. When a series of normal faults run parallel to one another with their downthrows all on the same side, the area involved is described as being step-faulted (Fig. 2.10). Horsts and rift structures (graben) are also illustrated in Figure 2.10.

An overthrust is a thrust fault which has an initial dip of 10° or less and its net slip measures several kilometres. Overthrusts may be folded or even overturned. Consequently, when they are subsequently eroded, remnants of the overthrust rocks may be left as outliers surrounded by rocks which lay beneath the thrust. These remnant areas are termed klippe and the area which separates them from the parent overthrust is referred to as a fenester or a window. The area which occurs in front of the overthrust is called the foreland.

(a)

(b)

Fig. 2.8 (a) Fault in strata of Limestone Group, Lower Carboniferous, near Howick, Northumberland, England; (b) fault zone formed from crushed flysch deposits (dominantly shales with some thin bands of sandstone), thrown against limestone; about 80 km northwest of Athens, Greece.

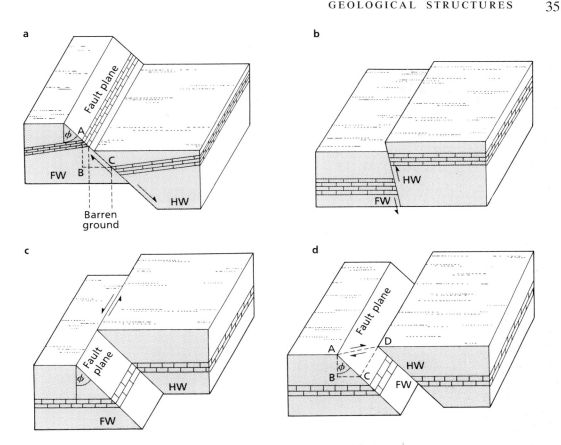

Fig. 2.9 Types of faults: (a) normal fault; (b) reverse fault; (c) wrench or strike-slip fault; (d) oblique-slip fault. FW, footwall; HW, hanging wall; AB, throw; BC, heave; φ, angle of hade. Arrows show the directions of relative displacement.

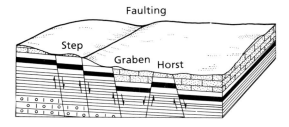

Fig. 2.10 Block diagram illustrating step-faulting and horst and graben structures.

2.2.2 Criteria for the recognition of faults

The abrupt ending of one group of strata against another may be caused by the presence of a fault, but abrupt changes also occur at unconformities and intrusive contacts. It is usually easy to distinguish

between these three relationships. Repetition of strata may be caused by faulting, that is, when the beds are repeated in the same order and dip in the same direction, whereas when they are repeated by folding they recur in the reverse order and may possess a different inclination (Fig. 2.11a). Omission of strata suggests that faulting has taken place, although such a feature could again occur as a result of unconformity (Fig. 2.11b).

Many features are characteristically associated with faulting and indicate the presence of a fault. Shear and tension joints are frequently associated with major faults and when formed along a fault are frequently referred to as feather joints because of their barblike appearance. Feather joints may be subdivided into pinnate shear joints and pinnate tension joints. Where pinnate shear planes are closely spaced and involve some displacement, then fracture cleavage is developed.

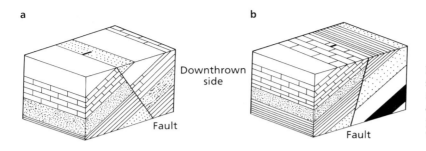

Fig. 2.11 (a) Repetition of bed at surface (fault parallel with strike and hading against dip); (b) omission of bed at surface (fault parallel with strike and hading with dip).

Slickensides are polished striated surfaces which occur on a fault plane and are caused by the frictional effects generated by its movement. Only slight movements are required to form slickensides and their presence has been noted along shear joints. The striations illustrate the general direction of movement. Very low scarps, sometimes less than a millimetre high, occur perpendicular to the striations and represent small accumulations of material formed as a consequence of the drag effect created by the movement of the opposing block. The shallow face of the scarp points in the direction in which the block moved. Sometimes two or more sets of slickensides, which usually intersect at an acute angle, may be observed indicating successive movements in slightly different directions or a sudden deviation in the movement during one displacement.

Intraformational shears, that is, zones of shearing parallel to bedding, are associated with faulting. They often occur in clays, mudstones, and shales at the contact with sandstones. Such shear zones tend to die out when traced away from the faults concerned and are probably formed as a result of flexuring of strata adjacent to faults. A shear zone may consist of a single, polished or slickensided shear plane. A more complex shear zone may be up to 300 mm in thickness. Intraformational shear zones are not restricted to argillaceous rocks, for example, they occur in chalk. Their presence means that the strength of the rock along the shear zone has been reduced to its residual value.

As a fault is approached, the strata involved frequently exhibit flexures which suggest that the beds have been dragged into the fault plane by the frictional resistance generated along it. Indeed, along some large dip-slip faults the beds may be inclined vertically. A related effect is seen in faulted gneisses and schists where a pre-existing foliation is strongly turned into the fault zone and a secondary foliation results.

If the movement along a fault has been severe, then the rocks involved may have been crushed (Fig. 2.8b), sheared, or pulverized (section 1.2.3). Where shales or clays have been faulted the fault zone may be occupied by clay gouge. Fault breccias (which consist of a jumbled mass of angular fragments containing a high proportion of voids), occur when more competent rocks are faulted. Crush breccias and conglomerates develop when rocks are sheared by a regular pattern of fractures. Movements of greater intensity are responsible for the occurrence of mylonite along a fault zone. The ultimate stage in the intensity of movements is reached with the formation of pseudotachylite.

Although a fault may not be observable, its effects may be reflected in the topography (Fig. 2.12), for example, if blocks are tilted by faulting, then a series of scarps are formed. If the rocks on either side of a fault are of different hardness, then a scarp may form along the fault as a result of differential erosion. Triangular facets may occur along a large fault scarp associated with an upland region. They represent the remnants left behind after swift-flowing streams have cut deep valleys into the scarp. Such streams initially deposit alluvial cones over the fault scarp. Scarplets are indicative of active faults and are found near the foot of mountains where they run parallel to the base of the range. Alternatively, natural escarpments may be offset by cross-faults. Stream profiles may be similarly interrupted by faults or, in a region of recent uplift, their courses may be relatively straight due to them following faults. Springs often occur along faults. A lake may form if a fault intersects the course of a river and the downthrown block is tilted upstream. Faults may be responsible for the formation of waterfalls in the path of a stream. Sag pools may be formed if the downthrown side settles different amounts along the strike of a fault. However, it must be emphasized that the physiographical features noted above may

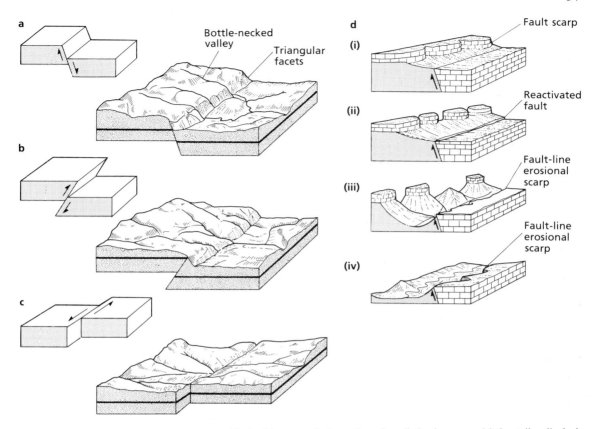

Fig. 2.12 (a) Fault scarp formed along normal fault; (b) reverse fault produces less distinctive scarp; (c) the strike-slip fault has produced a crush zone which is exploited by a stream. Drainage which once crossed the fault is now offset; (d) the formation of a fault-line scarp by differential erosion. The scarp ultimately (iv) does not reflect the original sense of displacement.

be developed without the aid of faulting and consequently they do not provide a foolproof indication of such stratal displacement.

Faults provide a path of escape and they are therefore frequently associated with mineralization, silicification, and igneous phenomena, for example, dykes are often injected along faults.

2.3 DISCONTINUITIES

A discontinuity represents a plane of weakness within a rock mass across which the rock material is structurally discontinuous. Although discontinuities are not necessarily planes of separation, most in fact are and they possess little or no tensile strength. Discontinuities vary in size from small fissures to huge faults. The most common discontinuities are

joints and bedding planes (Fig. 2.13). Other important discontinuities are planes of cleavage and schistosity.

2.3.1 Nomenclature of joints

Joints are fractures along which little or no displacement has occurred and are present within all types of rock. At the ground surface, joints may open as a consequence of denudation, especially weathering, and/or the dissipation of residual stress.

A group of joints which run parallel to each other are termed a joint set whilst two or more joint sets which intersect at a more or less constant angle are referred to as a joint system. If one set of joints is dominant, they are known as primary joints, the other set(s) being termed secondary. If joints are

Fig. 2.13 Discontinuities (i.e. joints and bedding planes) in Lincolnshire Limestone, Jurassic, near Ancaster, Lincolnshire, England.

planar and parallel or subparallel they are described as systematic. Conversely, when they are irregular they are termed non-systematic.

On the basis of size, joints can be divided into:
1 master joints, which penetrate several rock horizons and persist for hundreds of metres;
2 major joints, which are smaller joints but are still well-defined structures;
3 minor joints, which do not transcend bedding planes;
4 micro-joints are minute fractures that occasionally occur in finely bedded sediments, and may only be a few millimetres in size.

Joints may be associated with folds and faults, having developed towards the end of an active tectonic phase or when such a phase has subsided. However, joints do not appear to form parallel to other planes of shear failure such as normal and thrust faults. The orientation of joint sets in relation to folds depends upon their size, the type and size of the fold, and the thickness and competence of the rocks involved. At times the orientation of the joint sets can be directly related to the folding and may be defined in terms of the *a*, *b* and *c* axes of the 'tectonic cross' (Fig. 2.14). Those joints which cut the fold at right angles to the axis are called *ac* or cross-joints. The *bc* or longitudinal joints are perpendicular to the latter joints and diagonal or oblique joints make an angle with both the *ac* and the *bc* joints. Diagonal joints are classified as shear joints whereas *ac* and *bc* joints are regarded as tension joints.

Joints are formed through failure in tension, in shear, or through some combination of both. Rupture surfaces formed by extension tend to be clean and rough with little detritus. They tend to follow minor lithological variations. Simple surfaces of shearing are generally smooth and contain appreciable detritus. They are unaffected by local lithological changes.

Price (1966) contended that the majority of joints are post-compressional structures formed as a result of the dissipation of residual stress after folding has occurred. Some spatially restricted small joints

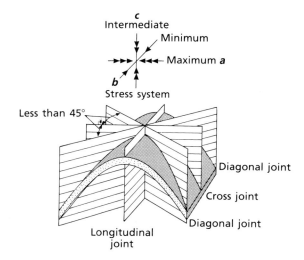

Fig. 2.14 Geometric orientation of longitudinal, cross-, and diagonal joints relative to fold axis and to principal stress axes, *a*, *b* and *c*.

associated with folds, such as radial tension joints, are probably initiated during folding. Such dissipation of the residual stresses occurs in the immediate neighbourhood of a joint plane so that a very large number of joints need to form in order to dissipate the stresses throughout a large volume.

Joints are also formed in other ways, for example, they develop within igneous rocks when they initially cool down, and in wet sediments when they dry out. The most familiar of these are the columnar joints in lava flows, sills, and some dykes. The cross-joints, longitudinal joints, diagonal joints, and flat-lying joints associated with large granitic intrusions have been described in Chapter 1. Sheet or mural joints have a similar orientation to flat-lying joints. When they are closely spaced and well developed they impart a pseudostratification to the host rock. It has been noted that the frequency of sheet jointing is related to the depth of overburden, that is, the thinner the rock cover the more pronounced the sheeting. This suggests a connection between removal of over-burden by denudation and the development of sheeting. For example, granitic intrusions may contain considerable residual strain energy and so with the gradual removal of load, residual stresses are dissipated by the formation of sheet joints.

2.3.2 Description of jointed rock masses

The shear strength of a rock mass and its deformability are very much influenced by the discontinuity pattern, its geometry, and how well it is developed. Observation of discontinuity spacing, whether in a field exposure or a core stick, aids appraisal of rock mass structure. In sedimentary rocks bedding planes are usually the dominant discontinuity and the rock mass can be described as shown in Table 2.3. The same boundaries can be used to describe the spacing of joints.

The mechanical behaviour of a rock mass is strongly influenced by the number of sets of discontinuities by which it is intersected, since this governs the amount of deformation that the rock mass will undergo. Systematic sets should be distinguished from non-systematic sets when recording discontinuities in the field. Barton (1978) suggested that the number of sets of discontinuities at any particular location could be described in the following manner:

1 Massive, occasional random joints
2 One discontinuity set
3 One discontinuity set plus random
4 Two discontinuity sets
5 Two discontinuity sets plus random
6 Three discontinuity sets
7 Three discontinuity sets plus random
8 Four or more discontinuity sets
9 Crushed rock, earth-like

As joints represent surfaces of weakness, the larger and more closely spaced they are, the more influential they become in reducing the effective strength of the rock mass. The persistence of a joint plane refers to its continuity. This is one of the most difficult properties to quantify since joints frequently continue beyond the rock exposure and consequently in such instances it is impossible to estimate their continuity. Nevertheless Barton (1978) suggested that the modal trace lengths measured for each discontinuity set can be described as follows:

Very low persistence	Less than 1 m
Low persistence	1–3 m
Medium persistence	3–10 m
High persistence	10–20 m
Very high persistence	Greater than 20 m

Block size provides an indication of how a rock mass is likely to behave, since block size and interblock shear strength determine the mechanical performance of a rock mass under given conditions of stress.

Table 2.3 Description of bedding plane and joint spacing

Description of bedding-plane spacing	Description of joint spacing	Limits of spacing
Very thickly bedded	Extremely wide	Over 2 m
Thickly bedded	Very wide	0.6–2 m
Medium bedded	Wide	0.2–0.6 m
Thinly bedded	Moderately wide	60 mm–0.2 m
Very thinly bedded	Moderately narrow	20–60 mm
Laminated	Narrow	6–20 mm
Thinly laminated	Very narrow	Under 6 mm

Table 2.4 Block size and equivalent discontinuity spacing

Term	Block size	Equivalent discontinuity spacings in blocky rock	Volumetric joint count $(J_V)^*$ (joints/m^3)
Very large	Over 8 m^3	Extremely wide	Less than 1
Large	0.2–8 m^3	Very wide	1–3
Medium	0.008–0.2 m^3	Wide	3–10
Small	0.0002–0.008 m^3	Moderately wide	10–30
Very small	Less than 0.0002 m^3	Less than moderately wide	Over 30

* After Barton (1978).

The following descriptive terms have been recommended for the description of rock masses in order to convey an impression of the shape and size of blocks of rock material (Barton, 1978):

1 Massive — few joints or very wide spacing
2 Blocky — approximately equidimensional
3 Tabular — one dimension considerably shorter than the other two
4 Columnar — one dimension considerably larger than the other two
5 Irregular — wide variations of block size and shape
6 Crushed — heavily jointed to 'sugar cube'

The block size may be described by using the terms given in Table 2.4 (Anon, 1977a).

Discontinuities, especially joints, may be open and the extent to which they are open (Table 2.5) is of importance in relation to the overall strength and permeability of a rock mass. Other joints may be partially or completely filled. The type and amount of filling not only influence the effectiveness with which the opposing joint surfaces are bound together,

thereby affecting the strength of the rock mass, but also influence permeability. If the infilling is sufficiently thick, the walls of the joint will not be in contact and hence the strength of the joint plane will be that of the infill material.

The nature of the opposing joint surfaces also influences rock mass behaviour as the smoother they are, the more easily can movement take place along them. However, joint surfaces are usually rough and may be slickensided. Hence, the nature of a joint surface may be considered in relation to its waviness, roughness, and the condition of the walls. Waviness and roughness differ in terms of scale and their effect on the shear strength of the joint. Waviness refers to first-order asperities which appear as undulations of the joint surface and are not likely to shear off during movement. Therefore the effects of waviness do not change with displacements along the joint surface. Waviness modifies the apparent angle of dip but not the frictional properties of the discontinuity. On the other hand, roughness refers to second-order asperities which are sufficiently small

Table 2.5 Description of the aperture of discontinuity surfaces

Anon (1977a)		Barton (1978)		
Description	Width of aperture	Description		Width of aperture
Tight	Zero	Closed	Very tight	Less than 0.1 mm
Extremely narrow	Less than 2 mm		Tight	0.1–0.25 mm
Very narrow	2–6 mm		Partly open	0.25–0.5 mm
Narrow	6–20 mm	Gapped	Open	0.5–2.5 mm
Moderately narrow	20–60 mm		Moderately wide	2.5–10 mm
Moderately wide	60–200 mm		Wide	Over 10 mm
Wide	Over 200 mm	Open	Very wide	10–100 mm
			Extremely wide	100–1000 mm
			Cavernous	Over 1 m

to be sheared off during movement. Increased roughness of the discontinuity walls results in an increased effective friction angle along the joint surface. These effects diminish or disappear when infill is present.

A set of terms to describe roughness has been suggested by Barton (1978) based upon two scales of observation:

1 small scale (several centimetres);
2 intermediate scale (several metres).

The intermediate scale of roughness is divided into stepped, undulating, planar, and the small scale of roughness, superimposed upon the former, includes rough (or irregular), smooth, and slickensided categories. The direction of the slickensides should be noted as shear strength may vary with direction. Barton recognized the classes shown in Figure 2.15.

The compressive strength of the rock comprising the walls of a discontinuity is a very important component of shear strength and deformability, especially if the walls are in direct rock-to-rock contact. Weathering (and alteration) frequently is concentrated along the walls of discontinuities, thereby reducing their strength. The weathered material can be assessed in terms of its grade and by index tests (Chapter 3).

Seepage of water through rock masses usually takes place via the discontinuities, although in some sedimentary rocks seepage through the pores may also play an important role. The prediction of groundwater level, probable seepage paths, and approximate water pressures frequently provides an indication of stability or construction problems. Barton (1978) suggested that seepage from open or filled discontinuities could be assessed according to the descriptive scheme shown in Table 2.6.

2.3.3 Strength of jointed rock masses and its assessment

Joints in a rock mass reduce its effective shear strength at least in a direction parallel with the discontinuities. Hence, the strength of jointed rocks is highly anisotropic. Joints offer no resistance to tension whereas they offer high resistance to compression. Nevertheless they may deform under compression if there are crushable asperities, compressible filling, or apertures along the joint or if the wall rock is altered.

Barton (1976) proposed the following empirical expression for deriving the shear strength (τ) along joint surfaces:

$$\tau = \sigma_n \tan (JRC \log_{10} (JCS/\sigma_n) + \phi_b) \qquad (2.1)$$

where σ_n is the effective normal stress, JRC is the joint roughness coefficient, JCS is the joint wall compressive strength, and ϕ_b is the basic friction angle. According to Barton, the values of the JRC range from 0 to 20, from the smoothest to the roughest surface (Fig. 2.16). The JCS is equal to the unconfined compressive strength of the rock if the joint is unweathered. This may be reduced by up to 75% when the walls of the joints are weathered. Both these factors are related as smooth-walled joints are less affected by the value of JCS, since failure of asperities plays a less important role. The smoother the walls of the joints, the more significant is the part played by its mineralogy (ϕ_b). The experience gained from rock mechanics indicates that under low, effective normal stress levels, such as occur in engineering, the shear strength of joints can vary within relatively wide limits. The maximum effective normal stress acting across joints considered critical for stability lies, according to Barton, in the range of 0.1 to 2.0 MPa.

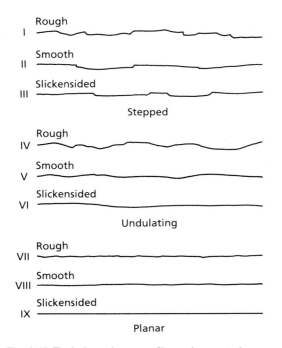

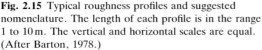

Fig. 2.15 Typical roughness profiles and suggested nomenclature. The length of each profile is in the range 1 to 10 m. The vertical and horizontal scales are equal. (After Barton, 1978.)

Table 2.6 Assessment of seepage from open and filled discontinuities. (After Barton, 1978.)

Open discontinuities		Filled discontinuities
Seepage rating	Description	Description
(1)	The discontinuity is very tight and dry, water flow along it does not appear possible	The filling material is heavily consolidated and dry, significant flow appears unlikely due to very low permeability
(2)	The discontinuity is dry with no evidence of water flow	The filling materials are damp but no free water is present
(3)	The discontinuity is dry but shows evidence of water flow, i.e. rust staining, etc.	The filling materials are wet, occasional drops of water
(4)	The discontinuity is damp but no free water is present	The filling materials show signs of outwash, continuous flow of water (estimate l/min)
(5)	The discontinuity shows seepage, occasional drops of water but no continuous flow	The filling materials are washed out locally, considerable water flow along outwash channels (estimate l/min and describe pressure, i.e. low, medium, high)
(6)	The discontinuity shows a continuous flow of water (estimate l/min and describe pressure, i.e. low, medium, high)	The filling materials are washed out completely, very high water pressures are experienced, especially on first exposure (estimate l/min and describe pressure)

Subsequently Hoek and Brown (1980) proposed that the peak triaxial compressive strengths of a wide range of rock materials could be described by the expression:

$$\sigma_1 = \sigma_3 + (m\sigma_c\sigma_3 + s\sigma_c^2)^{1/2} \qquad (2.2)$$

where σ_1 is the major principal stress at failure, σ_3 is the minor principal stress (the confining pressure in the case of a triaxial test), σ_c is the uniaxial compressive strength of the intact rock, and m and s are dimensionless constants which are approximately analogous to the angle of friction and cohesion. The constant, m, varies with rock type, ranging from about 0.001 for rock masses containing heavily weathered joint sets to about 25 for certain hard intact rock (Table 2.7). For intact rock, s is 1. In 1983 Hoek suggested that the shear strength, τ, along a discontinuous failure could be obtained from

$$\tau = (\cot \phi_i - \cos \phi_i) \frac{m\sigma_c}{8} \qquad (2.3)$$

where ϕ_i is the instantaneous angle of friction at given values of τ and σ^1.

2.3.4 Discontinuities and rock quality indices

Several attempts have been made to relate the numerical intensity of fractures to the quality of unweathered rock masses and to quantify their effect on deformability. For example, the concept of rock quality designation (RQD) was introduced by Deere (1964). It is based on the percentage core recovery when drilling rock with NX (57.2 mm) or larger-diameter diamond core drills. Assuming that a consistent standard of drilling can be maintained, the percentage of solid core obtained depends on the strength and degree of discontinuities in the rock mass concerned. The RQD is the sum of the core sticks in excess of 100 mm expressed as a percentage of the total length of core drilled. However, the RQD does not take account of the joint opening

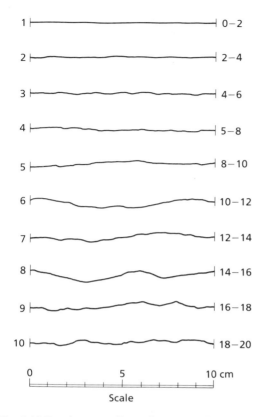

1	0–2
2	2–4
3	4–6
4	5–8
5	8–10
6	10–12
7	12–14
8	14–16
9	16–18
10	18–20

0   5   10 cm

Scale

Fig. 2.16 Roughness profiles and corresponding range of JRC values associated with each one. (After Barton, 1976.)

and condition, a further disadvantage being that with fracture spacings greater than 100 mm the quality is excellent irrespective of the actual spacing (Table 2.8). This particular difficulty can be overcome by using the fracture spacing index as suggested by Franklin *et al.* (1971). This simply refers to the frequency with which fractures occur per metre within a rock mass (Table 2.8).

The effect of discontinuities in a rock mass can be estimated by comparing the *in situ* compressional wave velocity with the laboratory sonic velocity of an intact core sample obtained from the rock mass. The difference in these two velocities is caused by the discontinuities which exist in the field. The velocity ratio, V_{cf}/V_{cl}, where V_{cf} and V_{cl} are the compressional wave velocities of the rock mass *in situ* and of the intact specimen respectively, was first proposed by Onodera (1963). For a high quality massive rock with only a few tight joints, the velocity ratio approaches unity. As the degree of jointing

and fracturing becomes more severe, the velocity ratio is reduced (Table 2.8). The sonic velocity is determined for the core sample in the laboratory under an axial stress equal to the computed overburden stress at the depth from which the rock material was taken, and at a moisture content equivalent to that assumed for the *in situ* rock. The field seismic velocity preferably is determined by uphole or crosshole seismic measurements in drillholes (Chapter 7) or test adits, since by using these measurements, it is possible to explore individual homogenous zones more precisely than by surface refraction surveys.

An estimate of the numerical value of the deformation modulus of a jointed rock mass can be obtained from various *in situ* tests. The values derived from such tests are always smaller than those determined in the laboratory from intact core specimens and the more heavily the rock mass is jointed, the larger the discrepancy between the two values. Thus, if the ratio between these two values of deformation modulus is obtained from a number of locations on a site, the engineer can evaluate the rock mass quality. Accordingly, the concept of the rock mass factor (j) was introduced by Hobbs (1975), who defined it as the ratio of the deformability of a rock mass to that of the deformability of the intact rock (Table 2.8).

2.3.5 Discontinuity surveys

One of the most widely used methods of collecting discontinuity data is simply by direct measurement on the ground. A direct survey can be carried out subjectively in that only those structures which appear to be important are measured and recorded. In a subjective survey the effort can be concentrated on the apparently significant joint sets. Nevertheless, there is a risk of overlooking sets which might be important. Conversely, in an objective survey all discontinuities intersecting a fixed line or area of the rock face are measured and recorded.

Several methods have been used for carrying out direct discontinuity surveys. In the fracture-set mapping technique all discontinuities occurring in 6 × 2 m zones, spaced at 30 m intervals along the face, are recorded. Alternatively, using a series of line scans provides a satisfactory method of joint surveying. The technique involves extending a metric tape across an exposure, levelling the tape, and then securing it to the face. Two other scan lines are set out as near as possible at right angles to the first, one

Table 2.7 Approximate relationship between rock mass quality and material constants. (After Hoek, 1983.)

Empirical failure criterion $\sigma'_1 = \sigma'_3 + (m\sigma_c\sigma'_3 + s\sigma_c^2)^{1/2}$ σ'_1 = major principal stress σ'_3 = minor principal stress σ_c = uniaxial compressive strength of intact rock m, s = empirical constants	Carbonate rocks with well developed crystal cleavage, e.g. dolostone, limestone, and marble	Lithified argillaceous rocks, e.g. mudstone, siltstone, shale, and slate (tested normal to cleavage)	Arenaceous rocks with strong crystals and poorly developed crystal cleavage, e.g. sandstone and quartzite	Fine-grained polyminerallic igneous crystalline rocks, e.g. andesite, dolerite, diabase, and rhyolite	Coarse-grained polyminerallic igneous and metamorphic crystalline rocks, e.g. amphibolite, gabbro, gneiss, granite, norite, and quartz diorite
Intact rock samples Laboratory size samples free from pre-existing fractures	$m = 7$ $s = 1$	$m = 10$ $s = 1$	$m = 15$ $s = 1$	$m = 17$ $s = 1$	$m = 25$ $s = 1$
Very good quality rock mass Tightly interlocking undisturbed rock with rough unweathered joints spaced at 1–3 m	$m = 3.5$ $s = 0.1$	$m = 5$ $s = 0.1$	$m = 7.5$ $s = 0.1$	$m = 8.5$ $s = 0.1$	$m = 12.5$ $s = 0.1$
Good quality rock mass Fresh to slightly weathered rock, slightly disturbed with joints spaced 1–3 m	$m = 0.7$ $s = 0.004$	$m = 1$ $s = 0.004$	$m = 1.5$ $s = 0.004$	$m = 1.7$ $s = 0.004$	$m = 2.5$ $s = 0.004$
Fair quality rock mass Several sets of moderately weathered joints spaced at 0.3–1 m disturbed	$m = 0.14$ $s = 0.0001$	$m = 0.20$ $s = 0.0001$	$m = 0.30$ $s = 0.0001$	$m = 0.34$ $s = 0.0001$	$m = 0.50$ $s = 0.001$
Poor quality rock mass Numerous weathered joints at 30–500 mm with some gouge. Clean, compacted rock fill	$m = 0.04$ $s = 0.00001$	$m = 0.05$ $s = 0.00001$	$m = 0.08$ $s = 0.00001$	$m = 0.09$ $s = 0.00001$	$m = 0.13$ $s = 0.00001$
Very poor quality rock mass Numerous heavily weathered joints spaced at 50 mm with gouge. Waste rock	$m = 0.007$ $s = 0$	$m = 0.010$ $s = 0$	$m = 0.015$ $s = 0$	$m = 0.017$ $s = 0$	$m = 0.025$ $s = 0$

Table 2.8 Classification of rock quality in relation to the incidence of discontinuities

Quality classification	RQD (%)	Fracture frequency per metre	Velocity ratio (V_{cf}/V_{cl})	Mass factor (j)
Very poor	0–25	Over 15	0.0–0.2	—
Poor	25–50	15–8	0.2–0.4	Less than 0.2
Fair	50–75	8–5	0.4–0.6	0.2–0.5
Good	75–90	5–1	0.6–0.8	0.5–0.8
Excellent	90–100	Less than 1	0.8–1.0	0.8–1.0

more or less vertical, the other horizontal. The distance along a tape at which each discontinuity intersects is noted, as is the direction of the pole to each discontinuity (this provides an indication of the dip direction). The dip of the pole from the vertical is recorded as this is equivalent to the dip of the plane from the horizontal. The strike and dip directions of discontinuities in the field can be measured with a compass and the amount of dip with a clinometer. Measurement of the length of a discontinuity provides information on its continuity. It has been suggested that measurements should be taken over distances of about 30 m, and to ensure that the survey is representative the measurements should be continuous over that distance. A minimum of at least 200 readings per locality is recommended to ensure statistical reliability. A summary of the other details which should be recorded about discontinuities is given in Figure 2.17.

The value of data on discontinuities gathered from orientated cores from drillholes depends in part on the quality of the rock concerned, in that poor quality rock is likely to be lost during drilling. It is impossible to assess the persistence, degree of separation, or the nature of the joint surfaces. What is more infill material, especially if it is soft, is not recovered by the drilling operations.

Core orientation can be achieved by using the Craelius core orientator or by integral sampling (Figs 2.18a and b). In the Craelius core orientator the teeth clamp the instrument in position inside of the core barrel. The housing contains a soft aluminium ring against which a ball bearing is indented by pressure from the cone when it reaches the bottom of the hole. The cone is released by pressure against the core stub and, when released, locks the probes in position and releases the clamping teeth to allow the instrument to ride up the barrel ahead of the core. In the first stage of integral sampling, a drillhole (diameter D') is drilled to a depth where the integral sample is to be obtained, then another hole (diameter D') is drilled coaxial with the former and with the same length as the required sample, into which a reinforcing bar is placed. The bar is then grouted to the rock mass. The integral sample is then obtained by overdrilling with diameter D. The method has been used with success in all types of rock masses, from massive to highly weathered varieties, and provides information on the spacing and orientation as well as the opening and infilling of discontinuities.

Drillhole inspection techniques include the use of drillhole periscopes, drillhole cameras, or closed-circuit television. The drillhole periscope affords direct inspection and can be orientated from outside the hole. However, its effective use is limited to about 30 m. The drillhole camera can also be orientated prior to photographing a section of the wall of a drillhole. The television camera provides a direct view of the drillhole and a recording can be made on videotape. These three systems are limited in that they require relatively clear conditions and so may be of little use below the water table, particularly if the water in the drillhole is murky. The televiewer produces an acoustic picture of the drillhole wall. One of its advantages is that drillholes need not be flushed prior to its use.

Many data relating to discontinuities can be obtained from photographs of exposures. Photographs may be taken looking horizontally at the rock mass from the ground or they may be taken from the air looking vertically, or occasionally obliquely, down at the outcrop. These photographs may or may not have survey control. Uncontrolled photographs are taken using hand-held cameras. Stereo-pairs are obtained by taking two photographs of the same face from positions about 5% of the distance of the face apart, along a line parallel to the face. Delineation of major discontinuity patterns and preliminary subdivision of the face into structural

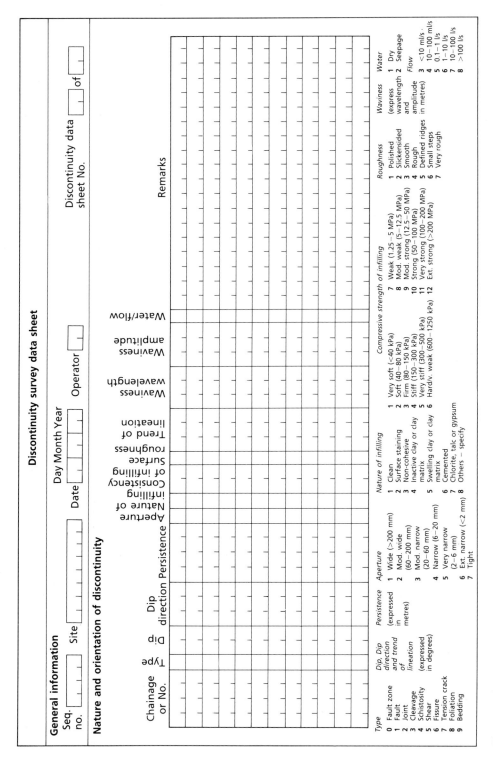

Fig. 2.17 Discontinuity survey data sheet. (After Anon, 1977a.)

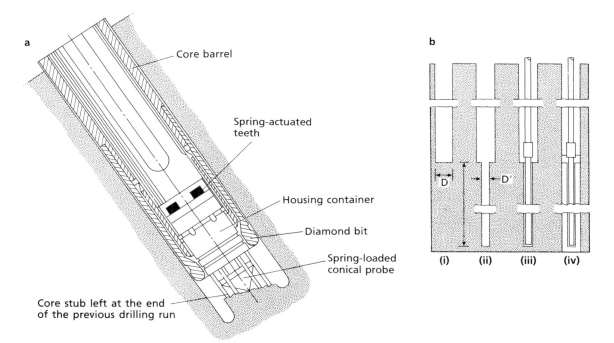

Fig. 2.18 (a) Details and method of operation of the Craelius core orientor. The probes take up the profile of the core stub left by the previous drilling run and are locked in position when the spring-loaded cone is released; (b) stages of the integral sampling method.

zones can be made from these photographs. Unfortunately, data cannot be transferred with accuracy from them onto maps and plans. Conversely, discontinuity data can be accurately located on maps and plans by using controlled photographs. Controlled photographs are obtained by aerial photography with complementary ground control or by ground-based phototheodolite surveys. Aerial photographs, with a suitable scale, have proved useful in the investigation of discontinuities. Photographs taken with a phototheodolite can also be used with a stereo-comparator which produces a stereoscopic model. Measurements of the locations or points in the model can be made with an accuracy of approximately 1 in 5000 of the mean object distance. As a consequence, a point on a face photographed from 50 m can be located to an accuracy of 10 mm. In this way the frequency, orientation, and continuity of discontinuities can be assessed. Such techniques prove particularly useful when faces which are inaccessible or unsafe have to be investigated.

2.3.6 Recording discontinuity data

Data from a discontinuity survey are usually plotted on a stereographic projection. The use of spherical projections, commonly the Schmidt or Wulf nets, means that traces of the planes on the surface of the 'reference sphere' can be used to define the dips and dip directions of discontinuity planes. In other words the inclination and orientation of a particular plane is represented by a great circle or a pole, normal to the plane, which are traced on an overlay placed over the stereonet. The method whereby great circles or poles are plotted on a stereogram has been explained by Hoek and Bray (1981). When recording field observations of the direction and amount of dip of discontinuities it is convenient to plot the poles rather than the great circles. The poles can then be contoured in order to provide an expression of orientation concentration. This affords a qualitative appraisal of the influence of the discontinuities on the engineering behaviour of the rock mass concerned (Fig. 2.19).

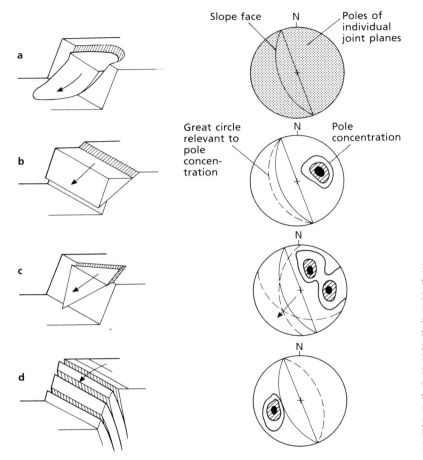

Fig. 2.19 Representation of
structural data concerning four
possible slope failure modes
plotted on equal-area stereonets
as poles, which are contoured to
show relative concentration, and
great circles: (a) Circular failure in
heavily jointed rock with no
identifiable structural pattern;
(b) plane failure in highly ordered
structure such as slate; (c) wedge
failure on two intersecting sets of
joints; (d) toppling failure caused
by steeply dipping joints. (After
Hock and Bray, 1981.)

3 · Surface processes

All landmasses are continually being worn away or denuded by weathering and erosion, the agents of erosion being the sea, rivers, wind, and ice. The detrital products resulting from denudation are transported by water, wind, ice, and the action of gravity, and are ultimately deposited. In this manner the surface features of the Earth are gradually, but constantly, changing. As landscapes are continually developing it is possible to distinguish successive stages in their evolution. These stages have been termed youth, maturity, and senility. However, the form of landscape which arises during any one of these stages is conditioned partly by the processes of denudation to which the area is subjected, and partly by the type of rocks and geological structure on which the landforms are being developed. Earth movements and type of climate also play a significant role in landscape development.

3.1 WEATHERING

Weathering of rocks is brought about by physical disintegration, chemical decomposition, and biological activity. The agents of weathering, unlike those of erosion, do not themselves provide for the transportation of debris from a rock surface. Therefore, unless this rock waste is otherwise removed it eventually acts as a protective cover, preventing further weathering taking place. If weathering is to be continuous, fresh rock exposures must be constantly revealed, which means that the weathered debris must be removed by the action of gravity, running water, wind, or moving ice.

The rate at which weathering takes place depends not only on the vigour of the weathering agent(s) but also on the durability of the rock itself (Fookes et al., 1988). The latter is governed by the mineralogical composition, texture and porosity of the rock, and incidence of discontinuities within the rock mass.

3.1.1 Mechanical weathering

Mechanical or physical weathering is particularly effective in climatic regions which experience significant diurnal changes of temperature. This does not necessarily imply a large range of temperature, as frost and thaw action can proceed where the range is limited.

As far as frost susceptibility is concerned the porosity, pore size, and degree of saturation all play an important role. When water turns to ice it increases in volume by up to 9% thus giving rise to an increase in pressure within the pores. This action is further enhanced by the displacement of pore water away from the developing ice front. Once ice has formed the ice pressures rapidly increase with decreasing temperature, so that at approximately −22°C ice can exert a pressure of up to 200 MPa. Usually coarse-grained rocks withstand freezing better than fine-grained types. The critical pore size for freeze−thaw durability appears to be about 0.005 mm. In other words rocks with larger mean pore diameters allow outward drainage and escape of fluid from the frontal advance of the ice line and are therefore less frost susceptible. Alternate freeze−thaw action causes cracks, fissures, joints, and some pore spaces to be widened. As the process advances, angular rock debris is gradually broken from the parent body.

The mechanical effects of weathering are well displayed in hot deserts, where wide diurnal ranges of temperature cause rocks to expand and contract. Because rocks are poor conductors of heat these effects are mainly localized in their outer layers where alternate expansion and contraction creates stresses which eventually rupture the rock. In this way flakes of rock break away from the parent material, the process being termed exfoliation. The effects of exfoliation are concentrated at the corners and edges of rocks so that their outcrops gradually become rounded (Fig. 3.1).

3.1.2 Chemical and biological weathering

Chemical weathering leads to mineral alteration and the solution of rocks. Alteration is principally effected by oxidation, hydration, hydrolysis, and carbonation whilst solution is brought about by acidified or alkalized waters. Chemical weathering also aids rock disintegration by weakening the rock fabric

Fig. 3.1 Weathering of granite, near Grunau, Namibia.

and by emphasizing any structural weaknesses, however slight, that it possesses. When decomposition occurs within a rock, the altered material frequently occupies a greater volume than that from which it was derived and in the process internal stresses are generated. If this swelling occurs in the outer layers of a rock, then it causes them to peel off from the parent body.

In dry air rocks decay very slowly. The presence of moisture hastens the rate tremendously, first, because water is itself an effective agent of weathering and, second, because it holds in solution substances which react with the component minerals of the rock. The most important of these substances are free oxygen, carbon dioxide, organic acids, and nitrogen acids.

Free oxygen is an important agent in the decay of all rocks which contain oxidizable substances, iron and sulphur being especially suspect. The rate of oxidation is quickened by the presence of water; indeed it may enter into the reaction itself, for example, in the formation of hydrates. However, its role is chiefly that of a catalyst. Carbonic acid is produced when carbon dioxide is dissolved in water and it may possess a pH value of about 5.7. The principal source of carbon dioxide is not the atmosphere but the air contained in the pore spaces in the soil where its proportion may be a hundred or so times greater than it is in the atmosphere. An abnormal concentration of carbon dioxide is released when organic matter decays. Furthermore, humic acids are formed by the decay of humus in soil

waters; they ordinarily have pH values between 4.5 and 5.0 but occasionally they may be less than 4.0.

The simplest reactions which take place on weathering are the solution of soluble minerals and the addition of water to substances to form hydrates. Solution commonly involves ionization, for example, when gypsum and carbonate rocks are weathered. Hydration and dehydration take place amongst some substances, a common example being gypsum and anhydrite:

$$CaSO_4 + 2H_2O = CaSO_4.2H_2O$$
(anhydrite)   (gypsum)

The above reaction produces an increase in volume of approximately 6% and accordingly causes the enclosing rocks to be wedged further apart. Iron oxides and hydrates are conspicuous products of weathering. Usually oxides are red and hydrates yellow to dark brown.

Sulphur compounds are readily oxidized by weathering. Because of the hydrolysis of the dissolved metal ion, solutions forming from the oxidation of sulphides are acidic. For instance, when pyrite is initially oxidized, ferrous sulphate and sulphuric acid are formed. Further oxidation leads to the formation of ferric sulphate. Very insoluble ferric oxide or hydrated oxide is formed if highly acidic conditions are produced.

Perhaps the most familiar example of a rock prone to chemical attack is limestone (Fookes and Hawkins, 1988). Limestones are chiefly composed of calcium carbonate and they are suspect to acid attack because CO_3 readily combines with H^+ to form the stable bicarbonate, HCO_3^-:

$$CaCO_3 + H_2CO_3 = Ca(HCO_3)_2$$

In water with a temperature of 25°C the solubility of calcium carbonate ranges from 0.01 to 0.05 g/l, depending upon the degree of saturation with carbon dioxide. Dolostone is somewhat less soluble than limestone. When a limestone is subject to dissolution any insoluble material present in it remains behind.

Weathering of the silicate minerals is primarily a process of hydrolysis. Much of the silica which is released by weathering forms silicic acid but where it is liberated in large quantities some of it may form colloidal or amorphous silica. Mafic silicates usually decay more rapidly than felsic silicates and in the process they release magnesium, iron, and lesser amounts of calcium and alkalis. Olivine is particularly unstable, decomposing to form serpentine, which on further weathering forms talc and carbonates. Chlorite is the commonest alteration product of augite (the principal pyroxene) and of hornblende (the principal amphibole).

When subjected to chemical weathering feldspars decompose to form clay minerals, the latter are consequently the most abundant residual products. The process is effected by the hydrolysing action of weakly carbonated water which leaches the bases out of the feldspars producing clays in colloidal form. The alkalis are removed in solution as carbonates from orthoclase (K_2CO_3) and albite (Na_2CO_3), and as bicarbonate from anorthite ($Ca(HCO_3)_2$). Some silica is hydrolysed to form silicic acid. Although the exact mechanism of the process is not fully understood the equation given below is an approximation towards the truth:

$$2KAlSi_3O_6 + 6H_2O + CO_2$$
(orthoclase)

$$= Al_2Si_2O_5(OH)_4 + 4H_2SiO_4 + K_2CO_3$$
(kaolinite)

The colloidal clay eventually crystallizes as an aggregate of minute clay minerals.

Clays are hydrated aluminium silicates and when they are subjected to severe chemical weathering in humid tropical regimes they break down to form laterite or bauxite. The process involves the removal of siliceous material and this is again brought about by the action of carbonated waters. Intensive leaching of soluble mineral matter from surface rocks takes place during the wet season. During the sub-sequent dry season groundwater is drawn to the surface by capillary action and minerals are precipitated there as the water evaporates. The minerals generally consist of hydrated peroxides of iron, and sometimes of aluminium, and very occasionally of manganese. The precipitation of these insoluble hydroxides gives rise to an impermeable lateritic soil. When this point is reached the formation of laterite ceases as no further leaching can occur. As a consequence, lateritic deposits are usually less than 7 m thick.

Plants and animals play an important role in the breakdown and decay of rocks, indeed their part in soil formation is of major significance. Tree roots penetrate cracks in rocks and gradually wedge the sides apart whilst the adventitious root system of grasses breaks down small rock fragments to particles of soil size. Burrowing rodents also bring about mechanical disintegration of weakened rocks. The action of bacteria and fungi is largely responsible for the decay of dead organic matter. Other bacteria are responsible, for the reduction of iron or sulphur compounds for example.

3.1.3 Tests of weatherability

The slake-durability test estimates the resistance to wetting and drying of a rock sample, particularly mudstones and rocks which exhibit a certain degree of alteration. In this test the sample, which consists of ten pieces of rock, each weighing about 40 g, is placed in a test drum, oven dried, and then weighed. After this, the drum, with sample, is half immersed in a tank of water and attached to a rotor arm which rotates the drum for a period of 10 min at 20 rev/min (Fig. 3.2). The cylindrical periphery of the drum is formed of 2 mm sieve mesh so that broken-down material can be lost whilst the test is in progress. After slaking, the drum and the material retained are dried and weighed. The slake-durability index is then obtained by dividing the weight of the sample retained by its original weight and expressing the answer as a percentage. The following scale is used:

Very low	Under 25%
Low	25−50%
Medium	50−75%
High	75−90%
Very high	90−95%
Extremely high	Over 95%

Taylor (1988) suggested that durable mudrocks were more easily distinguished from non-durable types on

Fig. 3.2 The slake-durability apparatus.

a basis of unconfined compressive strength and three-cycle slake-durability index (i.e. those mudrocks with a compressive strength exceeding 3.6 MPa and a three-cycle slake-durability index above 60% were regarded as durable).

Failure of consolidated and poorly cemented rocks occurs during saturation when the swelling pressure (or internal saturation swelling stress, σ_s), developed by capillary suction pressures exceeds their tensile strength. An estimate of σ_s can be obtained from the modulus of deformation (E):

$$E = \sigma_s/\epsilon_D \qquad (3.1)$$

where ϵ_D is the free-swelling coefficient. The latter is determined by a sensitive dial gauge recording the amount of swelling of an oven-dried core specimen along the vertical axis during saturation in water for 12 h, ϵ_D being obtained as follows:

$$\epsilon_D = \frac{\text{Change in length after swelling}}{\text{Initial length}} \qquad (3.2)$$

Olivier (1979) proposed the geodurability classification which is based on the free-swelling coefficient and uniaxial compressive strength (Fig. 3.3).

3.1.4 Engineering classification of weathering

Several attempts have been made to devise an engineering classification of weathered rock. The problem can be tackled in a number of ways. One method is to attempt to assess the grade of weathering by reference to some simple index test. Such methods provide a quantitative, rather than qualitative, assessment. When coupled with a grading system, this means that the disadvantages inherent in these simple index tests are largely overcome.

For instance, Iliev (1967) developed a coefficient of weathering (K) for granitic rock based upon the ultrasonic velocities of the rock material according to the expression:

$$K = (V_u - V_w)/V_u \qquad (3.3)$$

where V_u and V_w are the ultrasonic velocities of the fresh and weathered rock, respectively. A quantitative index indicating the grade of weathering as determined from the ultrasonic velocity and the corresponding coefficient of weathering is shown in Table 3.1.

After an extensive testing programme Irfan and Dearman (1978a) concluded that quick absorption, bulk density, point-load strength, and unconfined compressive strength prove reliable indices for the determination of a quantitative weathering index for granite (Table 3.2). This index can be related to the

Table 3.1 Ultrasonic velocity and grade of weathering

Grade of weathering	Ultrasonic velocity (m/s)	Coefficient of weathering
Fresh	Over 5000	0
Slightly weathered	4000–5000	0–0.2
Moderately weathered	3000–4000	0.2–0.4
Strongly weathered	2000–3000	0.4–0.6
Very strongly weathered	Under 2000	0.6–1.0

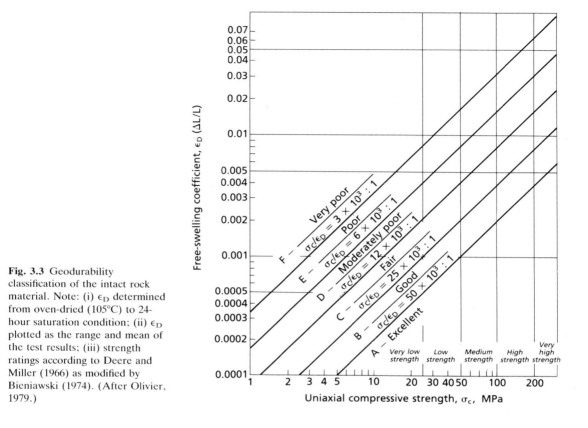

Fig. 3.3 Geodurability classification of the intact rock material. Note: (i) ϵ_D determined from oven-dried (105°C) to 24-hour saturation condition; (ii) ϵ_D plotted as the range and mean of the test results; (iii) strength ratings according to Deere and Miller (1966) as modified by Bieniawski (1974). (After Olivier, 1979.)

Table 3.2 Weathering indices for granite. (After Irfan and Dearman, 1978a.)

Type of weathering	Quick absorption (%)	Bulk density (t/m³)	Point-load strength (MPa)	Unconfined compressive strength (MPa)
Fresh	Less than 0.2	Over 2.61	Over 10	Over 250
Partially stained*	0.2–1.0	2.56–2.61	6–10	150–250
Completely stained*	1.0–2.0	2.51–2.56	4–6	100–150
Moderately weathered	2.0–10.0	2.05–2.51	0.1–4	2.5–100
Highly/completely weathered	Over 10	Less than 2.05	Less than 0.1	Less than 2.5

* Slightly weathered.

various grades of weathering recognized by visual determination and given in Table 3.3.

As mineral composition and texture influence the physical properties of a rock, petrographic techniques can be used to evaluate successive stages in mineralogical and textural changes brought about by weathering. Accordingly Irfan and Dearman (1978b)

also developed a quantitative method of assessing the grade of weathering of granite in terms of its megascopic and microscopic petrography. The megascopic factors included an evaluation of the amount of discoloration, decomposition, and disintegration shown by the rock. The microscopic analysis involved assessment of mineral composition

Table 3.3 Engineering grade classifications of weathered rock and their relation to engineering behaviour

Grade	Degree of decomposition	Field recognition (After Little, 1969; Fookes et al., 1972; Dearman, 1974)			Engineering behaviour		
		Rocks (mainly chemical decomposition)	Rocks (physical disintegration)*	Carbonate rocks (solution)	After Little (1969)	After Hobbs (1975)	After Martin and Hencher (1986)*
VI	Residual soil	The rock is discoloured and is completely changed to a soil in which the original fabric of the rock is completely destroyed. There is a large volume change	The rock is changed to a soil by granular disintegration and/or grain fracture. The structure of the rock is destroyed and the soil is a residuum of minerals unaltered from the original rock	Grades V and VI cannot occur. These grades can be applied to interbedded soluble and insoluble rocks. Void size should be recorded	Unsuitable for important foundations. Unstable on slopes when vegetation cover is destroyed and may erode easily unless hard cap is present. Requires selection before use as fill	In completely weathered rock and residual soil it may be possible to obtain fair quality samples depending upon the parent rock type and the consistency of the product. Generally the samples will tend to be less disturbed than when taken in the same rock in the highly weathered state. The bearing capacity and settlement characteristics of rock in these extreme states can be assessed using the usual methods for testing soils	A soil mixture with the original texture of the rock completely destroyed
V	Extremely weathered	The rock is discoloured and is wholly decomposed and friable, but the original fabric is mainly preserved. The properties of the rock mass depend in part on the nature of the parent rock. In granite rocks feldspars are completely kaolinized	The rock is changed to a soil by granular disintegration and/or fracture. The structure of the rock is preserved		Cannot be recovered as cores by ordinary rotary drilling methods. Can be excavated by hand or ripping without the use of explosives. Unsuitable for foundations of concrete dams or large structures. May be suitable for foundations of earth dams and for fill. Unstable in high cuttings at steep angles. New joint patterns may have formed. Requires erosion protection		No rebound from N. Schmidt hammer; slakes readily in water; geological pick easily indents when pushed into surface; rock is wholly decomposed but rock texture preserved

| IV | Highly weathered† | The rock is discoloured; discontinuities may be open and have discoloured surfaces (e.g. stained by limonite) and the original fabric of the rock near the discontinuities is altered; alteration penetrates deeply inwards, but corestones are still present. The rock mass is partially friable. Less than 50% rock | More than 50% and less than 100% of the rock is disintegrated by open discontinuities or spheroidal scaling spaced at 60 mm or less and/or by granular disintegration. The structure of the rock is preserved | More than 50% of the rock has been removed by solution. A small residuum may be present in the voids | Similar to Grade V. Sometimes recovered as core by careful rotary drilling. Unlikely to be suitable for foundations of concrete dams. Erratic presence of boulders makes it an unreliable foundation for large structures | In highly weathered rock difficulties will generally be encountered in obtaining undisturbed samples for testing. If samples are obtained the strength and modulus will generally be underestimated, frequently by large margins, even with apparently undisturbed samples. In such rocks in situ tests with either the Menard pressure meter or the plate should be carried out to determine the bearing capacity and settlement characteristics. The greatest difficulties in assessing bearing capacity and settlement are likely to be encountered in highly weathered rocks, in which the rock fabric becomes increasingly disintegrated or increasingly more plastic | N. Schmidt hammer rebound value 0 to 25; does not slake readily in water; geological pick cannot be pushed into surface; hand penetrometer strength index >250 kPa; rock weakened so that large pieces broken by hand; individual grains plucked from surface |

Continued on p. 56

Table 3.3 Continued

Grade	Degree of decomposition	Field recognition (After Little, 1969; Fookes et al., 1972; Dearman, 1974)			Engineering behaviour		
		Rocks (mainly chemical decomposition)	Rocks (physical disintegration)*	Carbonate rocks (solution)	After Little (1969)	After Hobbs (1975)	After Martin and Hencher (1986)*
III	Moderately weathered†	The rock is discoloured: discontinuities may be open and have greater discoloration with the alteration penetrating inwards; the intact rock is noticeably weaker, as determined in the field, than the fresh rock. The rock mass is not friable. 50–90% rock	Up to 50% of the rock is disintegrated by open discontinuities or by spheroidal scaling spaced at 60 mm or less and/or by granular disintegration. The structure of the rock is preserved	Up to 50% of the rock has been removed by solution. A small residuum may be present in the voids. The structure of the rock is preserved	Possessing some strength – large pieces (e.g. NX drill core) cannot be broken by hand. Excavated with difficulty without the use of explosives. Mostly crushes under bulldozer tracks. Suitable for foundations of small concrete structures and rock fill dams. May be suitable for semipervious fill. Stability in cuttings depends on structural features especially joint attitudes	In moderately weathered rock the intact modulus and strength can be very much lower than in the fresh rock and thus the j-value will be higher than in the fresh state, unless the joints and fractures have been opened by erosion or softened by the accumulation of weathering products. The intact modulus and strength can be measured in the laboratory and the bearing capacity assessed, in the same way as for fresh rock. Triaxial tests may be more appropriate than uniaxial tests, and it would be advisable to adopt conservative values for the factor of safety	N. Schmidt rebound value 25 to 45; considerably weathered but possessing strength such that pieces 55mm diameter cannot be broken by hand; rock material not friable

II	Slightly weathered	The rock may be slightly discoloured, particularly adjacent to discontinuities which may be open and have slightly discoloured surfaces; the intact rock is not noticeably weaker than the fresh rock. Some decomposed feldspar in granites. Over 90% rock	100% rock; discontinuities open and spaced at more than 60 mm	100% rock; discontinuity surfaces open. Very slight solution etching of discontinuity surfaces may be present	Requires explosives for excavation. Suitable for concrete dam foundations. Highly permeable through open joints. Often more permeable than the zones above or below. Questionable as concrete aggregate	In faintly and slightly weathered rock it is possible that the j-value, owing to the reduction in stiffness of the joints as a result of penetrative weathering alone, will show a fairly sharp decrease compared with that of the same rock in the fresh state. The intact modulus, by definition, is unaffected by penetrative weathering. The	N. Schmidt rebound value >45; more than one blow of geological hammer to break specimen; strength approaches that of fresh rock
I	Fresh rock	The parent rock shows no discoloration, loss of strength or other effects due to weathering	100% rock; discontinuities closed	100% rock; discontinuities closed	Staining indicates water percolation along joints; individual pieces may be loosened by blasting or stress relief and support may be required in tunnels and shafts	safe bearing capacity is not therefore affected by faint weathering, and may be only slightly affected by slight weathering	No visible signs of weathering

* Discontinuity spacing should be recorded.
† The ratio of the original rock to altered material should be estimated where possible.

Fig. 3.4 Excavation in highly weathered granite, Hong Kong.

and degree of alteration by modal analysis and a microfracture analysis.

Another method of assessing the grade of weathering is based on a simple description of the geological character of the rock concerned as seen in the field (Fig. 3.4), the description embodying different grades of weathering which are related to engineering performance (Table 3.3). This approach was first developed by Moye (1955), who proposed a grading system for the degree of weathering found in granite at the Snowy Mountains scheme in Australia.

Similar classifications were subsequently advanced and were mainly based on the degree of chemical decomposition exhibited by a rock mass. They were primarily directed towards weathering in granitic rocks. Dearman (1974) suggested descriptions which could be used to establish the grade of mechanical weathering, and that of solution weathering of relatively pure carbonate rock (Table 3.3). Others, working on different rock types, have proposed modified classifications of weathering grade, for example, Lovegrove and Fookes (1972) made slight variations in their identification of grades of weathering of volcanic tuffs and associated sediments in Fiji, as did Lee and De Freitus (1989) for weathered granites from Korea. Classifications of weathered chalk and weathered marl have been developed by Ward *et al.* (1968; see Table 5.30, p. 178) and Chandler (1969). Usually the grades will lie one above the other in a weathered profile developed from a single rock type, the highest grade being at

the surface. But this is not necessarily the case in complex geological conditions. Even so the concept of grade of weathering can still be applied. Such a classification can be used to produce maps showing the distribution of the grade of weathering at particular engineering sites. The dramatic effect of weathering upon the strength of rock is illustrated, according to grade of weathering, in Figure 3.5.

3.2 MOVEMENT OF SLOPES

3.2.1 Soil creep and valley bulging

Sharpe (1938) defined creep as the slow downslope movement of superficial rock or soil debris, which is usually imperceptible except by observations of long duration. Creep is a more or less continuous process, which is a distinctly surface phenomenon and occurs on slopes with gradients somewhat in excess of the angle of repose of the material involved. Like landslip, its principal cause is gravity although it may be influenced by seasonal changes in temperature and by swelling and shrinkage in surface rocks. Evidence of soil creep may be found on many soil-covered slopes. For example, it occurs in the form of small terracettes, downslope tilting of poles, the curving downslope of trees, and soil accumulation on the uphill sides of walls.

Solifluction is a form of creep which occurs in cold climates or high altitudes where masses of saturated rock waste move downslope. Generally, the bulk of the moving mass consists of fine debris but blocks of

Weathering description

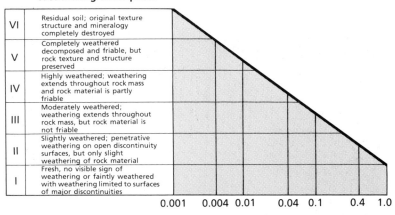

VI	Residual soil; original texture structure and mineralogy completely destroyed
V	Completely weathered decomposed and friable, but rock texture and structure preserved
IV	Highly weathered; weathering extends throughout rock mass and rock material is partly friable
III	Moderately weathered; weathering extends throughout rock mass, but rock material is not friable
II	Slightly weathered; penetrative weathering on open discontinuity surfaces, but only slight weathering of rock material
I	Fresh, no visible sign of weathering or faintly weathered with weathering limited to surfaces of major discontinuities

0.001   0.004 0.01   0.04 0.1   0.4   1.0

Strength reduction factor

Fig. 3.5 Strength reduction as a function of weathering. (After Stacey and Page, 1986.)

appreciable size may also be moved. Saturation may be due to either water from rain or melting snow. Moreover in periglacial regions water commonly cannot drain into the ground since it is frozen permanently. Solifluction differs from mudflow in that it moves much more slowly, the movement is continuous and it occurs over the whole slope.

Valley bulges consist of folds formed by mass movement of argillaceous material in valley bottoms (Fig. 3.6), the argillaceous material being overlain by thick, competent strata. The amplitude of the fold can reach 30 m in those instances where a single anticline occurs along the line of the valley. Alternatively, the valley floor may be bordered by a pair of reverse faults or a belt of small-scale folding. These features have been explained as a stress-relief phenomenon, that is, as stream erosion proceeded in the valley the excess loading on the sides caused the argillaceous material to squeeze out towards the area of minimum loading. This caused the rocks in the valley to bulge upwards. However, other factors may also be involved in the development of valley bulging such as high piezometric pressures, swelling clays or shales, and rebound adjustments of the stress field due to valley loading and excavation by ice.

The valleyward movement of argillaceous material results in cambering of the overlying competent strata, blocks of which become detached, and move down the hillside. Fracturing of cambered strata produces deep debris-filled cracks or 'gulls' which run parallel to the trend of the valley. Some gulls may be several metres wide. Small gulls are some-times found in relatively flat areas away from the slopes with which they are associated.

3.2.2 Landslides

Landsliding is one of the most effective and widespread mechanisms by which landscape is developed. It is of great interest to the engineer since an understanding of the causes of landsliding should help provide answers relating to the control of slopes, either natural or man-made. An engineer faced with a landslide is primarily interested in curing the harmful effects of the slide. In many instances the principal cause cannot be removed so that it may be more economical to alleviate the effects continually. In most landslides a number of causes contribute towards movement and any attempt to decide which one finally produced the failure is not only difficult but pointless. Often the final factor is nothing more than a trigger mechanism that sets in motion a mass which was already on the verge of failure.

Landslides represent the rapid downward and outward movement of slope-forming materials, the movement taking place by falling, sliding, or flowing, or some combination of these factors. Obviously this movement involves the development of a slip surface(s) between the separating and remaining masses. The majority of stresses found in most slopes are the sum of the gravitational stress from the self-weight of the material plus the residual stress.

Landslides occur because the forces creating movement, the disturbing forces (M_D), exceed those resisting it, the resisting forces (M_R), that is, the

Fig. 3.6 Valley bulging in interbedded shales and thin sandstones of Namurian age revealed during the excavation for the dam for Howden reservoir in 1933, South Yorkshire, England. (Courtesy of the Severn-Trent Water Authority.)

shear strength of the material concerned. In general terms the stability of a slope may be defined by a factor of safety (F) where:

$$F = M_R/M_D \qquad (3.4)$$

If the factor of safety exceeds one, then the slope is stable, whereas if it is less than one, the slope is unstable.

The common force tending to generate movements on slopes is, of course, gravity. Over and above this a number of causes of landslides can be recognized. These were grouped into two categories by Terzaghi (1950), namely internal causes and external causes. The former included those mechanisms within the mass which brought about a reduction of its shear strength to a point below the external forces imposed on the mass by its environment, thus inducing failure (Grainger and Harris, 1986). External mechanisms were those outside the mass involved, which were

responsible for overcoming its internal shear strength, thereby causing it to fail.

An increase in the weight of slope material means that shearing stresses are increased, leading to a decrease in the stability of a slope, which may ultimately give rise to a slide. This can be brought about by natural or artificial (man-made) activity. For example, removal of support from the toe of a slope, either by erosion or excavation, is a frequent cause of slides, as is overloading the top of a slope. Such slides are external slides in that an external factor causes failure. Other external mechanisms include earthquakes or other shocks and vibrations.

Internal slides are generally caused by an increase of pore water pressures within the slope material, which causes a reduction in the effective shear strength. It is generally agreed that in most landslides groundwater constitutes the most important single contributory cause. An increase in water content also means an increase in the weight of the slope material or its bulk density, which can induce slope failure. Significant volume changes may occur in some materials, notably clays, on wetting and drying out. Not only does this weaken the clay by developing desiccation cracks within it, but the enclosing strata may also be adversely affected.

Weathering can effect a reduction in strength of slope material, leading to sliding. The necessary breakdown of equilibrium to initiate sliding may take decades. In relatively impermeable cohesive soils the swelling process is probably the most important factor leading to a loss of strength and therefore to delayed failure.

A slope in dry frictional soil should be stable provided its inclination is less than the angle of repose. Slope failure tends to be caused by the influence of water, for example, seepage of groundwater through a deposit of sand in which slopes exist can cause them to fail. Failure on a slope composed of granular soil involves the translational movement of a shallow surface layer. The slip is often appreciably longer than it is in depth. This is because the strength of granular soils increases rapidly with depth. If, as is generally the case, there is a reduction in the density of the granular soil along the slip surface, then the peak strength is reduced ultimately to the residual strength. The soil will continue shearing without further change in volume once it has reached its residual strength. Although shallow slips are common, deep-seated shear slides can occur in granular soils.

In cohesive soils slope and height are interdependent and can be determined when the shear characteristics of the material are known. Because of their water-retaining capacity, due to their low permeability, excess pore water pressures are developed in cohesive soils. These pore pressures reduce the strength of the soil. Thus, in order to derive the strength of an element of the failure surface within a slope in cohesive soil, the pore water pressure at that point needs to be determined to obtain the total and effective stress. This effective stress is then used as the normal stress in a shear box or triaxial test to assess the shear strength of the clay concerned. Skempton (1964) showed that on a stable slope in clay the resistance offered along a slip surface, that is, its shear strength (s), is given by:

$$s = c' + (\sigma - u) \tan \phi' \qquad (3.5)$$

where c' is the cohesion intercept, ϕ' is the angle of shearing resistance (these are average values around the slip surface and are expressed in terms of effective stress), σ is the total overburden pressure and, u is the pore water pressure. In a stable slope only part of the total available shear resistance along a potential slip surface will be mobilized to balance the total shear force (τ), hence

$$\Sigma \tau = \Sigma c'/F + \Sigma (\sigma - u) \tan \phi'/F \qquad (3.6)$$

where F is the factor of safety. If the total shear force equals the total shear strength, then $F = 1$. Sliding occurs when F is below 1.

The factors which determine the degree of stability of steep slopes in hard unweathered crystalline rock (defined as rock with an unconfined compressive strength of 35 MPa and over) have been examined by Terzaghi (1962). Terzaghi contended that landsliding in such rock is largely dependent on the incidence, orientation, and nature of the discontinuities present. The value of the angle of shearing resistance (ϕ), required for a stability analysis, depends on the type and degree of interlock between the blocks on either side of the surface of sliding. Terzaghi concluded that the critical slope angle for slopes underlain by hard massive rocks with a random joint pattern is about 70°, provided the walls of the joints are not acted on by seepage pressures.

In a bedded and jointed rock mass, if the bedding planes are inclined, the critical slope angle depends upon their orientation in relation to the slope and the orientation of the joints (Hoek and Bray, 1981; Matheson, 1989). The relation between the angle of shearing resistance (ϕ) along a discontinuity, at which sliding will occur under gravity, and the inclination of the discontinuity (α) is important. If $\alpha < \phi$ the slope is stable at any angle, whilst if $\phi < \alpha$ then gravity will induce movement along the discontinuity surface and the slope cannot exceed the critical angle, which has a maximum value equal to the inclination of the discontinuities. It must be borne in mind, however, that rock masses are generally interrupted by more than one set of discontinuities.

3.2.3 Classification of landslides

Varnes (1978) classified landslides according to the type of movement undergone on the one hand and the type of materials involved on the other (Fig. 3.7). Types of movement were grouped into falls, slides, and flows; one can, of course, merge into another. The materials concerned were simply grouped as rocks and soils.

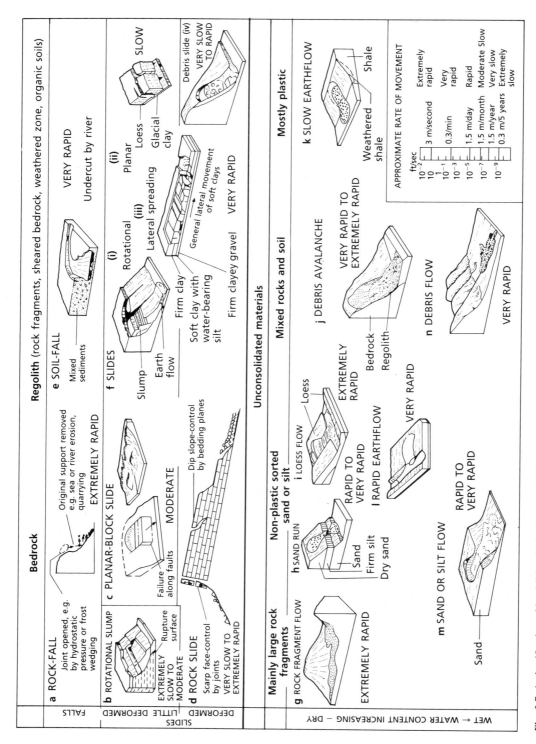

Fig. 3.7 A classification of landslides. (After Varnes, 1958.)

Falls are very common. The moving mass travels mostly through the air by free fall, by saltation, or rolling, with little or no interaction between the moving fragments. Movements are very rapid and may not be preceded by minor movements (Culshaw and Bell, 1992). In rockfalls the fragments are of various sizes and are generally broken in the fall (Fig. 3.8). They accumulate at the bottom of a slope as scree deposit. If a rockfall is active or very recent, then the slope from which it was derived is scarped. Frost−thaw action is one of the major causes of rockfall.

Toppling failure is a special type of rockfall which can involve considerable volumes of rock. The condition for toppling is defined by the position of the weight vector in relation to the base of the block. If the weight vector, which passes through the centre of gravity of the block, falls outside the base of the block, toppling will occur (Fig. 3.9). The danger of a slope toppling increases with increasing discontinuity angle and steep slopes in vertically jointed rocks

Fig. 3.8 Rockfall on slopes of Table Mountain, Cape Town, South Africa.

frequently exhibit signs of toppling failure.

In true slides the movement results from shear failure along one or several surfaces, such surfaces offering the least resistance to movement. The mass involved may or may not experience considerable deformation. One of the most common types of slide occurs in clay soils where the slip surface is approximately spoon-shaped. Such slides are referred to as rotational slides. They are commonly deep-seated (0.15 depth/length < 0.33). Although the slip surface is concave upwards it seldom approximates to a circular arc of uniform curvature. For example, if the shear strength of the soil is less in the horizontal than vertical direction, the arc may flatten out; if the soil conditions are reversed, then the converse may apply. Also, the shape of the slip surface is very much influenced by the existing discontinuity pattern.

Rotational slides usually develop from tension scars in the upper part of a slope, the movement being more or less rotational about an axis located above the slope. The tension cracks at the head of a rotational slide are generally concentric and parallel to the main scar. When the scar at the head of a rotational slide is almost vertical and unsupported, then further failure is usually just a matter of time. As a consequence successive rotational slides occur until the slope is stabilized (Fig. 3.10). These are retrogressive slides and they develop in a headward direction. All multiple retrogressive slides have a common basal shear surface in which the individual planes of failure are combined.

Translational slides occur in inclined stratified deposits, the movement occurring along a planar surface, frequently a bedding plane (Fig. 3.11). The mass involved in the movement becomes dislodged because the force of gravity overcomes the frictional resistance along the potential slip surface, the mass having been detached from the parent rock by a prominent discontinuity such as a major joint. Slab slides, in which the slip surface is roughly parallel to the ground surface, are a common type of translational slide. Such a slide may progress almost indefinitely if the slip surface is sufficiently inclined and the resistance along it is less than the driving force, whereas rotational sliding usually brings equilibrium to an unstable mass. Slab slides can occur on gentler surfaces than rotational slides and may be more extensive. Wedge failure is a type of translational failure in which two planar discontinuities intersect, the wedge as formed daylighting into the slope (Fig. 2.19c, p. 48).

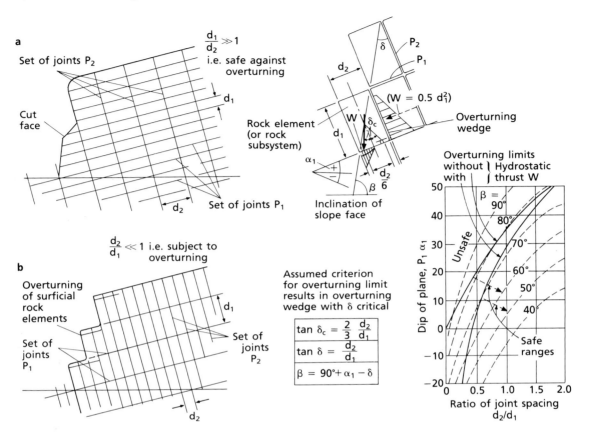

Fig. 3.9 Criteria for toppling failure in two dimensions. The diagram summarizes the criteria required for stability against overturning for two-dimensional conditions in terms of mean spacing and orientation of joint sets, slope angle, and hydrostatic forces (if present).

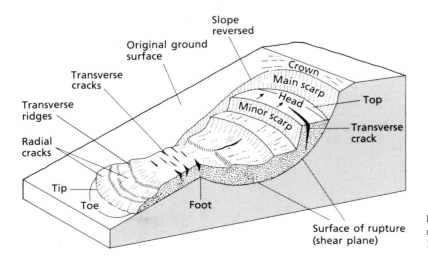

Fig. 3.10 The main features of a rotational slide. (After Varnes, 1978.)

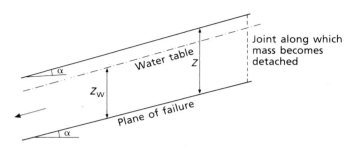

Fig. 3.11 A translational slide. Z, depth of plane of failure below surface; Z_w, depth of plane of failure below water table; α, angle of inclination of plane of failure and surface. In a translational slide it is assumed that the potential plane of failure lies near to and parallel to the surface. The water table also is inclined parallel to the surface. If the water table creates a hydrostatic component of pressure on the slip surface with flow out of the slope, then it has been shown that the factor of safety (F) is as follows: $F = c' + (\gamma Z \cos^2\alpha - \gamma_w Z_w) Z \tan \phi'/\gamma Z \sin \alpha \cos \alpha$ where γ_w is the unit weight of water and c' and ϕ' are the effective cohesion and angle of shearing resistance, respectively.

Rock slides and debris slides are usually the result of a gradual weakening of the bonds within a rock mass and are generally translational in character. Most rock slides are controlled by the discontinuity patterns within the parent rock. Water is seldom an important direct factor in causing rock slides although it may weaken bonding along joints and bedding planes. Freeze—thaw action, however, is an important cause. Rock slides commonly occur on steep slopes and most of them are of single rather than multiple occurrence. They are composed of rock boulders. Individual fragments may be very large and may move great distances from their source. Debris slides are usually restricted to the weathered zone or to surficial talus. With increasing water content debris slides grade into mudflows. These slides are often limited by the contact between loose material and the underlying firm bedrock.

In a flow the movement resembles that of a viscous fluid. Slip surfaces are usually not visible or are short lived and the boundary between the flow and the material over which it moves may be sharp or may be represented by a zone of plastic flow. Some content of water is necessary for most types of flow movement but dry flows can occur. Dry flows, which consist predominantly of rock fragments, are simply referred to as rock-fragment flows or rock avalanches and generally result from a rock slide or rockfall turning into a flow. They are generally very rapid and short-lived, and frequently are composed mainly of silt or sand. As would be expected they are of frequent occurrence in rugged mountainous regions where they usually involve the movement of many millions of tonnes of material. Wet flows occur when fine-grained soils, with or without coarse debris,

become mobilized by an excess of water. They may be of great length.

Progressive failure is rapid in debris avalanches and the whole mass, either because it is quite wet or is on a steep slope, moves downwards, often along a stream channel, and advances well beyond the foot of a slope. Debris avalanches are generally long and narrow and frequently leave V-shaped scars tapering headwards. These gulleys often become the sites of further movement.

Debris flows are distinguished from mudflows on a basis of particle size, the former containing a high percentage of coarse fragments, whilst the latter consist of at least 50% sand-size or less. Almost invariably debris flows follow unusually heavy rainfall or sudden thaw of frozen ground. These flows are of high density, perhaps 60 to 70% solids by weight, and are capable of carrying large boulders. Like debris avalanches they commonly cut V-shaped channels, at the sides of which coarser material may accumulate as the more fluid central area moves down channel. Both debris flows and mudflows may move over many kilometres.

Mudflows may develop when a rapidly moving stream of storm water mixes with a sufficient quantity of debris to form a pasty mass (Fig. 3.12). Because such mudflows frequently occur along the same courses, they should be kept under observation when significant damage is likely to result. Mudflows frequently move at rates ranging between 10 and 100 m/min and can travel over slopes inclined at 1° or less. They usually develop on slopes with shallow inclinations, that is, between 5 and 15°. Movement involves the development of forward thrusts due to undrained loading of the rear part of the mudflow

Fig. 3.12 Mudflow across a road near Pinetown, Natal, South Africa.

where the basal shear surface is inclined steeply downwards. A mudflow continues to move down shallow slopes due to this undrained loading, which is implemented by frequent small falls or slips of material from a steep rear scarp on the head of the moving mass. This not only aids instability by loading, but it also raises the water pressures along the back part of the slip surface.

An earthflow involves mostly cohesive or fine-grained material which may move slowly or rapidly (Pyles *et al.*, 1987) The speed of movement is to some extent dependent on water content in that the higher the content, the faster the movement. Slowly moving earthflows may continue to move for several years. These flows generally develop as a result of a build-up of pore water pressure, so that part of the weight of the material is supported by interstitial water with consequent decrease in shearing resistance. If the material is saturated a bulging frontal lobe is formed and this may split into a number of tongues which advance with a steady rolling motion. Earthflows frequently form the spreading toes of rotational slides due to the material being softened by the ingress of water.

3.3 FLUVIAL PROCESSES

3.3.1 The development of drainage systems

It is assumed that the initial drainage pattern which develops on a new surface consists of a series of subparallel rills flowing down the steepest slopes.

The drainage pattern then becomes integrated by micropiracy (the beheading of the drainage system of a small rill by that of a larger) and cross-grading. Micropiracy occurs when the ridges which separate the initial rills are overtopped and broken down. When the divides are overtopped the water tends to move towards those rills at a slightly lower elevation and in the process the divides are eroded. Eventually water drains from rills of higher elevation into adjacent ones of lower elevation (Fig. 3.13). The flow towards the master rill steadily increases and its development across the main gradient is termed cross-grading. The tributaries which flow into the master stream are subsequently subjected to cross-grading and so a dendritic system is developed.

The texture of the drainage system is influenced by rock type and structure, the nature of the vegetation cover and the type of climate. The drainage density affords a measure of comparison between the development of one drainage system and another. It is calculated by dividing the total length of a stream by the area it drains, and is generally expressed in kilometres per square kilometre.

Horton (1945) classified streams into orders. First-order streams are unbranched, and when two such streams become confluent they form a second-order stream. It is only when streams of the same order meet that they produce one of the higher rank, for example, a second-order stream flowing into a third-order stream does not alter its rank. The frequency with which streams of a certain order flow into those of the next order above them is referred to as the

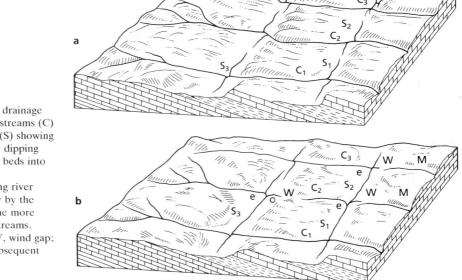

Fig. 3.13 (a) Trellised drainage pattern of consequent streams (C) and their subsequents (S) showing the erosion of a gently dipping series of hard and soft beds into escarpments; (b) later development illustrating river capture or micropiracy by the headward growth of the more vigorous subsequent streams. e, elbow of capture; W, wind gap; M, misfit stream; o, obsequent stream.

bifurcation ratio. The bifurcation ratio for any consecutive pair of orders is obtained by dividing the total number of streams of the lower order by the total number in the next higher order. Similarly the stream-length ratio is found by dividing the total length of streams of the lower order by the total length of those in the next higher order. Values of stream-length ratio depend mainly on drainage density and stream-entrance angles, and increase somewhat with increasing order. A river system is also assigned an order, which is defined numerically by the highest stream order it contains.

In the early stages of development in particular, rivers tend to accommodate themselves to the local geology, for example, tributaries may develop along fault zones. What is more, rock type has a strong influence on the drainage texture or channel spacing, that is, a low drainage density tends to form on resistant or permeable rocks whereas weak highly erodible rocks are characterized by a high drainage density.

The initial dominant action of master streams is vertical downcutting which is accomplished by the formation of pot-holes, which ultimately coalesce, and by the abrasive action of the load. Hence in the early stages of river development the cross-profile of the valley is sharply V-shaped. As time proceeds valley widening due to soil creep, slippage, rainwash,

and gullying becomes progressively more important and eventually lateral corrasion replaces vertical erosion as the dominant process. A river possesses few tributaries in the early stages, but as the valley widens their numbers increase, which affords a growing increment of rock waste to the master stream thereby enhancing its corrasive power.

During valley widening the stream erodes the valley sides by causing undermining and slumping to occur on the outer concave curves of meanders where steep cliffs or bluffs are formed. These are most marked on the upstream side of each spur. Deposition usually takes place on the convex side of meanders. They migrate both laterally and downstream, and their amplitude is progressively increased. In this manner spurs are continually eroded, first becoming more asymmetrical until they are eventually truncated (Fig. 3.14). The slow deposition which occurs on the convex side of a meander will, as lateral migration proceeds, produce a gently sloping area of alluvial ground called the flood plain. The flood plain gradually grows wider as the river bluffs recede until it is as broad as the amplitude of the meanders. It was at this period in river development that Davis (1909) regarded maturity as having been reached (Fig. 3.15). From now onwards the continual migration of meanders slowly reduces the valley floor to an almost flat plain which slopes

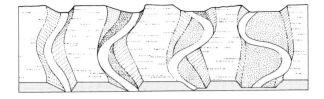

Fig. 3.14 Widening of valley floor by lateral corrasion.

gently downstream and is bounded by shallow valley sides.

Throughout its length a river channel has to adjust to several factors which change independently of the channel itself. These include the different rock types and structures across which it flows. The tributaries and inflow of water from underground sources affect the long profile but are independent of the channel. Other factors which bring about adjustment of a river channel are: flow resistance which is a function of particle size and the shape of transitory deposits such as bars, the method of load transport, and the channel pattern, including meanders and islands. Lastly, the river channel must also adjust itself to the river slope, width, depth, and velocity.

As the longitudinal profile or thalweg of a river is developed, the differences between the upper, middle, and lower sections of its course become more clearly defined until three distinctive tracts are observed. These are the upper or torrent, the middle or valley, and the lower or plain stage. The torrent stage includes the headstreams of a river where small fast-flowing streams are engaged principally in active downwards and headwards erosion. They possess steep-sided cross-profiles and irregular thalwegs. The initial longitudinal profile of a river reflects the irregularities which occur in its path. For instance, it may exhibit waterfalls or rapids where it flows across resistant rocks. However, such features are transient in the life of a river. In the valley tract the predominant activity is lateral corrasion. The shape of the valley sides depends upon the nature of the rocks being excavated, the type of climate, the rate of rock wastage, and meander development (Fig. 3.16). Some reaches in the valley tract may approximate to grade and there the meanders may have developed alluvial flats, whilst other stretches may be steep-sided with irregular longitudinal profiles. The plain tract is formed by the migration of meanders and deposition is the principal river activity.

Meanders, although not confined to, are characteristic of, flood plains. The consolidated veneer of alluvium, spread over a flood plain, offers little resistance to continual meander development, so the loops become more and more accentuated. As time proceeds the swelling loops approach one another. During flood the river may cut through the neck separating two adjacent loops and thereby straightens its course. As it is much easier for the river to flow through this new course, the meander loop is silted off and abandoned as an oxbow lake (Fig. 3.17).

Meander lengths vary from 7 to 10 times the width of the channel whilst crossovers occur at about every 5 to 7 channel widths. The amplitude of a meander

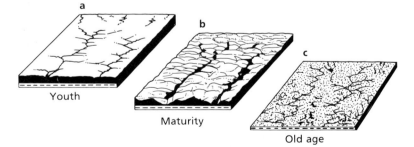

Fig. 3.15 Diagram illustrating the three main periods in the denudation of an uplifted land surface according to the Davisian interpretation of the 'normal' cycle of erosion: (a) in youth, parts of the initial surface survive; (b) in maturity, most or all of the initial surface has vanished and the landscape is mainly slopes, apart from valley floors; (c) in old age or senility, the landscape becomes subdued and gently undulating, rising only to residual hills representing the divides between adjoining drainage basins. Eventually such hills are worn down and the region becomes a peneplain.

Fig. 3.16 Fish River canyon, Namibia. The canyon owes its shape to a number of factors which include strata of differing hardness and attitude (horizontally lying rocks overlying steeply inclined strata), erosion in an arid region being largely restricted to infrequent periods after heavy rainfall, and rejuvenation.

bears little relation to its length but is largely determined by the erosion characteristics of the river bed and local factors. For example, in uniform material the amplitude of meanders does not increase progressively nor do meanders form oxbow lakes during the downstream migration of bends. Higher sinuosity is associated with small width relative to depth and a larger percentage of silt and clay, which affords greater cohesiveness, in the river banks. Relatively sinuous channels with a low width−depth ratio are developed by rivers transporting large quantities of suspended sediment. By contrast, the channel tends to be wide and shallow, and less sinuous, when the amount of bedload discharge is high.

A river is described as being braided if it splits into a number of separate channels or anabranches to adjust to a broad valley. The areas between the anabranches are occupied by islands built of gravel and sand. For the islands to remain stable the river banks must be more erodable so that they give way

rather than the islands. Braided channels occur on steeper slopes than do meanders.

Climatic changes and earth movements alter the base level to which a river grades. When a land surface is elevated the downcutting activity of rivers flowing over it is accelerated. The rivers begin to regrade their courses from their base level and as time proceeds their newly graded profiles are extended upstream until they are fully adjusted to the new conditions. Until this time the old longitudinal profile intersects with the new to form a knick point. The upstream migration of knick points tends to be retarded by outcrops of resistant rock. Consequently, after an interval of time they are usually located at hard rock exposures. The acceleration of downcutting consequent upon uplift frequently produces a new valley within the old, the new valley extending upstream to the knick point.

River terraces are also developed by rejuvenation. In the lower course of a river uplift leads to the river

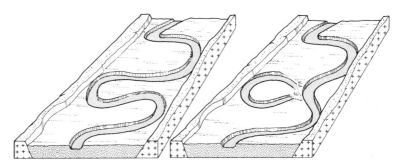

Fig. 3.17 Formation of oxbow lake.

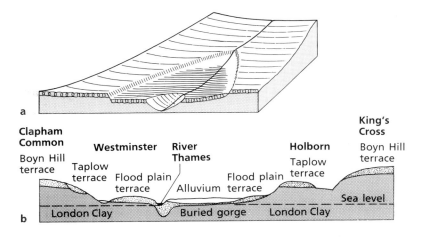

Fig. 3.18 (a) Paired river terraces due to rejuvenation, note valley in valley and knick points; (b) section across London to show the paired terraces and one of the buried 'gorges' of the Thames valley.

cutting into its alluvial plain. The lateral and downstream migration of meanders means that a new flood plain is formed but very often paired alluvial terraces, representing the remnants of the former flood plain, are left at its sides (Fig. 3.18).

Incised meanders are also associated with rejuvenation and are often found together with river terraces. When uplift occurs the downcutting action of meanders is accelerated and they carve themselves into the terrain over which they flow. The landforms which are then produced depend upon the character of the terrain and the relative rates of downcutting and meander migration. If vertical erosion is rapid, meander shift has little opportunity to develop and consequently the loops are not greatly enlarged. The resulting incised meanders are described as entrenched. However, when time is afforded for meander migration, they incise themselves by oblique erosion and the loops are enlarged; they are then referred to as ingrown meanders.

When incision occurs in the alluvium of a river plain the meanders migrate back and forth across the floor. On each successive occasion that a meander swings back to the same side, it does so at a lower level, thus small remnant terraces may be left above the newly formed plain. These terraces are not paired across the valley and their position and preservation depends on the swing of meanders over the valley. If downcutting is very slow, then erosion terraces are unlikely to be preserved.

Conversely, when sea level rises rivers again have to regrade their courses to the new base level, for example, during glacial times the sea level was much lower and rivers carved out valleys accordingly. As the last glaciation retreated, so the sea level rose and the rivers had to adjust to these changing conditions. This frequently meant that their valleys were filled with sediments. Hence buried channels are associated with the low-land sections of many rivers.

3.3.2 The work of rivers

The velocity of a river depends upon channel gradient, volume and configuration. One of the most commonly used equations applicable to open channel hydraulics is the Chezy formula:

$$v = C\sqrt{(Rs)} \qquad (3.7)$$

where v is the mean velocity, C is a coefficient which varies with the characteristics of the channel, R is the hydraulic radius, and s is the slope. Numerous attempts have been made to find a generally acceptable expression for C.

The Manning formula, based on field and experimental determinations of the value of the resistance coefficient (n) is also widely used:

$$v = \frac{1.49}{n}R^{2/3}s^{1/2} \qquad (3.8)$$

The experimental values of n vary from approximately 0.01 for smooth metal surfaces to 0.06 for natural, irregular channels containing large stones. Some values of n used in the Manning formula are as follows:

1 clean straight channel with no pools: 0.025−0.033
2 channel containing weeds and stones: 0.030−0.040
3 channel containing large stones: 0.045−0.060

The ratio between the cross-sectional area of a river channel and the length of its wetted perimeter determines the efficiency of the channel. This ratio is termed the hydraulic radius, and the higher its value, the more efficient is the river. The most efficient forms of channel are those with approximately circular or rectangular sections with widths approaching twice their depths; the most inefficient channel forms are very broad and shallow with wide wetted perimeters.

The quantity of flow can be estimated from measurements of cross-sectional areas and current speed of a river. Generally channels become wider relative to their depth and adjusted to larger flows with increasing distance downstream. Bankfull discharges also increase downstream in proportion to the square of the width of the channel or of the length of individual meanders, and in proportion to the 0.75 power of the total drainage area focused at the point in question.

Statistical methods are used to predict river flow and assume that recurrence intervals of extreme events bear a consistent relationship to their magnitudes. A recurrence interval, generally of 50 or 100 years, is chosen in accordance with given hydrological requirements. Sherman's (1932) concept of unit hydrograph postulates that the most important hydrological characteristics of any basin can be seen from the direct runoff hydrograph resulting from 25 mm of rainfall evenly distributed over 24 h. This is produced by drawing a graph of the total stream flow at a chosen point as it changes with time after such a storm, from which the normal baseflow caused by groundwater is subtracted (Fig. 3.19).

There is a highly significant relationship between mean annual flood discharge per unit area and drainage density. Peak discharge and the lag time of discharge (the time which elapses between maximum precipitation and maximum runoff) are also influenced by drainage density, as well as by the shape and slope of the drainage basin. Stream flow is generally most variable and flood discharges at a maximum per unit area in small basins. This is because storms tend to be local in occurrence. An average relationship exists between drainage density and baseflow or groundwater discharge. This is related to the permeability of the rocks present in a drainage basin, that is, the greater the quantity of water which moves on the surface of the drainage system, the higher the drainage density, which in turn means that the baseflow is lower. As pointed out previously, in areas of high drainage density the soils and rocks are relatively impermeable and water runs off rapidly. The amount of infiltration is reduced accordingly.

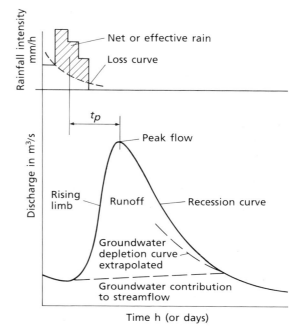

Fig. 3.19 Component parts of a hydrograph. When rainfall commences there is an initial period of interception and infiltration before any measurable runoff reaches the stream channels. During the period of rainfall these losses continue in a reduced form so that the rainfall graph has to be adjusted to show effective rain. When the initial losses are met, surface runoff begins and continues to a peak value which occurs at time tp, measured from the centre of gravity of the effective rain on the graph. Thereafter surface runoff declines along the recession limb until it disappears. Baseflow represents the groundwater contribution along the banks of the river.

The work undertaken by a river is threefold: it erodes rocks, transports the products thereof, and eventually deposits them (Fig. 3.20). Erosion occurs when the force provided by the river flow exceeds the resistance of the material over which it runs. The velocity needed to initiate movement is appreciably higher than that required to maintain movement. Four types of fluvial erosion have been distinguished: (i) hydraulic action, (ii) attrition, (iii) corrasion, and (iv) corrosion. Hydraulic action is the force of the water itself. Attrition is the disintegration which occurs when two or more particles which are suspended in water collide. Corrasion is the abrasive action of the load carried by a river on its channel. Most of the erosion done by a river is attributable to corrosive action. Hence a river carrying coarse, re-

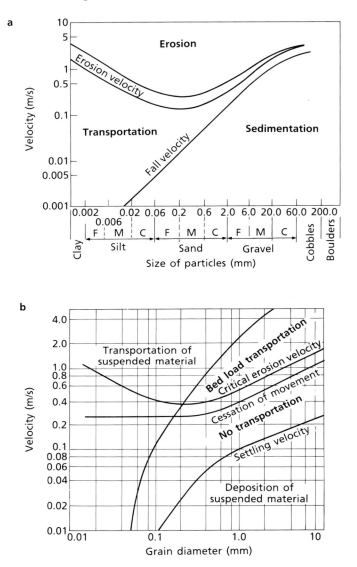

Fig. 3.20 (a) Curves for erosion, transportation, and deposition of uniform sediment. F, fine; M, medium; C, coarse. Note that fine sand is the most easily eroded material. (After Hjulstrom, 1935.) (b) Relation between flow velocity, grain size, entrainment, and deposition for uniform grains with a specific gravity of 2.65. The velocities are those 1.0 m above the bottom of the body of water. The curves are not valid for high sediment concentrations, which increase the fluid viscosity. (After Sundberg, 1956.)

sistant, angular rock debris possesses a greater ability to erode than does one transporting fine particles in suspension. Corrosion is the solvent action of river water.

In the early stages of river development erosion tends to be greatest in the lower part of the drainage basin. However, as the basin develops the zone of maximum erosion moves upstream and it is concentrated along the divides in the later stages. The amount of erosion accomplished by a river in a given time depends upon its volume and velocity of flow, the character and size of its load, the rock type and geological structure over which it flows, the infiltration capacity of the area it drains, the vegetation (which affects soil stability), and the permeability of the soil. The volume and velocity of a river influences the quantity of energy it possesses. When flooding occurs the volume of a river is greatly increased which leads to an increase in its velocity and competence. However, much energy is spent in overcoming the friction between the river and its channel so that energy losses increase with any increase in channel roughness. Obstructions, changing forms on a river bed such as sand bars, and vegetation offer added resistance to flow. Bends in a river also increase friction. Each of these factors causes deflection of the flow which dissipates energy by creating eddies, secondary circulation, and increased shear rate.

The load which a river carries is transported in four different ways:

1 traction, i.e. rolling of the coarsest fragments along the river bed;

2 smaller material, when it is caught in turbulent upward-moving eddies, proceeds downstream in a jumping motion referred to as saltation;

3 fine sand, silt, and mud may be transported in suspension;

4 soluble material is carried in solution.

Sediment yield may be determined by sampling both the suspended load and the bedload. It can also be derived from the amount of deposition which takes place when a river enters a relatively still body of water such as a lake or a reservoir.

The competence of a river to transport its load is demonstrated by the largest boulder it is capable of moving; it varies according to the velocity of a river and its volume, being at a maximum during flood. It has been calculated that the competence of a river varies as the sixth power of its velocity. The capacity of a river refers to the total amount of sediment which it carries. It varies according to the size of the rock fragments which form the load, and the velocity of the river. When the load consists of fine particles, the capacity is greater than when it is composed of coarse material. Usually the capacity of a river varies as the third power of its velocity.

Both the competence and capacity of a river are influenced by changes in the weather, the lithology and structure of the rocks over which it flows, and by vegetative cover and land use. Because the discharge of a river varies, sediments are not transported continously, for example, boulders may be moved only a few metres during a single flood. Sediments which are deposited over a flood plain may be regarded as being stored there temporarily.

Deposition occurs where turbulence is at a minimum or where the region of turbulence is near the surface of a river. For example, lateral accretion occurs, with deposition of a point bar, on the inside of a meander bend. The settling velocity for small grains in water is roughly proportional to the square of the grain diameter, for larger particles it is proportional to the square root of the grain diameter. An individual point bar grows as a meander migrates downstream and new ones are formed as a river changes its course during flood. Indeed, old meander scars are a common feature of flood plains. The combination of point bar and filled slough or oxbow lake gives rise to ridge and swale topography. The ridges consist of sandbars and the swales are sloughs filled with silt and clay.

An alluvial flood plain is the most common depositional feature of a river. The alluvium is made up of many kinds of deposits, laid down both in the channel and outside it (Marsland, 1986). Vertical accretion of a flood plain is accomplished by in-channel filling and the growth of overbank deposits during and immediately after floods. Gravels and coarse sands are moved chiefly at flood stages and deposited in the deeper parts of a river. As the river overtops its banks, its ability to transport material is lessened so that coarser particles are deposited near the banks to form levees. Levees stand above the general level of the adjoining plain so that the latter is usually poorly drained and marshy. This is particularly the case when levees have formed across the confluences of minor tributaries, so forcing them to wander over the flood plain until they find another entrance to the main river. Finer material is carried farther and laid down as backswamp deposits (Fig. 3.21). At this point a river sometimes aggrades its bed, eventually raising it above the level of the surrounding plain. Consequently, when levees are breached by flood water, hundreds of square kilometres may be inundated.

A delta develops when a river enters a quiet body of water such as a lake, or at its mouth when the tidal and current action of the sea is incapable of removing the sediment it carries. The speed at which the delta grows depends upon:

1 the rate of sediment supply;

2 the size of the basin;

3 the rate at which the basin is subsiding or being uplifted; and

4 the rate at which sediments are being removed by marine action.

3.3.3 Limestone (karst) topography and underground drainage

Limestones are well-jointed and bedded carbonate rocks which are susceptible to chemical attack by weakly acidified water. The degree of aggressiveness of water to limestone can be assessed on the basis of the relationship between the dissolved carbonate content, the pH value and the temperature of the water. At any given pH value, the cooler the water

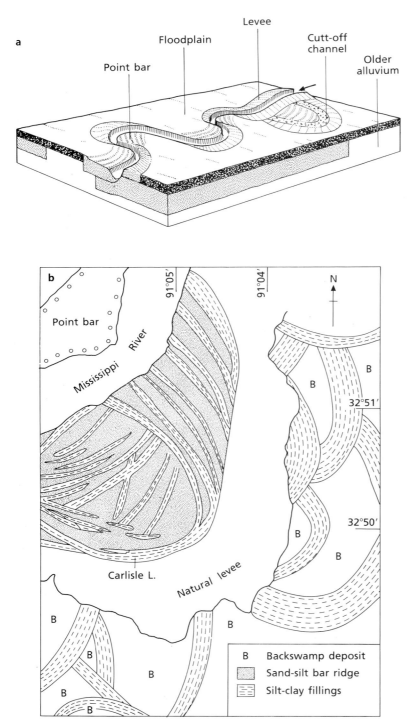

Fig. 3.21 (a) The main depositional features of a meandering channel; (b) map of a portion of the Mississippi River floodplain, showing various kinds of deposits. (After Fisk, 1944.)

Table 3.4 Solution of soluble rocks. (After James and Kirkpatrick, 1980.)

Rock	(a) Solubility (c_s) in pure water c_s (kg/ m³ at 10°C)*	(b) Solution rate constants (flow velocity — 0.05 m/s) (K) at 10°C m/s × 10⁵†	m^4 (kg/s × 10⁶)
		$\theta = 1$	$\theta = 2$
Halite	360.0	0.3	
Gypsum	2.5	0.2	
Anhydrite	2.0		0.8
Limestone	0.015	0.4	

* c_s is dependent upon temperature and the presence of other dissolved salts.

† K is dependent on temperature, flow velocity and other dissolved salts.

θ = order of dissolution reaction.

the more aggressive it is. If solution continues, its rate slackens and it eventually ceases when saturation is reached. Hence solution is greatest when the bicarbonate saturation is low. This occurs when water is circulating so that fresh supplies with low lime saturation are continually made available. As James and Kirkpatrick (1980) noted, a material dissolves at a rate and in a manner which is influenced by its solubility (c_s) and specific solution-rate constant (K) (Table 3.4). Not only is the rate of flow significant but the area of material exposed to flowing water is also important. Non-saline water can dissolve up to 400 mg/l of calcium carbonate.

The solution of limestone etches out and gradually enlarges joints to form clints, the blocks of limestone being termed grykes (Fookes and Hawkins, 1988). Limestone is frequently removed along bedding planes to form limestone pavements (Fig. 3.22a). Sometimes dissolution produces a highly irregular pinnacled surface on limestone pavements. The continued enlargement of joints, particularly where they intersect, eventually leads to the formation of funnel-shaped hollows known as sinkholes or swallow holes (Fig. 3.22b). Surface streams disappear underground via swallow holes. Some of the larger swallow holes are connected near the surface by irregular, inclined shafts known as ponors, to underground integrated systems of caverns and galleries. Larger surface depressions form when enlarged swallow holes coalesce to form uvalas or dolines. These features may range up to 1 km in diameter. Although some of these depressions are formed essentially by solution, others are at least partially created by the collapse of underground caverns. Still larger depressions which occur, for example, in the limestone region of western Yugoslavia, are called poljes. However, certain authors have suggested that these depressions are of tectonic origin. Any residual masses of limestone which, after a lengthy period of continuous erosion, remain as isolated hills are known as hums. These are invariably honeycombed with galleries, shafts, and caverns. However, the nature of karst landforms is influenced by climate, for example, tower karst (Fig. 3.23) is developed in tropical and subtropical areas.

Underground river systems in limestone deepen and widen their courses by mechanical erosion as well as by solution. During flood they usually erode the roofs as well as the sides and floors of the caverns and galleries through which they flow. In this way caverns are enlarged and their roofs are gradually thinned until they become so unstable that they wholly or partly collapse to form gorges or natural arches, respectively.

Surface drainage is usually sparse in areas of thick limestone. Dry valleys are common although they may be occupied by streams during periods of heavy rainfall. Underground streams may appear as vaclusian springs where the water table meets the surface. Occasionally, streams which rise on impervious strata may traverse a broad limestone outcrop without disappearing.

3.4 GLACIATION

A glacier may be defined as a mass of ice which is formed from recrystallized snow and refrozen meltwater and which moves under the influence of gravity. Glaciers develop above the snow line, that is, in regions of the world which are cold enough to allow snow to remain on the surface throughout the year. The snow line varies in altitude from sea level in polar regions to above 5000 m in equatorial regions. As the area of a glacier which is exposed to wastage is small compared with its volume, this accounts for the fact that glaciers penetrate into the warmer zones below the snow line.

Glaciers can be grouped into three types:
1 valley glaciers;
2 piedmont glaciers; and
3 ice sheets and ice caps.

Valley glaciers flow down pre-existing valleys from mountains where snow has collected and formed into ice. They disappear where the rate of melting exceeds the rate of supply of ice. When a number of

(a)

(b)

Fig. 3.22 (a) Limestone pavement above Malham Cove, Malham, North Yorkshire, England; (b) Gaping Ghyll sinkhole which descends some 150 m in Carboniferous Limestone, near Clapham, North Yorkshire, England.

valley glaciers emerge from a mountain region onto a plain, where they coalesce, they then form a piedmont glacier. Ice sheets are huge masses of ice which extend over areas which may be of continental size. Ice caps are of smaller dimensions. Today there are two ice sheets in the world, one extending over the Antarctic continent, the other covering most of Greenland.

3.4.1 Glacial erosion

Although pure ice is a comparatively ineffective agent as far as eroding massive rocks is concerned, it does acquire rock debris, which enhances its abrasive power. The larger fragments of rock embedded in the sole of a glacier tend to carve grooves in the path over which it travels whilst the finer material smooths and polishes rock surfaces. Ice also erodes by a quarrying process, whereby fragments are plucked from rock surfaces. Generally 'quarrying' is a more effective form of glacial erosion than abrasion.

The rate of glacial erosion is extremely variable and depends upon the velocity of the glacier, the weight of the ice, the abundance and physical

Fig. 3.23 Limestone towers, south of Guilin, China.

character of the rock debris carried at the bottom of the glacier, and the resistance offered by the rocks of the glacier channel. The erodibility of the surface over which a glacier travels will vary with depth and hence with time. Once the weathered overburden and open-jointed bedrock have been removed, the rate of glacial erosion slackens. This is because 'quarrying' becomes less effective and hence the quantity of rock fragments contributed to abrasive action is gradually reduced.

Continental ice sheets move very slowly and may be effective agents of erosion only temporarily, removing the weathered mantle from, and smoothing off the irregularities of, a landscape. The pre-glacial relief features are consequently afforded protection by the overlying ice against denudation, although the surface is somewhat modified by the formation of hollows and hummocks.

The most common features produced by glacial abrasion are striations on rock surfaces which were formed by rock fragments embedded in the base of the glacier. Many glaciated slopes formed of resistant rocks which are well-jointed display evidence of erosion in the form of ice-moulded hummocks, which are known as *roches moutonnées*. Large, highly resistant obstructions, like volcanic plugs, which lie in the path of advancing ice give rise to features called crag and tail (Fig. 3.24). The resistant obstruction forms the crag and offers protection to the softer rocks which occur on its lee side and form the tail.

Drumlins are mounds which are rather similar in shape to the inverted bowl of a spoon (Fig. 3.25). They vary in composition ranging from 100% bedrock to 100% glacial deposit. Obviously, those types formed of bedrock originated as a consequence of glacial erosion. However, even those composed of glacial debris were, at least in part, moulded by glacial action. Drumlins range up to 1 km in length and some may be over 70 m in height. Usually they do not occur singly but in scores or even hundreds in drumlin fields.

Corries are located at the head of glaciated valleys, being the features in which ice accumulated. Hence they formed at or close to the snow line. Corries are

Fig. 3.24 Crag and tail, Edinburgh. The Castle Rock probably represents an early phase of volcanic activity associated with the ancient volcano, Arthur's Seat.

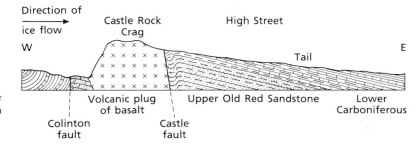

Fig. 3.25 Drumlin along the Tweed valley near Kelso, Scotland.

frequently arranged in tiers and in such instances give rise to corrie stairways up a mountain side. Because of their shape, corries are often likened to amphitheatres in that they are characterized by steep backwalls and steep sides (Fig. 3.26). Their floors are generally rock basins. Corries vary in size, some of the largest being about 1 km across. The dominant factor influencing their size is the nature of the rock in which they were excavated.

The cross-profile of a glaciated valley is typically steep-sided with a comparatively broad, flat bottom and it is commonly referred to as U-shaped (Fig. 3.27). Most glaciated valleys are straighter than those of rivers since their spurs have been truncated

by ice. In some glaciated valleys a pronounced bench or shoulder occurs above the steep walls of the trough. Tributary streams of ice flow across the shoulders to the main glacier. When the ice disappears the tributary valleys are left hanging above the level of the trough floor. The valleys are then occupied by streams. Those in the hanging valleys cascade down the slopes of the main trough as waterfalls. An alluvial cone may be deposited at the base of the waterfall.

Generally glaciated valleys have a scalloped or stepped long profile and sometimes the head of the valley is terminated by a major rock step known as a trough's end. Such rock steps develop where a

Fig. 3.26 Corrie, near Mt Cook, South Island, New Zealand.

Fig. 3.27 Glaciated Teleki Valley, Mount Kenya, Kenya.

number of tributary glaciers, descending from corries at the head of the valley, converge and thereby effectively increase erosive power. A simple explanation of a scalloped valley floor can be found in the character of the rock type. Not only is a glaciated valley stepped but reversed gradients are also encountered within its path. The reversed gradients are located in rock-floored basins which occur along the valley. Rock basins appear to be formed by localized ice action.

Fjords are found along the coasts of glaciated highland regions which have suffered recent submergence, they represent the drowned part of a glaciated valley. Frequently, a terminal rock barrier, the threshold, occurs near the entrance of a fjord. Some thresholds rise very close to sea level and some may be uncovered at low tide. However, water landward of the threshold very often is deeper than the known post-glacial rise in sea level, for example, depths in excess of 1200 m have been recorded in some Norwegian fjords.

3.4.2 Glacial deposits: unstratified drift

Glacial deposits form a more significant element of the landscape in lowland areas than they do in highlands. Two kinds of glacial deposits are distinguished — unstratified drift (or till), and stratified drift. However, one type commonly grades into the other. Till is usually regarded as being synonymous with boulder clay and is deposited directly by ice, whilst stratified drift is deposited by meltwaters.

Till consists of a variable assortment of rock debris which ranges in size from very fine rock flour to boulders and is characteristically unsorted. The compactness of a till varies according to the degree of consolidation undergone, the amount of cementation, and grain size. Tills which contain less than 10% clay fraction are usually friable whilst those with over 10% clay tend to be massive and compact.

Distinction has been made between tills derived from rock debris carried along at the base of a glacier (lodgement till) and those deposits which were transported within or at the terminus of the ice (ablation till). Lodgement till is commonly compact and fissile, and the fragments of rock it contains are frequently orientated in the path of ice movement. Ablation till accumulates as the ice, in which the parent material is entombed, melts. Hence, it is usually uncompacted and non-fissile, and the boulders present display no particular orientation. Since ablation till consists only of the load carried at the time of ablation, it usually forms a thinner deposit than does lodgement till.

A moraine is an accumulation of drift deposited directly from a glacier. There are six types of moraines deposited by valley glaciers. Rock debris which a glacier wears from its valley sides and which is supplemented by material that falls from the valley slopes above the ice forms a lateral moraine. When two glaciers become confluent a medial moraine develops from the merger of the two inner lateral moraines. Material which falls onto the surface of a glacier and then makes its way via crevasses into the centre, where it becomes entombed, is termed an englacial moraine. Some of this debris, however, eventually reaches the base of the glacier and there enhances the material eroded from the valley floor, constituting a subglacial moraine. A ground moraine is often irregularly distributed since it is formed when basal ice becomes overloaded with rock debris and is forced to deposit some of it. The material which is deposited at the snout of a glacier when the rate of wastage is balanced by the rate of outward flow of ice is known as a terminal moraine (Fig. 3.28). Terminal moraines possess a curved outline impressed upon them by the lobate nature of the snout of the ice. They are usually discontinuous, being interrupted where streams of meltwater issue from the glacier. Frequently, a series of terminal moraines may be found traversing a valley, the farthest down-valley marking the point of maximum extension of the ice, the others indicating pauses in glacial retreat. The latter types are called recessional moraines.

Ground moraines and terminal moraines are the two principal types of moraines deposited by ice sheets which spread over lowland areas, where the terminal moraines of ice sheets may rise to heights of some 60 m. In plan they commonly form a series of crescents, each crescent corresponding to a lobe at the snout of the ice. If copious amounts of meltwater drained from the ice front, then morainic material was washed away, hence a terminal moraine either did not develop or, if it did, was of inconspicuous dimension.

3.4.3 Fluvioglacial deposits; stratified drift

Stratified deposits of drift are often subdivided into two categories:

1 those deposits which accumulate beyond the limits of the ice, forming in streams, lakes, or seas (proglacial deposits); and

2 those deposits which develop in contact with the ice (ice-contact deposits).

Most meltwater streams which deposit outwash fans do not originate at the snout of a glacier but

Fig. 3.28 Terminal moraine, Fox glacier, west coast of South Island, New Zealand.

from within or upon the ice. Many of the streams which flow through a glacier have steep gradients and are therefore efficient transporting agents, but when they emerge at the snout, they do so on to a shallower incline and deposition results. Outwash deposits are typically cross-bedded and range in size from boulders to coarse sand. When first deposited the porosity of these sediments varies from 25 to 50%. They are therefore very permeable and hence can resist erosion by local runoff. The finer silt–clay fraction is transported further downstream. Also in this direction an increasing amount of stream alluvium is contributed by tributaries so that eventually the fluvioglacial deposits cannot be distinguished. Most outwash masses are terraced.

Five different types of stratified drift deposited in glacial lakes have been recognized:
1 terminal moraines;
2 deltas;
3 bottom deposits;
4 ice-rafted erratics; and
5 beach deposits.
Terminal moraines that formed in glacial lakes differ from those which arose on land in that lacustrine deposits are interstratified with drift. Glacial lake deltas are usually composed of sands and gravels which are typically cross-bedded. By contrast, those sediments which accumulated on the floors of glacial

lakes are fine-grained, consisting of silts and clays. These fine-grained sediments are sometimes composed of alternating laminae of finer and coarser grain size. Each couplet has been termed a varve. Large boulders which occur on the floors of glacial lakes were transported on rafts of ice and were deposited when the ice melted. Usually the larger the glacial lake, the larger were the beach deposits which developed about it. If changes in lake level took place, then these may be represented by a terraced series of beach deposits.

Deposition which takes place at the contact of a body of ice is frequently sporadic and irregular. Locally, the sediments possess a wide range of grain size, shape, and sorting. Most are granular and variations in their engineering properties reflect differences in particle size distribution and shape. Deposits often display abrupt changes in lithology and consequently in relative density. They are characteristically deformed since they sag, slump, or collapse as the ice supporting them melts.

Kame terraces are deposited by meltwater streams which flow along the contact between the ice and the valley side (Fig. 3.29). The drift is principally derived from the glacier although some is supplied by tributary streams. They occur in pairs, one each side of the valley. If a series of kame terraces occur on the valley slopes, then each pair represents a pause in

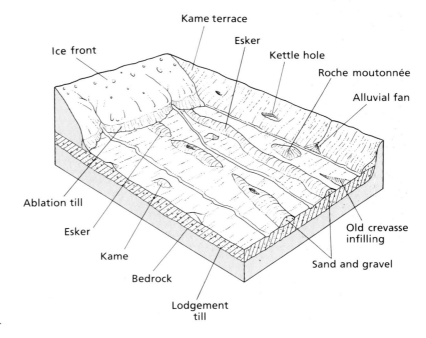

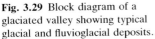
Fig. 3.29 Block diagram of a glaciated valley showing typical glacial and fluvioglacial deposits.

the process of glacier thinning. The surfaces of these terraces are often pitted with kettle holes (these are depressions where large blocks of ice remained unmelted whilst material accumulated around them). Narrow kame terraces are usually discontinuous, spurs having impeded deposition. Kames are mounds of stratified drift which originate as small deltas or fans built against the snout of a glacier where a tunnel in the ice, along which meltwater travels, emerged (Fig. 3.29). Other small ridge-like kames accumulate in crevasses in stagnant or near-stagnant ice. Many kames do not survive deglaciation for any appreciable period of time.

Eskers are long, narrow, sinuous, ridge-like masses of stratified drift which are unrelated to surface topography (Fig. 3.29), for example, eskers may climb up valley sides and cross low watersheds. They represent sediments deposited by streams which flowed within channels in a glacier. Although eskers may be interrupted their general continuity is easily discernible and some may extend lengthwise for several hundred kilometres. Eskers may reach up to 50 m in height, whilst they range up to 200 m wide. Their sides are often steep. Eskers are composed principally of sands and gravels, although silts and boulders are found within them. These deposits are generally cross-bedded.

3.4.4 Other glacial effects

Ice sheets have caused diversions of drainage in areas of low relief. In some areas which were completely covered with glacial deposits the post-glacial drainage pattern may bear no relationship to the surface beneath the drift, indeed moraines and eskers may form minor water divides. As would be expected notable changes occurred at or near the margin of the ice. There lakes were formed which were drained by streams whose paths disregarded pre-glacial relief. Evidence of the existence of pro-glacial lakes is to be found in the lacustrine deposits, strandlines, and overflow channels which they leave behind.

Where valley glaciers extend below the snow line they frequently pond back streams which flow down the valley sides and thereby give rise to lakes. If any col between two valleys is lower than the surface of the glacier occupying one of them, then the water from an adjacent lake dammed by this glacier eventually spills into the adjoining valley and in so doing erodes an overflow channel. Marginal spillways may develop along the side of a valley at the contact with the ice.

The enormous weight of an overlying ice sheet causes the Earth's crust beneath it to sag. Once the ice sheet disappears the land slowly rises to recover its former position and thereby restore isostatic equilibrium. Consequently, the areas of northern Europe and North America presently affected by isostatic uplift more or less correspond with those areas which were formerly covered with ice. At the present day the rate of isostatic recovery in, for example, the centre of Scandinavia is approximately 1 m per century. Isostatic uplift is neither regular nor continuous, thus the rise in the land surface has at times been overtaken by a rise in sea level. The latter was caused by meltwater from the retreating ice sheets.

With the advance and retreat of ice sheets in Pleistocene times the level of the sea fluctuated. Marine terraces (strandlines) were produced during interglacial periods when the sea was at a much higher level. The post-glacial rise in sea level has given rise to drowned coastlines such as rias and fjords, young developing cliff-lines, aggraded lower stretches of river valleys, buried channels, submerged forests, marshlands, shelf seas, straits, and the formation of numerous islands.

3.4.5 Frozen ground phenomena in periglacial environments

Frozen ground phenomena are found in regions which experience a tundra climate, that is, in those regions where the summer temperatures are only warm enough to cause thawing in the upper metre or so of the soil. Beneath the upper or active zone the subsoil is permanently frozen and so is known as the permafrost layer. Because of this layer summer meltwater cannot seep into the ground, the active zone then becomes waterlogged and the soils on gentle slopes are liable to flow. Layers or lenses of unfrozen ground termed taliks may occur, often temporarily, in the permafrost (Fig. 3.30).

Permafrost is an important characteristic, although it is not essential to the definition of periglacial conditions, the latter referring to conditions under which frost action is the predominant weathering agent. Permafrost covers 20% of the Earth's land surface and during Pleistocene times it was developed over an even larger area. Ground cover, surface water, topography, and surface materials all influence the distribution of permafrost. The temperature of perennially frozen ground below the depth of seasonal change ranges from slightly less than 0° to

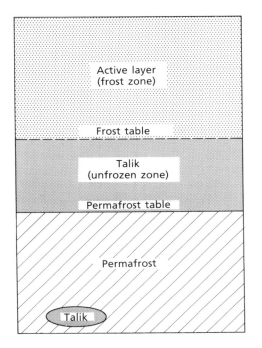

Fig. 3.30 Terminology of some features associated with permafrost.

and sand to a maximum in gravel where it may extend to 2 m in depth.

Prolonged freezing gives rise to shattering in the frozen layer, fracturing taking place along joints and cracks. Frost shattering, due to ice action in Pleistocene times, has been found to extend to depths of 30 m in the Chalk and to 12 m in the Borrowdale Volcanic Series in England. In this way the rock concerned suffers a reduction in bulk density and an increase in deformability and permeability. Fretting and spalling are particularly rapid where the rock is closely fractured. Frost shattering may be concentrated along certain preferred planes, if joint patterns are suitably oriented. Preferential opening takes place most frequently in those joints which run more or less parallel with the ground surface. Silt and clay frequently occupy the cracks in frost-shattered ground, down to appreciable depth, having been deposited by meltwater. Their presence may cause stability problems.

Stress relief following the disappearance of ice on melting may cause the enlargement of joints. This may aid failure on those slopes which were oversteepened by glaciation.

Frost wedging is one of the chief factors of mechanical weathering in tundra regimes (Fig. 3.31). Frozen soils often display a polygonal pattern of cracks. Individual cracks may be 1.2 m wide at their top, may penetrate to depths of 10 m, and may be some 12 m apart. They form when, because of

−12°C. Generally the depth of thaw is less, the higher the latitude. It is at a minimum in peat or highly organic sediments and increases in clay, silt,

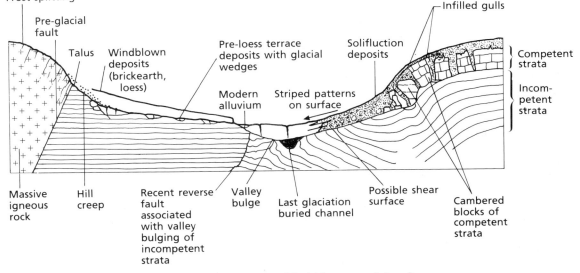

Fig. 3.31 Idealized section of valley showing important periglacial features and deposits.

exceptionally low temperatures, shrinkage of the ground occurs. Ice wedges occupy these cracks and cause them to expand. When the ice disappears an ice-wedge pseudomorph is formed by sediment, frequently sand, filling the crack.

Ground may undergo notable disturbance as a result of mutual interference of growing bodies of ice or from excess pore pressures developed in confined water-bearing lenses. Involutions are plugs, pockets, or tongues of highly disturbed material, generally possessing inferior geotechnical properties, which have been intruded into overlying layers. They are formed as a result of hydrostatic uplift in water trapped under a refreezing surface layer. They usually are confined to the active layer.

The movement downslope as a viscous flow of saturated rock waste is referred to as solifluction. It is probably the most significant process of mass wastage in tundra regions. Such movement can take place down slopes with gradients as low as 2°. The movement is extremely slow, most measurements showing rates ranging between 10 and 300 mm per year. Solifluction deposits commonly consist of gravels, which are characteristically poorly sorted, sometimes gap-graded, and poorly bedded. These gravels consist of fresh, poorly worn, locally derived material. Individual deposits are rarely more than 3 m thick and frequently display flow structures.

Periglacial action accelerates hill creep, the latter being particularly well developed on thinly bedded or cleaved rocks. Creep material may give way to solifluction deposits on approaching the surface. These deposits consist mainly of flat rock fragments oriented parallel with the hillside and are interrupted by numerous shallow slips.

Sheets and lobes of solifluction debris, transported by mudflow activity, are commonly found at the foot of slopes. These materials may be reactivated by changes in drainage, by stream erosion, by sediment overloading, or during construction operations. Solifluction sheets may be underlain by slip surfaces, the residual strength of which controls their stability.

Oversteepening of glaciated valleys and meltwater channels occur when ground is stabilized by deep permafrost or supported by ice masses. Frost sapping at the bottom of scarp features also causes oversteepening. When the support disappears, the oversteepened slopes become potentially unstable. Meltwater in a shattered rock mass gives rise to an increase in pore pressures which in turn leads to movement or instability along bedding planes and joints. Increase in the moisture content of cohesive material brings about a reduction in its strength and may cause it to swell, thereby aggravating the instability, due to oversteepening, in the near-surface zone. As a result landsliding on a large scale is associated with such oversteepened slopes.

The solubility of carbon dioxide in water varies inversely with temperature, for example, it is 1.7 times greater at 0°C than at 15°C. Accordingly cold meltwaters frequently have had a strong leaching effect on calcareous rocks. Some pipes and swallow holes in chalk may have been produced by such meltwaters. The problem of pipes and swallow holes in chalk is aggravated by their localized character and the frequent absence of surface evidence. They are often undetected by a conventional site investigation.

3.5 WIND ACTION AND DESERT LANDSCAPES

By itself wind can only remove uncemented rock debris, which it can perform more effectively if the debris is dry rather than wet. But once armed with rock particles, the wind becomes a noteworthy agent of abrasion. The size of individual rock particles which the wind can transport depends on the strength of the wind, and particle shape and weight. The distance which the wind, given that its velocity remains constant, can carry rock particles depends principally upon their size.

3.5.1 Wind action

Wind erosion takes place when air pressure overcomes the force of gravity on surface particles. At first particles are moved by saltation. The impact of saltating particles on others may cause them to move by creep, saltation, or suspension. Saltation accounts for three-quarters of the grains transported by wind, most of the remainder being carried in suspension, the rest are moved by creep or traction. Saltating grains may rise to a height of up to 2 m, their trajectory then being flattened by faster moving air and tailing off as the grains fall to the ground. The length of the trajectory is roughly ten times the height.

One of the most important factors in wind erosion is its velocity. Its turbulence, frequency, duration, and direction are also important. As far as the mobility of particles is concerned the important factors are their size, shape, and density. It would appear that particles less than 0.1 mm in diameter are usually transported in suspension; those between 0.1 and 0.5 mm are normally transported in saltation; and those larger than 0.5 mm tend to be moved by traction or creep. Grains with a specific gravity of 2.65, such as quartz sand, are most suspect to wind erosion in the size range 0.1 to 0.15 mm. A wind blowing at 12 km/h will move grains of 0.2 mm diameter — a lesser velocity will keep the grains moving.

Because wind can only remove particles of a limited size range, if erosion is to proceed beyond the removal of existing loose particles, then remaining material must be sufficiently broken down by other agents of erosion or weathering. Material which is not sufficiently reduced in size will seriously inhibit further wind erosion. Removal of fine material leads to a proportionate increase of coarser material that cannot be removed. The latter affords increasing protection against continuing erosion and eventually a wind-stable surface is created. Binding agents, such as silt, clay, and organic matter hold particles together and so make wind erosion more difficult. Soil moisture also contributes to cohesion between particles.

Generally, a rough surface tends to reduce the velocity of the wind immediately above it. Consequently, particles of a certain size are not as likely to be blown away as they would be on a smooth surface. The longer the surface distance over which a wind can blow without being interrupted, the more likely it is to attain optimum efficiency.

There are three types of wind erosion: (i) deflation; (ii) attrition; and (iii) abrasion. Deflation results in the lowering of land surfaces by loose, unconsolidated rock waste being blown away by the wind. The effects of deflation are most acutely seen in arid and semi-arid regions. Basin-like depressions are formed by deflation in the Sahara and Kalahari deserts. However, downward lowering is almost invariably arrested when the water table is reached since the wind cannot readily remove moist rock particles. Also, deflation of sedimentary material, particularly alluvium, creates a protective covering if the material contains pebbles. The fine particles are removed by the wind, leaving a surface formed of pebbles. The suspended load carried by the wind is further comminuted by attrition, turbulence causing the particles to collide vigorously with one another.

When the wind is armed with grains of sand it possesses great erosive force, the effects of which are best displayed in rock deserts. Accordingly, any surface subjected to prolonged attack by wind-blown sand is polished, etched, or fluted. Abrasion has a selective action, picking out the weaknesses in rocks, for example, discontinuities are opened and rock pinnacles developed. Since the heaviest rock particles are transported near to the ground, abrasion there is at its maximum and rock pedestals are formed. In deserts flat, smoothed surfaces produced by wind erosion are termed desert pavements.

The differential effects of wind erosion are well illustrated in areas where alternating beds of hard and soft rock are exposed. If strata are steeply tilted, then because soft rocks are more readily worn away than hard rocks, a ridge-and-furrow relief develops. Such ridges are called yardangs. Conversely, when an alternating series of hard and soft rocks are more or less horizontally bedded, features known as zeugens are formed. In such cases the beds of hard rock act as resistant caps affording protection to the soft rocks beneath. Nevertheless, any weaknesses in the hard caps are picked out by weathering and the caps are eventually breached exposing the underlying soft rocks. Wind erosion rapidly eats into the latter and in the process the hard cap is undermined. As the action continues tabular masses (known as mesas and buttes), are left in isolation (Fig. 3.32).

3.5.2 Desert dunes

About one-fifth of the land surface of the Earth is desert. Approximately four-fifths of this desert area consists of exposed bedrock or weathered rock waste. The rest is mainly covered with deposits of sand. Bagnold (1941) recognized five main types of sand accumulations:
1 sand drifts and sand shadows;
2 whalebacks;
3 low-scale undulations;
4 sand sheets; and
5 true dunes.
He further distinguished two kinds of true dunes: the barchan and the seif (Fig. 3.33).

Several factors control the form which an accumulation of sand adopts: (i) the rate at which sand is supplied; (ii) wind speed, frequency, and constancy of direction; (iii) the size and shape of the sand grains; (iv) the nature of the surface across which the sand is moved. Sand drifts accumulate at the exits of the gaps through which wind is channelled and are extended downwind. However, such drifts, unlike true dunes, are dispersed if they are moved downwind. Whalebacks are large mounds of com-

Fig. 3.32 Mesa and denuded mesa, near Goageb, Namibia.

(a)   (b)

Fig. 3.33 Sand dunes: (a) barchans, near Luderitz, Namibia; (b) seif dune, near Sossusvlei, Namibia.

paratively coarse sand which are thought to represent the relics of seif dunes. Presumably the coarse sand is derived from the lower parts of seifs, where accumulations of coarse sand are known to exist. These features develop in regions devoid of vegetation. By contrast, undulating mounds are found in the peripheral areas of deserts where the patchy cover of vegetation slows the wind and creates sand traps. Large undulating mounds are composed of fine sand. Sand sheets are also developed in the marginal areas of deserts. These sheets consist of fine sand which is well sorted; often presenting a smooth surface which is capable of resisting wind erosion. A barchan is crescentic in outline and is orientated at right angles to the prevailing wind direction, whilst a seif is a long ridge-shaped dune running parallel to the direction of the wind. Seif dunes are much larger than barchans and may extend lengthwise for up to 90 km and reach heights up to 100 m. Barchans are rarely more than 30 m in height and their width is usually about twelve times their height. Seifs occur in great numbers running approximately equidistant from each other, each crest being separated from another by anything from 30 to 500 m.

It is commonly believed that sand dunes come into being where some obstacle prevents the free flow of sand, sand piling up on the windward side of the obstacle to form a dune. But in areas where there is an exceptionally low rainfall and therefore little vegetation to impede the movement of sand, observation has revealed that dunes develop most readily on flat surfaces devoid of large obstacles. It would seem that where the size of the sand grains varies or where a rocky surface is covered with pebbles, dunes grow over areas of greater width than 5 m. Such patches exert a frictional drag on the wind causing eddies to blow sand towards them. Sand is trapped between the larger grains or pebbles and an accumulation results. If a surface is strewn with patches of sand and pebbles, deposition takes place over the pebbles. However, patches of sand exert a greater frictional drag on strong winds than do patches of pebbles and so deposition under such conditions takes place over the former. When strong winds sweep over a rough surface they become transversely unstable and barchans may develop.

Longitudinal dunes may develop from barchans. Suppose that the tails of a barchan for some reason become fixed, for example, by vegetation or by the water table rising to the surface, then the wind continues to move the central part until the barchan eventually loses its convex shape becoming concave towards the prevailing wind. As the central area becomes further extended, the barchan may split. The two separated halves are gradually rotated by the eddying action of the wind until they run parallel to one another, in line with the prevailing wind direction. Dunes that develop in this manner are often referred to as blowouts.

Seif dunes appear to form where winds blow in two directions, that is, where the prevailing winds are gentle and carry sand into an area, the sand then being driven by occasional strong winds into seif-like forms. Seifs may also develop along the bisectrix between two diagonally opposed winds of roughly equal strength. Because of their size, seif dunes can trap coarse sand much more easily than can barchans. This material collects along the lower flanks of the dune; and barchans sometimes occur in the troughs of seif dunes. On the other hand, the trough may be floored by bare rock.

3.5.3 Stream action in arid and semi-arid regions

It must not be imagined that stream activity plays an insignificant role in the evolution of landscape in arid and semi-arid regions. Admittedly the amount of rainfall occurring in arid regions is small and falls irregularly whilst that of semi-arid regions is markedly seasonal; but in both instances it frequently falls as heavy and often violent showers. The result is that the river channels cannot cope with the amount of rain water and extensive flooding takes place. These floods develop with remarkable suddenness and either form raging torrents, which tear their way down slopes excavating gullies as they go, or they may form sheet floods. Dry wadis are rapidly filled with swirling water and are thereby enlarged. Such floods are short-lived since the water soon becomes choked with sediment and the consistency of the resultant mudflow eventually reaches a point when further movement is checked. Much water is lost by percolation, and mudflows are also checked where there is an appreciable slackening in gradient.

Some of the most notable features produced by stream action in arid and semi-arid regions are found in intermontane basins, that is, where mountains circumscribe a basin of inland drainage. The rain which falls on the mountains causes flooding and active erosion. Mechanical weathering plays a significant role in the mountain zone. Boulders, 2 m or more in diameter, are found in gullies which cut the mountain slopes whilst finer gravels, sands, and muds are washed downstream.

Alluvial cones or fans, which consist of irregularly assorted sediment, are found along the foot of the mountain belt where it borders the pediment — the marked change in gradient accounting for the rapid deposition. The particles composing the cones are almost all angular in shape, boulders and cobbles being more frequent upslope, grading downslope into fine gravels. The cones have a fairly high permeability. When these alluvial cones merge into one another they form a bahada. The streams which descend from the mountains rarely reach the centre of the basin since they become choked by their own deposits and split into numerous distributaries which spread a thin veneer of gravels over the pediment.

Hydrocompaction may occur on alluvial cones, particularly if they are irrigated. The dried surface layer of these cones may contain many voids. Percolating water frequently reduces the strength of the material which, in turn, reduces the void space. This gives rise to settlement or hydrocompaction.

Pediments in semi-arid regions are graded plains cut by the lateral erosion of ephemeral streams. King (1963) described the pediments found in the semi-arid areas of South Africa as smooth, flat zones across which a maximum of water is discharged (that from the mountains together with that of the pediments). Thus, pediments are adjusted to dispose of water in the most efficient way and the heavy rainfall characteristic of semi-arid regions means that this is often in the form of sheet wash. Although true laminar flow occurs during sheet wash, as the flowing water deepens laminar flow yields to turbulent flow. The latter possesses much greater erosive power and occurs during and immediately after heavy rainfall. This is why pediments carry only a thin veneer of rock debris. With less rainfall there is insufficient water to form sheets and it is confined to rills and gullies. According to King the rock waste transported across the pediment is relatively fine and is deposited in hollows thereby smoothing the slope. The abrupt change in the slope at the top of the pediment is caused by a change in the principal processes of earth sculpture, the nature of the pediment being governed by sheet erosion whilst that of the steep hillsides is controlled by the downward movement of rock debris.

Aeolian and fluvial deposits, notably sand, also may be laid down in the intermediate zone between the pediment and the central depression or playa. However, if deflation is active this zone may be barren of sediments. Sands are commonly swept into dunes and the resultant deposits are cross-bedded.

The central area of a basin is referred to as the playa and it sometimes contains a lake (Fig. 3.34). This area is covered with deposits of sand, silt, clay, and evaporites. The silts and clays often contain crystals of salt whose development comminutes their host. Silts usually exhibit ripple marks whilst clays are frequently laminated. Desiccation structures such as mudcracks are developed on an extensive scale in these fine-grained sediments. If the playa lake has contracted to leave a highly saline tract, then this area is termed a salina. The capillary rise generally extends to the surface, leading to the formation of a salt crust. Where the capillary rise is near to, but does not normally reach, the surface, desiccation ground patterns provide an indication of its closeness.

Duricrust is a surface, or near-surface, hardened accumulation or encrusting layer, formed by precipitation of salts on evaporation of saline groundwater. It may be composed of calcium or magnesium carbonate, gypsum, silica, alumina, or iron oxide (or even halite), in varying proportions. It may occur in

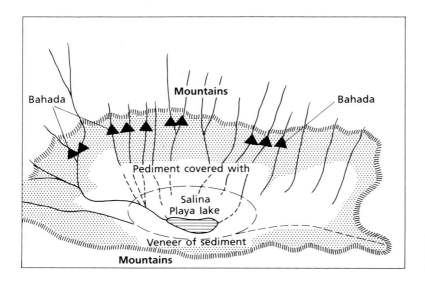

Fig. 3.34 Intermontane basin showing bahadas, pediment, salina, and playa lake.

Fig. 3.35 An inselberg, north of Bulawayo, Zimbabwe.

a variety of forms, ranging from a few millimetres in thickness to over a metre. A leached cavernous, porous, or friable zone is frequently found beneath the duricrust. When describing duricrusts those terms ending in crete (e.g. calcrete), refer to hardened surfaces usually occurring on hard rock; those ending in crust (e.g. gypcrust) represent softer accumulations which usually are found in salt playas, salinas, or sabkhas (coastal salt marshes). Locally, especially near the coast, sands may be cemented with calcrete to form caprock or miliolite. Desert fill often consists of mixtures of nodular calcrete, calcrete fragments, and drifted sand. Aggressive salty ground occurs, for example, in sabkhas, salinas, salt playas, and some duricrusts. Salt weathering brings about the disintegration of rock through crystallization, hydration, and thermal expansion.

The extension of pediments on opposing sides of a mountain mass means that the mountains are slowly reduced until a pediplain is formed. The pediments are first connected through the mountain mass by way of pediment passes. The latter become progressively enlarged forming pediment gaps. Finally, opposing pediments meet to form a pediplain on which there are residual hills. Such isolated, steep-sided residual hills have been termed inselbergs or bornhardts (Fig. 3.35). They are characteristically developed in the semi-arid regions of Africa, where they are usually composed of granite or gneiss, that is, of more resistant rock than that which forms the surrounding pediplain.

3.6 COASTS AND SHORELINES

Johnson (1919) distinguished three elements in the shore zone: (i) the coast; (ii) the shore; and (iii) the

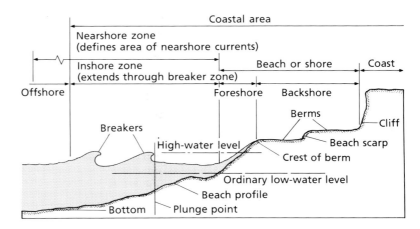

Fig. 3.36 Terminology of beach features.

offshore (Fig. 3.36). The coast was defined as the land immediately behind the cliffs, whilst the shore was regarded as that area between the base of the cliffs and low-water mark. That area which extended seawards from the low-water mark was termed the offshore. The shore was further divided into fore-shore and backshore, the former embracing the intertidal zone whilst the latter extended from the foreshore to the cliffs.

3.6.1 Wave action

When wind blows across the surface of deep water it causes an orbital motion of the water particles in the plane normal to the wind direction. Because adjacent particles are at different stages of their circular course a wave is produced. The motion is transmitted to the water beneath the surface but the orbitals are rapidly reduced in size with increasing depth and the motion dies out at a depth equal to that of the wave length (Fig. 3.37). There is no progressive forward motion of the water particles in such a wave although the form of the wave profile moves rapidly in the direction in which the wind is blowing. Such waves are described as oscillatory waves.

The parameters of a wave are defined as follows: the wave length (L) is the horizontal distance between each crest; the wave height (H) is the vertical distance between the crest and the trough; and the wave period (T) is the time interval between the passage of successive wave crests. The rate of propagation of the wave form is the wave length divided by the wave period. The height and period of waves are functions of the wind velocity, the fetch (the distance over which the wind blows), and the length of time for which the wind blows.

Fetch is the most important factor determining wave size and efficiency of transport. Winds of moderate force which blow over a wide stretch of water generate larger waves than do strong winds which blow across a short reach. Where the fetch is less than 32 km the wave height increases directly as, and the wave period increases as the square root of, the wind velocity. Long waves only develop where the fetch is large, for example, the largest waves are generated in the southern oceans where their lengths may exceed 600 m and their periods be greater than 20 s. Usually wave lengths in the open sea are less than 100 m and the speed of propagation is approximately 50 km/h.

Those waves which are developed in storm centres in the centre of an ocean may journey to its limits. This explains why large waves may occur along a coast during fine weather.

Waves frequently approach a coastline from different areas of generation. If they are in opposition, then their height is decreased whilst their height is increased if they are in phase.

Four types of waves have been distinguished: (i) forced; (ii) swell; (iii) surf; and (iv) forerunners. Forced waves are those formed by the wind in the generating area, they are usually irregular. On moving out of the area of generation the waves become long and regular. They are then referred to as swell or free waves. As these waves approach a shoreline they feel bottom, which disrupts their pattern of motion, changing them from oscillation to translation waves, that is, they break into surf. The longest and lowest waves are commonly termed forerunners or swell.

The breaking of a wave is influenced by its steepness, the slope of the sea floor, and the presence of an opposing or supplementary wind. When waves enter water equal in depth to their wave length they begin to feel bottom, and their length decreases whilst their height increases. Their velocity of travel

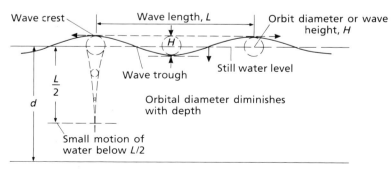

Direction of travel of wave profile →

Wave crest — Wave length, L — Orbit diameter or wave height, H
H
Wave trough — Still water level
$\frac{L}{2}$
d
Orbital diameter diminishes with depth
Small motion of water below $L/2$

Fig. 3.37 Orbital motion of water particles during the passage of an idealized sinusoidal wave in deep water. The orbital diameter decreases with depth and disappears at a depth of approximately one half the wave length.

or celerity (c) is also reduced in accordance with the expression:

$$c = \left[\frac{gL}{2\pi} \tanh \left(\frac{2\pi Z}{L} \right) \right]^{1/2} \qquad (3.9)$$

where L is the wave length, Z is the depth of the water, and g is acceleration due to gravity. As a result, the wave steepens until the wave train consists of peaked crests separated by relatively flat troughs. The wave period, however, remains constant. Steepening accelerates towards the breaker zone and the wave height grows to several times what it was in deep water. Three types of breaking waves are distinguished: (i) plunging; (ii) spilling; and (iii) surging or swash. Plunging breakers collapse when their wave height is approximately equal to the depth of the water. They topple suddenly and fall with a crash. They are usually a consequence of long, low swell and their formation is favoured by opposing winds. Spilling breakers begin to break when the wave height is just over one-half of the water depth and they do so gradually over some distance. Generally, they result from steep wind waves and they commonly occur when the wind is blowing in the direction of wave propagation. Surging breakers or swash rush up the beach and are usually encountered on beaches with steep profiles. The term backwash is used to describe the water which subsequently descends the beach slope.

Four dynamic zones have been recognized within the nearshore current system of the beach environment:
1 the breaker zone;
2 the surf zone;
3 the transition zone; and
4 the swash zone (Fig. 3.38).
The breaker zone is that in which waves break. The surf zone refers to that region between the breaker zone and the effective seaward limit of the backwash. The presence and width of a surf zone is primarily a function of the beach slope and tidal phase. Those beaches which have gentle foreshore slopes are often characterized by wide surf zones during all tidal phases whereas steep beaches seldom possess this zone. The transition zone includes that region where backwash interferes with the water at the leading edge of the surf zone and it is characterized by high turbulence. That region where water moves up and down the beach is termed the swash zone.

Swash tends to pile water against the shore and thereby gives rise to currents which move along the shore termed longshore currents. After flowing parallel to the beach the water runs back to the sea in narrow flows called rip currents. In the neighbourhood of the breaker zone rip currents extend from the surface to the floor whilst in the deeper reaches they override the bottom water which

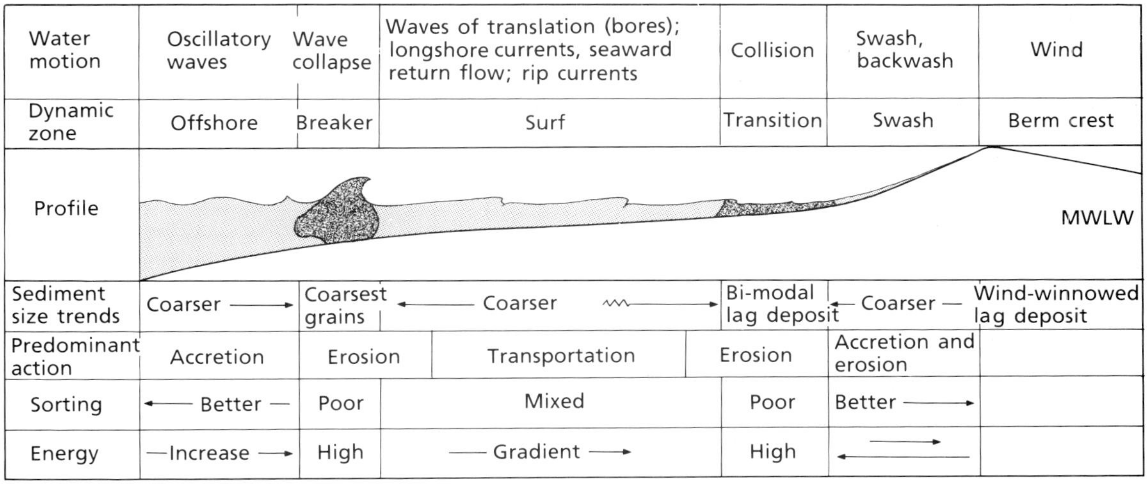

Water motion	Oscillatory waves	Wave collapse	Waves of translation (bores); longshore currents, seaward return flow; rip currents	Collision	Swash, backwash	Wind
Dynamic zone	Offshore	Breaker	Surf	Transition	Swash	Berm crest
Profile						MWLW
Sediment size trends	Coarser →	Coarsest grains	← Coarser ⌇ →	Bi-modal lag deposit	← Coarser —	Wind-winnowed lag deposit
Predominant action	Accretion	Erosion	Transportation	Erosion	Accretion and erosion	
Sorting	← Better —	Poor	Mixed	Poor	Better →	
Energy	—Increase →	High	— Gradient →	High	← →	

Fig. 3.38 Summary diagram schematically illustrating the effect of the four major dynamic zones in the beach environment. Hatched areas represent zones of high concentrations of suspended grains. The surf zone is bounded by two high-energy zones — the breaker and the transition zones. MWLW, mean water low water. (After Ingle, 1966.)

still maintains an overall onshore motion. The positions of rip currents are governed by submarine topography, coastal configuration, and the height and period of the waves. They frequently occur on the upcurrent sides of points and on either side of convergences where the water moves away from the centre of convergence and turns seawards. The nearshore circulation system comprises:

1 onshore movement of water by wave action in the breaker zone;

2 lateral transport by longshore currents in the breaker zone;

3 seaward return of flow as rip currents through the surface zone; and

4 longshore movement in the expanding head of a rip current.

Tides may play an important part in beach processes. In particular the tidal range is responsible for the area of the foreshore over which waves are active. Tidal streams are especially important where a residual movement resulting from differences between ebb and flood occurs, and where there is abundant loose sediment for the tidal streams to transport. They are frequently fast enough to carry coarse sediment but the forms, notably bars, normally associated with tidal streams in the offshore zone or in tidal estuaries usually consist of sand. These features therefore only occur where sufficient sand is available. In quieter areas tidal mud flats and salt marshes are developed, where the tide ebbs and floods over large flat expanses depositing muddy material. Mud also accumulates in runnels landward of high ridges, in lagoons, and on the lower foreshore where shelter is provided by offshore banks.

3.6.2 Marine erosion

Coasts undergoing erosion display two basic elements of the coastal profile: the cliff, and the bench or platform. In any theoretical consideration of the evolution of a coastal profile it is assumed that the coast is newly uplifted above sea level. After a time a wave-cut notch may be excavated and its formation intensifies marine erosion in this narrow zone. The development of a notch varies according to the nature of the rock in which excavation is proceeding. It may be present if the rocks are unconsolidated or if the bedding planes dip seawards. Where a notch develops it gives rise to a bench and the rock above is undermined and collapses to form a cliff face.

Potholes are common features on most wave-cut benches. They are excavated by pebbles and boulders being swilled around in depressions. As they increase in diameter, they coalesce and so lower the surface of the bench. The debris produced by cliff recession gives rise to a rudimentary beach. In tidal seas the base of a cliff is generally at high-tide level whilst in non-tidal seas it is usually above still-water level.

As erosion continues the cliff increases in height and the bench widens. The slope adopted by the bench below sea level is determined by the ratio of the rate of erosion of the slope to the recession of the cliff. A submarine accumulation terrace forms in front of the bench and is extended out to sea. Because of the decline in wave energy, consequent upon the formation of a wide, flat bench, the submarine terrace deposits may be spread over the lower part of the bench in the final stages of its development. The rate of cliff recession is correspondingly retarded and the cliff becomes gently sloping and moribund.

If the relationship between the land and sea remains constant, then erosion and consequently the recession of land beneath the sea is limited. Although sand can be transported at a depth of half the length of storm waves, bedrock is abraded at only half this depth or less. As soon as a submarine bench slope of 0.01 to 0.05 is attained, bottom abrasion generally ceases and any further deepening is brought about by organisms or chemical solution. Such rock destruction can occur at any depth but the floor can only be lowered where currents remove the altered material.

The nature of the impact of a wave upon a coastline depends to some extent on the depth of the water and partly on the size of the wave. The vigour of marine action drops sharply with increasing depth from the water surface, at approximately the same rate as the decline in the intensity of wave motion. Erosion is unlikely to take place at a depth of more than 60 m along the coast of an open sea and at less than that in closed seas.

If deep water occurs alongside cliffs, then waves may be reflected without breaking and in so doing they may interfere with incoming waves. In this way clapotis (standing waves which do not migrate) are formed. It is claimed that the oscillation of standing waves causes an alternate increase and decrease of pressure along discontinuities in rocks which occur in that part of the cliff face below the water line. Also, when waves break, a jet of water is thrown against the cliff at approximately twice the velocity of the wave and, for a few seconds, this increases the pressure within the discontinuities. Such action gradually dislodges blocks of rock.

Those waves with a period of approximately 4 s are usually destructive whilst those with a lower frequency, that is, a period of about 7 s or over are constructive. When high-frequency waves collapse they form plunging breakers and the mass of water is accordingly directed downwards at the beach. In such instances swash action is weak and, because of the high frequency of the waves, is impeded by the backwash. Thus material is removed from the top of the beach. The motion within waves which have a lower frequency is more elliptical and produces a strong swash which drives material up the beach. In this case the backwash is reduced in strength because water percolates into the beach deposits and therefore little material is returned down the beach. Although large waves may throw material above high-water level and thus act as constructive agents, they have an overall tendency to erode the beach, whilst small waves are constructive.

Swash is relatively ineffective compared with backwash on some shingle beaches. This action frequently leads to very rapid removal of the shingle from the foreshore into the deeper water beyond the breakpoint. Storm waves on such beaches, however, may throw some pebbles to considerable elevations above mean sea level creating a storm-beach ridge and, because of rapid percolation of water through the shingle, backwash does not remove these pebbles. On the other hand, when steep storm waves attack a sand beach they are usually destructive and the coarser the sand, the greater the quantity which is removed. Some of this sand may form a submarine bar at the breakpoint, whilst some is carried into deeper water offshore. It is by no means a rarity for the whole beach to be removed by storm waves.

Waves usually leave little trace on massive smooth rocks except to polish them. However, where there are irregularities or projections on a cliff face the upward spray of breaking waves quickly removes them (it has been noted that the force of upward spray along a sea wall can be as much as twelve times that of the horizontal impact of the wave).

The degree to which rocks are traversed by discontinuities affects the rate at which they are removed by marine erosion. The attitude of joints and bedding planes is especially important. Where the bedding planes are vertical or dip inland, then the cliff recedes vertically under marine attack. But if beds dip seawards blocks of rock are more readily dislodged since the removal of material from the base of the cliff means that the rock above lacks support and tends to slide into the sea. Joints may be enlarged into deep, narrow inlets. Marine erosion is also concentrated along fault planes.

The height of a cliff is another factor which influences the rate at which coastal erosion takes place. The higher the cliff the more material that falls when its base is undermined. This in turn means that a greater amount of debris has to be broken down and removed before the cliff is once more attacked with the same vigour.

Erosive forms of local relief include such features as wave-cut notches, caves, blowholes, marine arches, and stacks (Fig. 3.39). Marine erosion is concentrated in areas along a coast where the rocks offer less resistance. Caves and small bays or coves are excavated where the rocks are softer or highly jointed. At the landward end of large caves there is often an opening to the surface, through which spray issues, which is known as a blowhole. Blowholes are formed by the collapse of jointed blocks loosened by wave-compressed air. A marine arch is developed when two caves on opposite sides of a headland unite. When the arch falls the isolated remnant of the headland is referred to as a stack.

Wave refraction is the process whereby the direction of wave travel changes because of changes in the topography of the nearshore sea floor. When waves approach a straight beach at an angle they tend to swing parallel to the shore due to the retarding effect of the shallowing water. At the breakpoint such waves seldom approach the coast at an angle exceeding 20° irrespective of the offshore angle to the beach. As the waves break they develop a longshore current; indeed wave refraction is often the major factor in dictating the magnitude and direction of longshore drift. Wave refraction is also responsible for the concentration of erosion on headlands which leads to a coast being smoothed gradually in outline. As waves approach an irregular shoreline refraction causes them to turn into shallower water so that the wave crests run roughly parallel to the depth contours. Along an indented coast shallower water is first met with off headlands. This results in wave convergence and an increase in wave length, with wave crests becoming concave towards headlands. Conversely, where waves move towards a depression in the sea floor they diverge, are decreased in height, and become convex towards the shoreline.

(a)

(b)

(c)

Fig. 3.39 Coastal features formed by erosion. (a) Stacks, the Apostles, south coast of Victoria, Australia; (b) marine arch, London Bridge, south coast of Victoria, Australia. Formerly there were two arches but the central one collapsed recently; (c) blowhole, 'pancake rocks' Punakaki, west coast of South Island, New Zealand.

3.6.3 Constructive action of the sea

Beaches may be supplied with sand which is almost entirely derived from the adjacent sea floor although in some areas a larger proportion is produced by cliff erosion. During periods of low waves the differential velocity between onshore and offshore motion is sufficient to move sand onshore except where rip currents are operational. Onshore movement is particularly notable when long-period waves approach a coast, whereas sand is removed from the foreshore during high waves of short period.

The beach slope is produced by the interaction of swash and backwash. It is also related to the grain size (Table 3.5) and permeability of the beach, for example, the loss of swash due to percolation into beaches composed of grains of 4 mm in median diameter is ten times greater than into those where the grains average 1 mm. As a result there is almost as much water in the backwash on fine beaches as there is in the swash, so the beach profile is gentle and the sand is hard-packed.

Those waves which produce the most conspicuous constructional features on a shingle beach are storm waves. A small foreshore ridge develops on a shingle beach at the limit of the swash when constructional waves are operative. Similar ridges or berms may form on a beach composed of coarse sand. Berms represent a marked change in slope and usually occur a small distance above the high-water mark. However, they may be overtopped by high spring tides. Berms are not such conspicuous features on beaches of fine sand. Greater accumulation occurs on coarse sandy beaches because their steeper gradient means that the wave energy is dissipated over a relatively narrow width of beach.

Sandbars are constructional features of small size which are characteristic of tideless seas. Their location is related to the break point which in turn is related to wave size. Consequently, more than one bar may form, the outermost being attributable to storm waves, the inner to normal waves.

When waves move parallel to the coast they simply move sand and shingle up and down the beach; but when they approach the coast at an angle, material is moved up the beach by the swash in the direction normal to that of wave approach and then is rolled down the steepest slope of the beach by the backwash. Consequently, material is moved in a zig-zag path along the beach. This is known as longshore drift. Such action can transport pebbles appreciable distances along a coast. The duration of movement along a coastline is dependent upon the direction of the dominant winds. An indication of the direction of longshore drift is provided by the orientation of spits along a coast.

Material may be supplied to the littoral sediment budget by coastal erosion, by feed from offshore or by contributions from rivers. After sediment has been distributed along the coast by longshore drift it may be deposited in a sediment reservoir and therefore lost from the active environment. Sediment reservoirs formed offshore take the form of bars where the material is in a state of dynamic equilibrium, but from which it may easily re-enter the system. Dunes are the commonest type of onshore reservoir, from which sediment is less likely to re-enter the system.

Bay-head beaches are one of the most common types of coastal deposits and they tend to straighten a coastline. Wave refraction causes longshore drift to move from headlands to bays where sediments are deposited. Marine deposition also helps straighten coastlines by building beach plains.

Spits are deposits which grow out from the coast. They are supplied with material chiefly by longshore

Table 3.5 Average beach face slopes compared to sediment diameters. (After Shepard, 1963.)

Type of beach sediment	Size (mm)	Average slope of the beach face
Very fine sand	0.0625−0.125	1°
Fine sand	0.125−0.25	3°
Medium sand	0.25−0.05	5°
Coarse sand	0. 5−1	7°
Very coarse sand	1−2	9°
Granules	2−4	11°
Pebbles	4−64	17°
Cobbles	64−256	24°

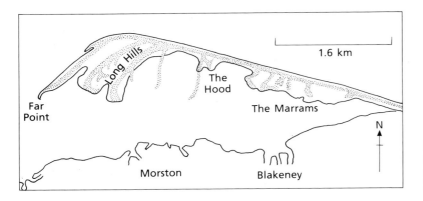

Fig. 3.40 Blakeney Point, north coast of Norfolk, a spit with recurved laterals. (After Steers, 1946.)

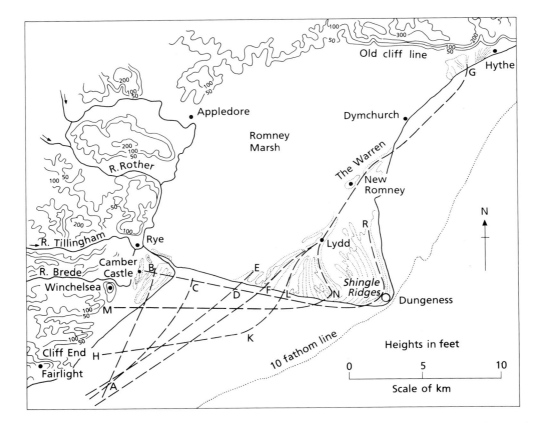

Fig. 3.41 Map of the evolution of Dungeness. Dungeness is conceived to have been a spit formed from the coast in the vicinity of Fairlight (A) and extending eastwards across a broad bay now occupied by Romney Marsh. In the early stages, the sea tended to build up a fan-shaped mass of shingle ridges represented by the phases AB, AC, and AD. Following the early stages of development the spit was extended right across the bay (AEFG). Immediately after this stage the ridge was breached, the most important break being near Fairlight. Robbed of its supply of shingle from the west, the spit began to swing round to face the dominant waves from the south-west. The ensuing series of stages, HKEFG, LME, and BMN, are consistent with the known orientation of shingle ridges on Dungeness. During this period the spit suffered erosion on its south-west-facing portion, the shingle being transported round the Ness and built up into new ridges on the east-facing side, where the dominant waves come from an easterly or north-easterly direction. (After Lewis, 1971.)

drift. Their growth is spasmodic and alternates with episodes of retreat. The stages in the development of many complex spits and forelands are marked by beach ridges which are frequently continuous over long distances. While longshore drift provides the material for construction their development results from spasmodic progradation, by frontal wave accretion during major storms. The distal end of a spit is frequently curved (Fig. 3.40). Those spits which extend from the mainland to link up with an island are known as tombolas.

Bay bars are constructed across the entrance to bays by the growth of a spit being continued from one headland to another. Bays may also be sealed off if spits, which grow from both headlands, merge. If two spits extending from an island meet, then they form a looped bar.

A cuspate bar arises either where a change in a direction of spit growth takes place so that it eventually joins the mainland again, or where two spits coalesce. If progradation occurs, then cuspate bars give rise to cuspate forelands (Fig. 3.41).

Offshore bars or barriers consist of ridges of shingle or sand. They usually project above sea level, extend for several kilometres, and are located a few kilometres offshore.

4   Groundwater conditions and supply

4.1 THE ORIGIN AND OCCURRENCE OF GROUNDWATER

The principal source of groundwater is meteoric water, that is, precipitation (rain, sleet, snow, and hail). However, two other sources are very occasionally of some consequence — juvenile water and connate water. The former is derived from magmatic sources whilst the latter represents the water in which sediments were deposited. Connate water has been trapped in the pore spaces of sedimentary rocks as they were formed and has never been expelled.

The amount of water that infiltrates into the ground depends upon how precipitation is dispersed, i.e. on what proportions are assigned to immediate runoff and to evapotranspiration, the remainder constituting the proportion allotted to infiltration/ percolation. Infiltration refers to the seepage of surface water into the ground, percolation being its subsequent movement, under the influence of gravity, to the zone of saturation. In reality one cannot be separated from the other. The infiltration capacity is influenced by the rate at which rainfall occurs (which also affects the quantity of water available), the vegetation cover, the porosity of the soils and rocks, their initial moisture content, and the position of the zone of saturation.

The retention of water in a soil depends upon the capillary force and the molecular attraction of the particles. As the pores in a soil become thoroughly wetted the capillary force declines so that gravity becomes more effective. In this way downward percolation can continue after infiltration has ceased but as the soil dries, so capillarity increases in importance. No further percolation occurs after the capillary and gravity forces are balanced. Thus water percolates into the zone of saturation when the retention capacity is satisfied. This means that the rains which occur after the deficiency of soil moisture has been catered for are those which count as far as supplementing groundwater is concerned.

4.2 THE WATER TABLE

The pores within the zone of saturation are filled with water, generally referred to as phreatic water. The upper surface of this zone is therefore known as the phreatic surface but is more commonly termed the water table. Above the zone of saturation is the zone of aeration in which both air and water occupy the pores. The water in the zone of aeration is commonly referred to as vadose water. Meinzer (1942) divided this zone into three subzones: (i) soil water; (ii) the intermediate belt; and (iii) the capillary fringe. The uppermost or soil water belt discharges water into the atmosphere in perceptible quantities by evapotranspiration. In the capillary fringe, which occurs immediately above the water table, water is held in the pores by capillary action. An intermediate belt occurs when the water table is far enough below the surface for the soil-water belt not to extend down to the capillary fringe. The degree of saturation decreases from the water table upwards, saturation occurring only in the immediate neighbourhood of the water table.

The geological factors which influence percolation not only vary from one soil or rock outcrop to another but may do so within the same one. This, together with the fact that rain does not fall evenly over a given area, means that the contribution to the zone of saturation is variable, which influences the position of the water table, as do the points of discharge. A rise in the water table as a response to percolation is partly controlled by the rate at which water can drain from the area of recharge. Accordingly it tends to be greatest in areas of low transmissivity (section 4.5.3). Mounds and ridges form in the water table under the areas of greatest recharge. Superimpose upon this the influence of water draining from lakes, streams, and wells and it can be appreciated that a water table is continually adjusting towards equilibrium. Because of the low flow rates in most rocks this equilibrium is rarely, if ever, attained before another disturbance occurs. By using measurements of groundwater levels obtained from wells and by observing the levels at which springs discharge, it is possible to construct groundwater

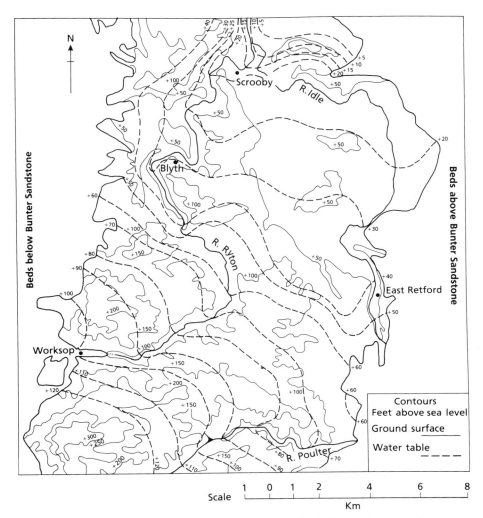

Fig. 4.1 Sketch map of part of Nottinghamshire showing the water table in the Bunter Sandstone (now Sherwood Sandstone).

contour maps showing the form and elevation of the water table (Fig. 4.1).

The water table fluctuates in position, particularly in those climates where there are marked seasonal changes in rainfall. Thus, permanent and intermittent water tables can be distinguished, the former marking the level beneath which the water table does not sink whilst the latter is an expression of the fluctuation. Usually water tables fluctuate within the lower and upper limits rather than between them, especially in humid regions, since the periods between successive recharges are small. The position at which the water table intersects the surface is termed the spring line. Intermittent and permanent springs similarly can be distinguished.

A perched water table is one which forms above a discontinuous impermeable layer such as a lens of clay in a formation of sand, the clay impounding a water mound.

4.3 AQUIFERS, AQUICLUDES, AND AQUITARDS

An aquifer is the term given to a rock or soil mass which not only contains water but from which water can be readily abstracted in significant quantities.

The ability of an aquifer to transmit water is governed by its permeability. Indeed, the permeability of an aquifer usually is in excess of 10^{-5} m/s.

By contrast, a formation with a permeability of less than 10^{-9} m/s is one which, in engineering terms, is regarded as impermeable and is referred to as an aquiclude, for example, clays and shales. Even when such rocks are saturated they tend to impede the flow of water through stratal sequences.

An aquitard is a formation which transmits water at a very slow rate but which, over a large area of contact, may permit the passage of large amounts of water between adjacent aquifers which it separates. Sandy clays provide an example.

An aquifer is described as unconfined when the water table is open to the atmosphere, that is, the aquifer is not overlain by material of lower permeability (Fig. 4.2a). Conversely a confined aquifer is one which is overlain by impermeable rocks. Confined aquifers may have relatively small recharge areas as compared with unconfined aquifers and therefore may yield less water. An aquifer which is overlain and/or underlain by aquitard(s) is described as a leaky aquifer (Fig. 4.2b). Even though these impervious bounding beds offer relatively high resistance to the flow of water through them, large amounts of water may flow from aquitard to aquifer or vice versa as a result of the extensive contact area between them.

Very often the water in a confined aquifer is under piezometric pressure, i.e. there is an excess of pressure sufficient to raise the water above the base of the overlying bed when the aquifer is penetrated by a well. Piezometric pressures are developed when the buried upper surface of a confined aquifer is lower than the water table in the aquifer at its recharge area. Where the piezometric surface is above ground level, then water overflows from a well. Such wells are described as artesian. A synclinal structure is the commonest cause of artesian conditions (Fig. 4.3a). Other geological structures which give rise to artesian conditions are illustrated in Figure 4.3b.

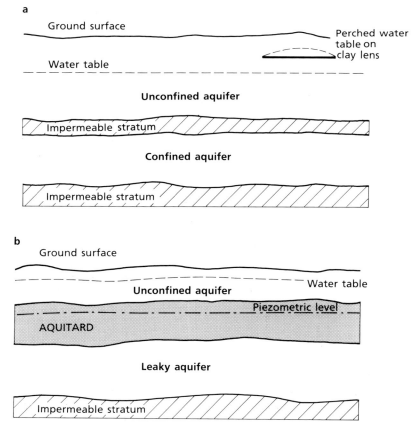

Fig. 4.2 (a) Diagram illustrating unconfined and confined aquifers, with a perched water table in the vadose zone; (b) diagram illustrating a leaky aquifer.

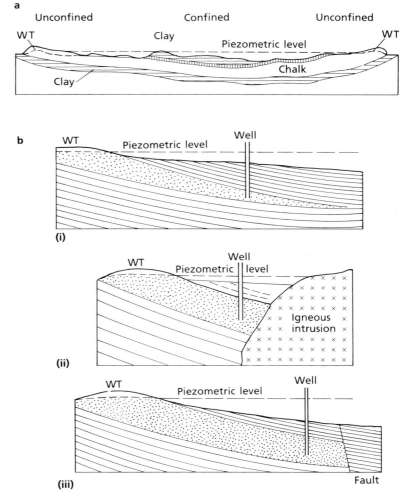

Fig. 4.3 (a) Section across an artesian basin; (b) other examples of artesian conditions (permeable layer, stippled, sandwiched between impermeable beds).

4.4 CAPILLARY MOVEMENT IN SOIL

Capillary movement in a soil refers to the movement of moisture (as a result of surface tension) through the minute pores between the soil particles, which act as capillaries. Therefore moisture can rise from the water table. This movement, however, can occur in any direction, not just vertically upwards. It occurs whenever evaporation takes place from the surface of the soil, thus exerting a 'surface tension pull' on the moisture, the forces of surface tension increasing as evaporation proceeds. Accordingly, capillary moisture is in hydraulic continuity with the water table and is raised against the force of gravity, the degree of saturation decreasing from the water

table upwards. Equilibrium is attained when the forces of gravity and surface tension are balanced.

The boundary separating capillary moisture from the gravitational water in the zone of saturation is ill-defined and cannot be determined accurately. That zone immediately above the water table which is saturated with capillary moisture is referred to as the closed capillary fringe, whilst above this, air and capillary moisture exist together in the pores of the open capillary fringe. The width of the capillary fringe is largely dependent upon the particle size distribution and density of the soil mass, which in turn influence pore size, that is, the smaller the pore size, the greater the width. For example, capillary moisture can rise to great heights in clay soils (Table

Table 4.1 Capillary rises and pressures in soils. (After Jumikis, 1968.)

Soil	Capillary rise (mm)	Capillary pressure (kPa)
Fine gravel	Up to 100	Up to 1.0
Coarse sand	100–150	1.0–1.5
Medium sand	150–300	1.5–3.0
Fine sand	300–1000	3.0–10.0
Silt	1000–10000	10.0–100.0
Clay	Over 10000	Over 100.0

Table 4.2 Soil suction pressure and pF value

pF value	mm water	Equivalent suction (kPa)
0	10	0.1
1	100	1.0
2	1000	10.0
3	10000	100.0
4	100000	1000.0
5	1000000	10000.0

4.1) but the movement is very slow. In soils which are poorly graded the height of the capillary fringe generally varies whereas in uniformly textured soils it attains roughly the same height. Where the water table is at shallow depth and the maximum capillary rise is large, moisture is continually attracted from the water table, due to evaporation from the soil surface, so that the uppermost soil is near saturation.

Below the water table the water contained in the pores is under normal hydrostatic load, the pressure increasing with depth. These pressures exceed atmospheric pressure and are designated positive pressures. The pressures existing in the capillary zone are less than atmospheric and are termed negative pressures. Thus, the water table is usually regarded as a datum of zero pressure between the positive pore pressure below and the negative above.

At each point where moisture menisci are in contact with soil particles the forces of surface tension are responsible for the development of capillary or suction pressure (Table 4.1). The air/water interfaces move into the smaller pores, the radii of curvature of the interfaces decrease, and the soil suction increases. Hence the drier the soil, the higher is the soil suction.

Soil suction is a negative pressure and indicates the height to which a column of water could rise due to such suction. Since this height or pressure may be very large, a logarithmic scale has been adopted to express the relationship between soil suction and moisture content. The latter is referred to as the pF value (Table 4.2).

Soil suction tends to compress soil particles, contributing towards the strength and stability of the soil. There is a particular suction pressure for a particular moisture content in a given soil, the magnitude of which is governed by whether it is becoming wetter or drier. As a clay soil dries out the soil suction may increase to the order of several thousands of kilopascals. However, the strength of a soil attributable to soil suction is only temporary and is destroyed upon saturation. At that point soil suction is zero.

4.5 POROSITY AND PERMEABILITY

Porosity and permeability are the two most important factors governing the accumulation, migration, and distribution of groundwater. Both may change within a rock or soil mass in the course of its geological evolution. And it is not uncommon to find variations in both porosity and permeability per metre of depth beneath the ground surface.

4.5.1 Porosity

The porosity, n, of a rock can be defined as the percentage pore space within a given volume and is expressed as follows:

$$n = V_v/V \times 100 \qquad (4.1)$$

where V_v is the volume of the voids and V is the total volume of the material concerned. A closely related property is the void ratio, e, that is, the ratio of the volume of the voids to the volume of the solids, V_s:

$$e = V_v/V_s \qquad (4.2)$$

Where the ground is fully saturated the void ratio can be derived from:

$$e = mG_s \qquad (4.3)$$

m being the moisture content and G_s the specific gravity. Both the porosity and the void ratio indicate the relative proportion of void volume in the material and the relationships between the two are as follows:

$$n = e/(1 + e) \qquad (4.4)$$

and

$$e = n/(1 - n) \qquad (4.5)$$

Total or absolute porosity is a measure of the total void volume and is the excess of bulk volume over grain volume per unit of bulk volume. It is usually determined as the excess of grain density (i.e. specific gravity), over dry density, per unit of grain density and can be obtained from the following expression:

$$\text{Absolute porosity} = \left\{ 1 - \frac{\text{Dry density}}{\text{Grain density}} \right\} \times 100 \tag{4.6}$$

The effective or net porosity is a measure of the effective void volume of a porous medium and is determined as the excess of bulk volume over grain volume and occluded pore (i.e. a pore with no connection to others) volume. It may be regarded as the pore space from which water can be removed.

The factors affecting the porosity of a rock include particle size distribution, sorting, grain shape, fabric, degree of compaction and cementation, solution effects, and mineralogical composition, particularly the presence of clay particles (Bell, 1978a). The highest porosity is commonly attained when all the grains are the same size. The addition of grains of different sizes lowers porosity and this is, within certain limits, directly proportional to the amount added. Irregularities in grain shape result in a larger possible range of porosity, as irregular forms may theoretically be packed either more tightly or more loosely than spheres. Similarly, angular grains may cause either an increase or a decrease in porosity.

After a sediment has been buried and indurated, several additional factors help determine its porosity. The chief amongst these are closer spacing of grains, deformation and granulation of grains, recrystallization, secondary growth of minerals, cementation, and, in some cases, dissolution. Hence diagenetic changes undergone by a rock may either increase or decrease its original porosity.

The porosity can be determined experimentally by using either the standard saturation method or an air porosimeter. Both tests give an effective value of porosity, although that obtained by the air porosimeter may be somewhat higher because air can penetrate pores more easily than can water.

The porosity of a deposit does not necessarily provide an indication of the amount of water that can be obtained therefrom. Nevertheless the water content of a soil or rock is related to its porosity. The water content, m, of a porous material is usually expressed as the percentage of the weight of the solid material, W_s, that is:

$$m = (W_w/W_s) \times 100 \tag{4.7}$$

where W_w is the weight of the water. The degree of saturation, S_r, refers to the relative volume of water, V_w, in the voids, V_v, and is expressed as a percentage:

$$S_r = (V_w/V_v) \times 100 \tag{4.8}$$

4.5.2 Specific retention and specific yield

The capacity of a material to yield water is of greater importance than is its capacity to hold water as far as supply is concerned. Even though a rock or soil may be saturated, only a certain proportion of water can be removed by drainage under gravity or pumping, the remainder being held in place by capillary or molecular forces. The ratio of the volume of water retained, V_{wr}, to that of the total volume of rock or soil, V, expressed as a percentage, is referred to as the specific retention, S_{re}:

$$S_{re} = V_{wr}/V \times 100 \tag{4.9}$$

The amount of water retained varies directly in accordance with the surface area of the pores and indirectly with regard to the pore space. The specific surface of a particle is governed by its size and shape. For example, particles of clay have far larger specific surfaces than do grains of sand. As an illustration, a grain of sand, 1 mm in diameter, has a specific surface of about $0.002 \, \text{m}^2/\text{g}$, compared with kaolinite, which varies from approximately 10 to $20 \, \text{m}^2/\text{g}$. Hence clays have much higher specific retention than sands (Fig. 4.4).

The specific yield, S_y, of a rock or soil refers to its water yielding capacity attributable to gravity drainage as occurs when the water table declines. It is the

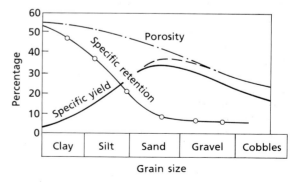

Fig. 4.4 Relationship between grain size, porosity, specific retention, and specific yield. Well-sorted material (– – – –). Average material (———).

Table 4.3 Some examples of specific yield

Material	Specific yield (%)
Gravel	15–30
Sand	10–30
Dune sand	25–35
Sand and gravel	15–25
Loess	15–20
Silt	5–10
Clay	1–5
Till (silty)	4–7
Till (sandy)	12–18
Sandstone	5–25
Limestone	0.5–10
Shale	0.5–5

ratio of the volume of water, after saturation, that can be drained by gravity, V_{wd}, to the total volume of the aquifer, expressed as a percentage; hence:

$$S_y = V_{wd}/V \times 100 \qquad (4.10)$$

The specific yield plus the specific retention is equal to the porosity of the material:

$$n = S_y + S_{re} \qquad (4.11)$$

when all the pores are interconnected. The relationship between the specific yield and particle size distribution is shown in Figure 4.4. In soils the specific yield tends to decrease as the coefficient of uniformity (Chapter 5) increases. Examples of the specific yield of certain common types of soil and rock are given in Table 4.3 (it must be appreciated that individual values of specific yield can vary considerably from those quoted).

4.5.3 Permeability

Permeability may be defined as the ability of a rock to allow the passage of fluids into or through it without impairing its structure. In ordinary hydraulic usage a substance is termed permeable when it permits the passage of a measurable quantity of fluid in a finite period of time and impermeable when the rate at which it transmits that fluid is slow enough to be negligible under existing temperature–pressure conditions (Table 4.4). The permeability of a particular material is defined by its coefficient of permeability or hydraulic conductivity (the term hydraulic conductivity is now frequently used in place of coefficient of permeability, it being the flow in m^3/s through a unit cross-sectional area of a material). The transmissivity or flow in m^3/day

through a section of aquifer 1 m wide under a hydraulic gradient of unity is sometimes used as a convenient quantity in the calculation of groundwater flow instead of the hydraulic conductivity. The transmissivity, T, and the coefficient of permeability, k, are related to each other as follows:

$$T = kH \qquad (4.12)$$

where H is the saturated thickness of the aquifer.

The flow through a unit cross-section of material is modified by temperature, hydraulic gradient, and the hydraulic conductivity. The latter is affected by the uniformity and range of grain size, shape of the grains, stratification, the amount of consolidation and cementation undergone, and the presence and nature of discontinuities. Temperature changes affect the flow rate of a fluid by changing its viscosity. The rate of flow is commonly assumed to be directly proportional to the hydraulic gradient but this is not always so in practice.

Permeability and porosity are not necessarily as closely related as would be expected. For example, very fine textured sandstones frequently have a higher porosity than coarser ones, though the latter are more permeable.

The permeability of a clastic material is also affected by the interconnections between the pore spaces. If these are highly tortuous, then the permeability is reduced. Consequently, tortuosity figures importantly in permeability, influencing the extent and rate of free-water saturation. It can be defined as the ratio of the total path covered by a current flowing in the pore channels between two given points to the straight-line distance between them. The sizes of the throat areas between pores are all important.

Stratification in a formation varies within limits both vertically and horizontally. Frequently it is difficult to predict what effect stratification has on the permeability of the beds. Nevertheless, in the great majority of cases where a directional difference in permeability exists, the greater permeability is parallel to the bedding, for example, ratios of 5:1 are not uncommon in sandstones.

The permeability of intact rock (primary permeability) is usually several orders less than *in situ* permeability (secondary permeability). Hence as far as the assessment of flow through rock masses is concerned, field tests provide more reliable results than can be obtained from testing intact samples in the laboratory. Although the secondary permeability is affected by the frequency, continuity and openness, and amount of infilling of discontinuities, a

Table 4.4 Relative values of permeabilities

Rock types	Porosity Primary (grain) %	Secondary (fracture)*	Permeability range (m/s)	Well yields	Type of water bearing unit
Sediments, unconsolidated			Very high (10^0) – Impermeable (10^{-10})	High / Medium / Low	
Gravel	30–40		Very high–High	High	Aquifer
Coarse sand	30–40		High–Medium	High	Aquifer
Medium to fine sand	25–35		Medium	Medium	Aquifer
Silt	40–50	Occasional	Very low	Low	Aquiclude
Clay, till	45–55	Often fissured	Very low–Impermeable	Low	Aquiclude
Sediments, consolidated					
Limestone, dolostone	1–50	Solutions joints, bedding planes	Very high–Low	High–Low	Aquifer or aquiclude
Coarse, medium sandstone	<20	Joints and bedding planes	Very high–Low	High–Low	Aquifer or aquiclude
Fine sandstone	<10	Joints and bedding planes	Medium–Very low	Medium–Low	Aquifer or aquiclude
Shale, siltstone	—	Joints and bedding planes	Very low–Impermeable	Low	Aquiclude or aquifer
Volcanic rocks, e.g. basalt	—	Joints and bedding planes	Very high–Low	High–Low	Aquifer or aquiclude
Plutonic and metamorphic rocks	—	Weathering and joints decreasing as depth increases	Very low–Impermeable	Low	Aquiclude or aquifer

Permeability range scale (m/s): 10^0 Very high · 10^{-2} High · 10^{-4} Medium · 10^{-6} Low · 10^{-8} Very low · 10^{-10} Impermeable

* Rarely exceeds 10%.

Table 4.5 Estimation of secondary permeability from discontinuity frequency

| Term | k (m/s) | Permeability | |
		Rock mass description	Interval (m)
Very closely to extremely closely spaced discontinuities	Less than 0.2	Highly permeable	$10^{-2}-1$
Closely to moderately widely spaced discontinuities	0.2−0.6	Moderately permeable	$10^{-5}-10^{-2}$
Widely to very widely spaced discontinuities	0.6−2.0	Slightly permeable	$10^{-9}-10^{-5}$
No discontinuities	Over 2.0	Effectively impermeable	Less than 10^{-9}

rough estimate of the permeability can be obtained from their frequency (Table 4.5). Admittedly such estimates must be treated with caution and cannot be applied to rocks which are susceptible to solution.

4.5.4 Storage coefficient

The storage coefficient or storativity, S, of an aquifer has been defined as the volume of water released from or taken into storage per unit surface area of the aquifer, per unit change in head normal to that surface (Fig. 4.5). It is a dimensionless quantity.

Changes in storage in an unconfined aquifer represent the product of the volume of the aquifer between the water table before and after a given period of time and the specific yield. The storage coefficient of an unconfined aquifer virtually corresponds to the specific yield as more or less all the water is released from storage by gravity drainage and only an extremely small part results from compression of the aquifer and the expansion of water. In confined aquifers water is not yielded simply by gravity drainage from the pore space because there is no falling water table and the material remains

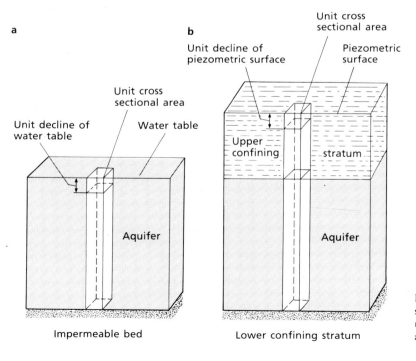

Fig. 4.5 Diagram illustrating the storage coefficient of: (a) an unconfined aquifer; (b) a confined aquifer.

saturated. Hence, other factors are involved regarding yield such as consolidation of the aquifer and expansion of water consequent upon lowering of the piezometric surface. Therefore much less water is yielded by confined than unconfined aquifers.

4.6 FLOW THROUGH SOILS AND ROCKS

Water possesses three forms of energy:
1 potential energy attributable to its height;
2 pressure energy owing to its pressure; and
3 kinetic energy due to its velocity.

The latter can usually be discounted in any assessment of flow through soils. Energy in water is usually expressed in terms of head. The head possessed by water in soils or rocks is manifested by the height to which water will rise in a standpipe above a given datum. This height is usually referred to as the piezometric level and provides a measure of the total energy of the water. If at two different points within a continuous area of water there are different amounts of energy, then there will be a flow towards the point of lesser energy and the difference in head is expended in maintaining that flow. Other things being equal, the velocity of flow between two points is directly proportional to the difference in head between them. The hydraulic gradient, i, refers to the loss of head or energy of water flowing through the ground. This loss of energy by the water is due to the friction resistance of the ground material and this is greater in fine than coarse-grained soils. Thus, there is no guarantee that the rate of flow will be uniform in a formation. Indeed, this is exceptional. However, if it is assumed that the resistance to flow is constant, then for a given difference in head the flow velocity is directly proportional to the flow path.

4.6.1 Darcy's law

Before any mathematical treatment of groundwater flow can be attempted certain simplifying assumptions have to be made:
1 that the material is isotropic and homogeneous;
2 that there is no capillary action; and
3 that a steady state of flow exists.

Since rocks and soils are anisotropic and heterogeneous, as they may be subject to capillary action and flow through them is characteristically unsteady, any mathematical assessment of flow must be treated with caution.

The basic law concerned with flow is that enunciated by Darcy (1856) which states that the rate of flow, v, per unit area is proportional to the gradient of the potential head, i, measured in the direction of flow:

$$v = ki \tag{4.13}$$

and for a particular rock or soil or part of it, of area, A:

$$Q = vA = Aki \tag{4.14}$$

where Q is the quantity in a given time. The ratio of the cross-sectional area of the pore spaces in a soil to that of the whole soil is given by $e/(1 + e)$, where e is the void ratio. Hence a truer velocity of flow, that is, the seepage velocity, v_s, is

$$v_s = [(1 + e)/e]ki \tag{4.15}$$

Darcy's law is valid as long as a laminar flow exists. Departures from this law therefore occur when the flow is turbulent such as when the velocity of flow is high. Such conditions exist in very permeable media, normally when the Reynolds number* can attain values above 4. Accordingly, it is usually accepted that this law can be applied to those soils which have finer textures than gravels. Furthermore, Darcy's law probably does not accurately represent the flow of water through a porous medium of extremely low permeability because of the influence of surface and ionic phenomena, and the presence of gases.

Apart from an increase in the mean velocity, the other factors which cause deviations from the linear laws of flow include:
1 the non-uniformity of pore spaces, since differing porosity gives rise to differences in the seepage rates through pore channels;
2 an absence of a running-in section where the velocity profile can establish a steady-state parabolic distribution;
3 deviations may be developed by perturbations due to jet separation from wall irregularities.

---

* Reynolds number, N_R, is commonly used to distinguish between laminar and turbulent flow and is expressed as follows

$$N_R = \rho \, \frac{vR}{\mu}$$

where ρ is density, v is mean velocity, R is hydraulic radius and μ is dynamic viscosity. Flow is laminar for small values of Reynolds number.

Darcy omitted to recognize that permeability also depends upon the density, ρ, and dynamic viscosity of the fluid, μ, involved, and the average size, D_n, and shape of the pores in a porous medium. In fact, permeability is directly proportional to the unit weight of the fluid concerned and is inversely proportional to its viscosity. The latter is very much influenced by temperature. The following expression attempts to take these factors into account:

$$k = CD_n^2 \, \rho/\mu \qquad (4.16)$$

where C is a dimensionless constant or shape factor which takes note of the effects of stratification, packing, particle size distribution, and porosity. It is assumed in this expression that both the porous medium and the water are mechanically and physically stable, but this may never be true. For example, ionic exchange on clay and colloid surfaces may bring about changes in mineral volume which, in turn, affect the shape and size of the pores. Moderate to high groundwater velocities tend to move colloids and clay particles. Solution and deposition may result from the pore fluids. Small changes in temperature and/or pressure may cause gas to come out of solution which may block pore spaces.

It has been argued that a more rational concept of permeability would be to express it in terms that are independent of the fluid properties. Thus the intrinsic permeability (k_i) characteristic of the medium alone has been defined as:

$$k_i = CD_n^2 \qquad (4.17)$$

However, it has proved impossible to relate C to the properties of the medium. Even with uniform spheres it is difficult to account for the variations in packing arrangement.

The Kozeny–Carmen equation for deriving the coefficient of permeability also takes the porosity, n, into account as well as the specific surface area of the porous medium, S_a, which is defined per unit volume of solid:

$$k = C_0 \frac{n^3}{(1-n)^2 S_a^2} \qquad (4.18)$$

where C_0 is a coefficient, the suggested value of which is 0.2.

4.6.2 General equation of flow

When considering the general case of flow in porous media, it is assumed that the media is isotropic and

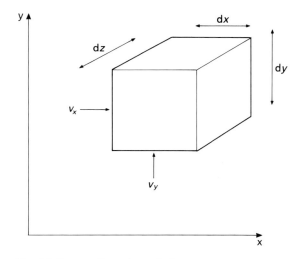

Fig. 4.6 Seepage through a soil element.

homogeneous as far as permeability is concerned. If an element of saturated material is taken, with the dimensions dx, dy, and dz (Fig. 4.6) and flow is taking place in the $x-y$ plane, then the generalized form of Darcy's law is:

$$v_x = k_x i_x \qquad (4.19)$$

$$v_x = k_x \frac{\delta h}{\delta x} \qquad (4.20)$$

and

$$v_y = k_y i_y \qquad (4.21)$$

$$v_y = k_y \frac{\delta h}{\delta y} \qquad (4.22)$$

where h is the total head under steady-state conditions and k_x, i_x and k_y, i_y are respectively, the coefficients of permeability and the hydraulic gradients in the x and y directions. Assuming that the fabric of the medium does not change and that the water is incompressible, then the volume of water entering the element is the same as that leaving in any given time, hence:

$$v_x \mathrm{d}y\mathrm{d}z + v_y \mathrm{d}x\mathrm{d}z = \left(v_x + \frac{\delta v_x}{\delta x}\mathrm{d}x \right)\mathrm{d}y\mathrm{d}z$$
$$+ \left(v_y + \frac{\delta v_y}{\delta y}\mathrm{d}y \right)\mathrm{d}x\mathrm{d}z \qquad (4.23)$$

In such a situation the difference in volume between the water entering and leaving the element is zero, therefore:

$$\frac{\delta v_x}{\delta x} + \frac{\delta v_y}{\delta y} = 0 \tag{4.24}$$

Equation (4.24) is referred to as the flow-continuity equation. If equations (4.20) and (4.22) are substituted in the continuity equation, then:

$$k_x \frac{\delta^2 h}{\delta x^2} + k_y \frac{\delta^2 h}{\delta y^2} = 0 \tag{4.25}$$

If there is a recharge or discharge to the aquifer ($-w$ and $+w$, respectively), then this term must be added to the right-hand side of equation (4.25). If it is assumed that the hydraulic conductivity is isotropic throughout the media so that $k_x = k_y$, then equation (4.25) becomes:

$$\frac{\delta^2 h}{\delta x^2} + \frac{\delta^2 h}{\delta y^2} = 0 \tag{4.26}$$

This is the two-dimensional Laplace equation for steady-state flow in an isotropic porous medium. The partial differential equation governing the two-dimensional unsteady flow of water in an anisotropic aquifer can be written as:

$$T_x \frac{\delta^2 h}{\delta x^2} + T_y \frac{\delta^2 h}{\delta y^2} = S \frac{\delta h}{\delta t} \tag{4.27}$$

where T and S are the coefficients of transmissivity and storage respectively.

4.6.3 Flow through stratified deposits

In a stratified sequence of deposits the individual beds will, no doubt, have different permeabilities, so that vertical permeability will differ from horizontal permeability. Consequently, in such situations, it may be necessary to determine the average values of the coefficient of permeability normal to (k_v) and parallel to (k_h) the bedding. If the total thickness of the sequence is H_T and the thicknesses of the individual layers are H_1, H_2, H_3, $\ldots$, H_n, with corresponding values of the coefficient of permeability k_1, k_2, k_3, $\ldots$, k_n, then k_v and k_h can be obtained as follows:

$$k_v = \frac{H_T}{H_1/k_1 + H_2/k_2 + H_3/k_3 + \ldots + H_n/k_n} \tag{4.28}$$

and

$$k_h = \frac{H_1 k_1 + H_2 k_2 + H_3 k_3 + \ldots + H_n k_n}{H_T}. \tag{4.29}$$

4.6.4 Fissure flow

Generally it is the interconnected systems of discontinuities which determine the permeability of a particular rock mass. The permeability of a jointed rock mass is usually several orders higher than that of intact rock. According to Serafim (1968), the following expression can be used to derive the filtration through a rock mass intersected by a system of parallel-sided joints with a given opening, e, separated by a given distance, d:

$$k = \frac{e^3 \gamma_w}{12 d \mu} \tag{4.30}$$

where γ_w is the unit weight of water and μ its viscosity. The velocity of flow, v, through a single joint of constant gap, e, is expressed by:

$$v = \left(\frac{e 2 \gamma_w}{12 \mu} \right)_i \tag{4.31}$$

where i is the hydraulic gradient.

Wittke (1973) suggested that where the spacing between discontinuities is small in comparison with the dimensions of the rock mass, it is often admissible to replace the fissured rock with regard to its permeability, by a continuous anisotropic medium, the permeability of which can be described by means of Darcy's law. He also provided a resumé of procedures by which three-dimensional problems of flow through rocks under complex boundary conditions could be solved.

4.7 PORE PRESSURES, TOTAL PRESSURES, AND EFFECTIVE PRESSURES

Subsurface water is normally under pressure which increases with increasing depth below the water table to very high values. Such water pressures have a significant influence on the engineering behaviour of most rock and soil masses and their variations are responsible for changes in the stresses in these masses, which affect their deformation characteristics and failure.

The efficiency of a soil in supporting a structure is influenced by the effective or intergranular pressure, that is, the pressure between the particles of the soil which develops resistance to applied load. Because the moisture in the pores offers no resistance to shear, it is effective or neutral and therefore pore pressure has also been referred to as neutral pressure. Since the pore or neutral pressure plus the effective pressure equals the total pressure, reduction in pore pressure increases the effective pressure. Reduction of the pore pressure by drainage consequently affords better conditions for carrying a proposed structure.

The effective pressure at a particular depth is simply obtained by multiplying the unit weight of the soil by the depth in question and subtracting the pore pressure for that depth. In a layered sequence the individual layers may have different unit weights. The unit weight of each should then be multiplied by its thickness and the pore pressure involved subtracted. The effective pressure for the total thickness involved is then obtained by summing the effective pressures of the individual layers (Fig. 4.7). Water held in the capillary fringe by soil suction does not affect the values of pore pressure below the water table. However, the weight of water held in the capillary fringe does increase the weight of overburden and so the effective pressure.

Volume changes brought about by loading compressive soils depend upon the level of effective stress and are not affected by the area of contact. The latter may also be neglected in saturated or near-saturated soils.

There is some evidence which suggests that the law of effective stress as used in soil mechanics, in which the pore pressure is subtracted from all direct stress components, holds true for some rocks. Those with low porosity may, at times, prove the exception. However, Serafim (1968) suggested it appeared that pore pressures have no influence on brittle rocks. This is probably because the strength of such rocks is mainly attributable to the strength of the bonds between the component crystals or grains.

The changes in stresses and the corresponding displacements due, for example, to construction work influence the permeability of a rock mass. For instance, with increasing effective shear stress the permeability increases along discontinuities orientated parallel to the direction of shear stress, whilst it is lowered along those running normal to the shear stress. Consequently, the imposition of shear stresses, and the corresponding strains, lead to an anisotropic permeability within joints.

Piezometers are installed in the ground in order to monitor and obtain accurate measurements of pore water pressures (Fig. 4.8). Observations should be made regularly so that changes due to external factors such as excessive precipitation, tides, the seasons, etc., are noted, it being most important to record the maximum pressures which have occurred. Standpipe piezometers allow the determination of the position of the water table and the permeability. For example, the water level can be measured with

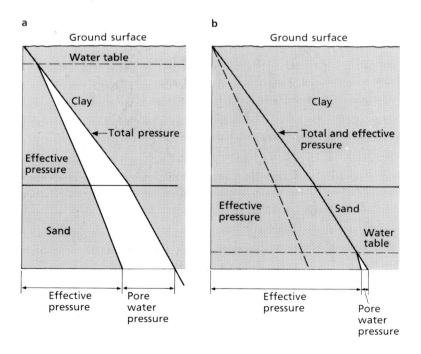

Fig. 4.7 Pressure diagrams to illustrate the influence of lowering the water table on effective pressure: (a) water table just below the ground surface; (b) water table has been lowered into the sand and effective pressure, σ', is increased accordingly. In the clay the effective pressure and total pressure, σ, are the same.

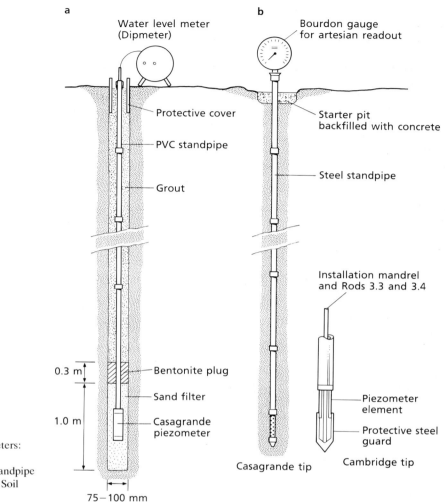

a

Water level meter
(Dipmeter)

Protective cover

PVC standpipe

Grout

0.3 m — Bentonite plug

Sand filter

1.0 m — Casagrande
piezometer

75 – 100 mm

b

Bourdon gauge
for artesian readout

Starter pit
backfilled with concrete

Steel standpipe

Installation mandrel
and Rods 3.3 and 3.4

Piezometer
element

Protective steel
guard

Casagrande tip   Cambridge tip

Fig. 4.8 Standard piezometers:
(a) borehole standpipe
piezometer; (b) drive-in standpipe
piezometers. (Courtesy of Soil
Instruments, Ltd.)

an electric dipmeter or with piezometer tips which have leads going to a constant head permeability unit enabling the rate of flow through the tip to be measured. Hydraulic piezometers can be installed at various depths in a borehole where it is required to determine water pressures. They are connected to a manometer board which records the changes in pore water pressure. Usually simpler types of piezometer are used in the more permeable soils. When a piezometer is installed in a borehole it should be surrounded with a filter of clean sand. The sand should be sealed both above and below the piezometer to enable the water pressures at that particular level to be measured with a minimum of influence from the surrounding strata, since the latter may contain

water at different pressures. The response to piezometers in rock masses can be very much influenced by the incidence and geometry of the discontinuities so that the values of water pressure obtained may be misleading if due regard is not given to these structures.

4.8 CRITICAL HYDRAULIC GRADIENT, QUICK CONDITIONS, AND HYDRAULIC UPLIFT PHENOMENA

As water flows through the soil and loses head, its energy is transferred to the particles past which it is moving, which in turn creates a drag effect on the particles. If the drag effect is in the same direction as the force of gravity, then the effective pressure is increased and the soil is stable. Indeed the soil tends to become more dense. Conversely, if water flows towards the surface, then the drag effect is counter to gravity thereby reducing the effective pressure between particles. If the velocity of upward flow is sufficient it can buoy up the particles so that the effective pressure is reduced to zero. This represents a critical condition where the weight of the submerged soil is balanced by the upward acting seepage force. The critical hydraulic gradient (i_c) can be calculated from the following expression:

$$i_c = \frac{G_s - 1}{1 + e} \qquad (4.32)$$

where G_s is the specific gravity of the particles and e is the void ratio. A critical condition sometimes occurs in silts or sands. If the upward velocity of flow increases beyond the critical hydraulic gradient a quick condition develops.

Quicksands, if subjected to deformation or disturbance, can undergo a spontaneous loss of strength. This loss of strength causes them to flow like viscous liquids. The quicksand phenomenon can be explained as follows:
1 the sand or silt concerned must be saturated and loosely packed;
2 on disturbance the constituent grains become more closely packed which leads to an increase in pore water pressure, reducing the forces acting between the grains. This brings about a reduction in strength. If the pore water can escape very rapidly the loss in strength is momentary;
3 the pore water cannot escape readily. This is fulfilled if the sand/silt has a low permeability and/or the seepage path is long.
Casagrande (1936) demonstrated that a critical porosity existed, above which a quick condition could be developed. He maintained that many coarse-grained sands, even when loosely packed, have porosities approximately equal to the critical condition whilst medium- and fine-grained sands, especially if uniformly graded, exist well above the critical porosity when loosely packed. Accordingly, fine sands tend to be potentially more unstable than coarse-grained varieties and also, finer sands have lower permeabilities.

Quick conditions brought about by seepage forces are frequently encountered in excavations made in fine sands which are below the water table, e.g. in cofferdam work. Because the velocity of the upward seepage force increases as the critical gradient is

exceeded, the soil begins to boil more and more violently. Structures then fail by sinking into the quicksand. Liquefaction of potential quicksands may be caused by sudden shocks such as the action of heavy machinery (notably pile driving), blasting, and earthquakes. Such shocks increase the stress carried by the water, the neutral stress, and give rise to a decrease in the effective stress and shear strength of the soil. A quick condition can also develop in a layered soil sequence where the individual beds have different permeabilities. Hydraulic conditions are particularly unfavourable where water initially flows through a very permeable horizon with little loss of head, which means that flow takes place under a high hydraulic gradient.

Piping refers to erosive action where sediments are removed by seepage forces, so forming subsurface cavities and tunnels. In order that erosion tunnels may form, the soil must have some cohesion — the greater the cohesion, the wider the tunnel. In fact, fine sands and silts, especially those with dispersive characteristics, are most susceptible to piping failures. The danger of piping occurs when the hydraulic gradient is high, that is, when there is a rapid loss of head over a short distance. For example, this has been associated with earth dams. As the pipe develops by backward erosion it nears the source of water supply (e.g. the reservoir) so that eventually the water breaks into and rushes through the pipe. Ultimately the hole, so produced, collapses from lack of support.

Hydraulic uplift phenomena are associated with artesian conditions, that is, where water flowing under pressure through the ground is confined between two impermeable horizons. This can cause serious trouble in excavations and both the position of the water table and the piezometric pressures should be determined before work commences. Otherwise excavations which extend close to strata under artesian pressure may heave or be severely damaged due to blowouts taking place in the floors. Slopes may also fail and such sites may have to be

4.9 GROUNDWATER EXPLORATION

A groundwater investigation requires a thorough appreciation of the hydrology and geology of the area concerned and a groundwater inventory needs to determine possible gains and losses affecting the subsurface reservoir. Of particular interest is the information concerning the lithology, stratigraphical sequence, and geological structure, as well as the hydrogeological characteristics of the subsurface materials. Also of importance are the positions of the water table and piezometric level, and their fluctuations.

In major groundwater investigations, records of precipitation, temperatures, wind movement, evaporation, and humidity may provide essential or useful supplementary information. Similarly, data relating to stream flow may be of value in helping to solve the groundwater equation since seepage into or from streams constitutes a major factor in the discharge or recharge of groundwater. The chemical and bacterial qualities of groundwater obviously require investigation.

Essentially, an assessment of groundwater resources involves the location of potential aquifers within economic drilling depths. Whether or not an aquifer will be able to supply the required amount of water depends on its thickness and spatial distribution, its porosity and permeability, whether it is fully or partially saturated, and whether or not the quality of the water is acceptable. Pumping lift and the effect of drawdown upon it also has to be considered.

The desk study involves a consideration of the hydrological, geological, hydrogeological, and geophysical data available concerning the area in question. Particular attention should be given to assessing the lateral and vertical extent of any potential aquifers, to their continuity and structure, to any possible variations in formation characteristics, and to possible areas of recharge and discharge. Additional information relating to groundwater chemistry, the outflow of springs and surface runoff, data from pumping tests, from mine workings, from waterworks, or meteorological data, should be considered. Information on vegetative cover, land utilization, topography, and drainage pattern can prove of value at times.

Aerial photographs may aid recognition of broad rock and soil types and thereby help locate potential aquifers. The combination of topographical and geological data may help identify areas of likely groundwater recharge and discharge. In particular, the nature and extent of superficial deposits may provide some indication of the distribution of areas of recharge and discharge. Aerial photographs allow the occurrence of springs to be recorded.

Variations in water content in soils and rocks which may not be readily apparent on black and white photographs are often clearly depicted by false colour. The specific heat of water is usually two to ten times greater than that of most rocks, which therefore facilitates its detection in the ground. Indeed, the specific heat of water can cause an aquifer to act as a heat sink which, in turn, influences near-surface temperatures.

Because the occurrence of groundwater is much influenced by the nature of the ground surface, aerial photographs can yield useful information. Also the vegetative cover may be identifiable from aerial photographs and as such may provide some clue as to the occurrence of groundwater. In arid and semi-arid regions, the presence of phreatophytes, that is, plants which have a high transpiration capacity and derive water directly from the water table, indicates that the water table is near the surface. By contrast, xerophytes can exist at low moisture contents in soil and their presence would suggest that the water table was at an appreciable depth. Thus, groundwater prediction maps can sometimes be made from aerial photographs. These can be used to help locate the sites of test wells.

Geological mapping frequently forms the initial phase of exploration and should identify potential aquifers such as sandstones and limestones, and distinguish them from aquicludes. Superficial deposits may perform a confining function in relation to the major aquifers they overlie, or because of their lithology they may play an important role in controlling recharge to major aquifers. Furthermore, geological mapping should locate igneous intrusions and major faults and it is important during the mapping programme to establish the geological structure. Direct subsurface exploration techniques are dealt with in Chapter 7.

4.9.1 Geophysical methods and groundwater investigation

As far as seismic methods are concerned generally velocities in crystalline rocks are high to very high (Table 7.3). Velocities in sedimentary rocks increase concomitantly with consolidation and with increase in the degree of cementation and diagenesis. Unconsolidated sedimentary accumulations have maximum velocities varying as a function of the mineralogy, the volume of voids (either air-filled or water-filled), and grain size.

The porosity tends to lower the velocity of shock waves through a material. Indeed it has been suggested that the compressional wave velocity (V_p) is related to the porosity (n) of a normally consolidated sediment as follows:

$$\frac{1}{V_p} = \frac{n}{V_{pf}} + \frac{1-n}{V_{pl}} \qquad (4.33)$$

where V_{pf} is the velocity in the pore fluid and V_{pl} is the compressional wave velocity for the intact material as determined in the laboratory. The compressional wave velocities may be raised appreciably by the presence of water. Because of the relationship between seismic velocity and porosity, seismic velocity is broadly related to the intergranular permeability of sandstone formations. However, in most sandstones fissure flow makes a more important contribution to groundwater movement than does intergranular or primary flow. A highly fractured or weathered rock mass will exhibit a lower compressional velocity than one which is sound. The effect of discontinuities within a rock mass can be estimated by comparing the *in situ* compressional wave velocity, V_{pf}, with the laboratory sonic velocity, V_{pl}, of an intact core obtained from the rock mass.

The resistivity method does not provide satisfactory quantitative results if the potential aquifer(s) being surveyed are thin, that is, 6 m or less in thickness, especially if they are separated by thick argillaceous horizons. In such situations either cumulative effects are obtained or anomalous resistivities are measured, the interpretation of which is extremely difficult. Also, the resistivity method is more successful when used to investigate a formation which is thicker than the one above it (Hawkins and Chadha, 1990). Most rocks and soils conduct an electric current only because they contain water. But the widely differing resistivity of the various types of pore water can cause variations in the resistivity of soil and rock formations ranging from a few tenths of an ohm-metre to hundreds of ohm-metres. In addition, the resistivity of water changes markedly with temperature and temperature increases with depth. Hence, for each bed under investigation the temperature of both rock and water must be determined or closely estimated and the calculated resistivity of the pore water at that temperature converted to its value at a standard temperature (i.e. 25°C).

As the amount of water present is influenced by the porosity of a rock, the resistivity provides a measure of its porosity. For example, in granular materials in which there are no clay minerals, the relationship between the resistivity, ρ, on the one hand and the density of the pore water, ρ_w, the porosity, n, and the degree of saturation, S_r, on the other is as follows:

$$\rho = a\rho_w n^{-x} S_r^{-y} \qquad (4.34)$$

where a, x, and y are variables (x ranges from 1.0 for sand to 2.5 for sandstone and y is approximately 2.0 when the degree of saturation is greater than 30%). If clay minerals do occur in sands or sandstones, then the resistivity of the pore water is significantly reduced by ion exchange between the latter and the clay minerals so that the above relationship becomes invalid. For those formations which occur below the water table and are therefore saturated, the above expression becomes

$$\rho = a\rho_w n^{-x} \qquad (4.35)$$

since $S_r = 1$ (i.e. 100%).

In a fully saturated sandstone a fundamental empirical relationship exists between the electrical and hydrogeological properties which involves the concept of the formation resistivity factor, F_a, defined as:

$$F_a = \frac{\rho_0}{\rho_w} \qquad (4.36)$$

where ρ_0 is the resistivity of the saturated sandstone and ρ_w is the resistivity of the saturating solution. In a clean sandstone, that is, one in which the electrical current passes through the interstitial electrolyte during testing with the rock mass acting as an insulator, the formation resistivity factor is closely related to the porosity. Worthington (1973) showed that the formation resistivity factor was related to the true formation factor, F, by the expression

$$F = \frac{\rho_A F_a}{\rho_A - F_a \rho_w} \qquad (4.37)$$

in which ρ_A is a measure of the effective resistivity of the rock matrix. In clean sandstone ρ_A is infinitely large, consequently $F = F_a$. Generally F is related to the porosity, n, by the equation:

$$F = a/n^m \qquad (4.38)$$

where a and m are constants for a given formation (e.g., in the Bunter Sandstone of the Fylde region, Lancashire, $a = 1.05$ and $m = 1.47$).

The true formation factor in certain formations has also been shown to be broadly related to intergranular permeability, k_g, by the expression:

$$F = b/k_g^n \qquad (4.39)$$

where b and n are constants for a given formation. In sandstones in which intergranular flow is important, the above expressions can be used to estimate hydraulic conductivity and thence, if the thickness of the aquifer is known, transmissivity. The techniques are not useful in highly indurated sandstones, where the intergranular permeabilities are less than 1.0×10^{-7} m/s, and the flow is controlled by fissures. Similarly these methods are of little value in multilayered aquifers except perhaps where the thicknesses of the different layers are fortuitously distributed so as to allow a complete and definite geophysical interpretation.

Generally speaking magnetic and gravity methods are not used in groundwater exploration except to derive a regional picture of the subsurface geology. They are referred to in Chapter 7, as is geophysical logging of drillholes.

4.9.2 Maps

Isopachyte maps can be drawn to shown the thickness of a particular aquifer and the depth below the surface of a particular bed. They can be used to estimate the positions and depths of drillholes. They also provide an indication of the distribution of potential aquifers.

Maps showing groundwater contours are compiled when there is a sufficient number of observation wells to determine the configuration of the water table. Data on surface water levels in reservoirs and streams that have free connection with the water table also should be used in the production of such maps. These maps are usually compiled for the periods of the maximum, minimum, and mean annual positions of the water table. A water table contour map is most useful for studies of unconfined groundwater.

As groundwater moves from areas of higher potential towards areas of lower potential and as the contours on groundwater contour maps represent lines of equal potential, the direction of groundwater flow moves from highs to lows at right angles to the contours. Analysis of conditions revealed by groundwater contours is made in accordance with Darcy's law. Accordingly spacing of contours is dependent on the flow rate and on aquifer thickness and permeability. If continuity of flow rate is assumed, then the spacing depends upon aquifer thickness and permeability. Hence areal changes in contour spacing may be indicative of changes in aquifer conditions. However, because of the heterogeneity of most aquifers, changes in gradient must be carefully interpreted in relation to all factors. The shape of the contours portraying the position of the water table helps to indicate where areas of recharge and discharge of groundwater occur. Groundwater

mounds can result from the downward seepage of surface water. In an ideal situation the gradient from the centre of such a recharge area will decrease radially and at a declining rate. An impermeable boundary or change in transmissivity will affect this pattern.

Depth-to-water table maps show the depth to water from the ground surface. They are prepared by overlaying a water table contour map on a topographical map of the same area (and scale) and recording the differences in values at the points where the two types of contours intersect. Depth-to-water contours are then interpolated in relation to these points. A map indicating the depth to the water table can also provide an indication of areas of recharge and discharge. Both are most likely to occur where the water table approaches the surface.

Water level change maps are constructed by plotting the change in the position of the water table recorded at wells during a given interval of time (Fig. 4.9). The effect of local recharge or discharge often shows as distinct anomalies on water level change maps, for example, it may indicate that the groundwater levels beneath a river have remained constant while falling everywhere else. This would suggest an influent relationship between the river and aquifer. Hence, such maps can help identify the locations where there are interconnections between surface water and groundwater. These maps also permit an estimation to be made of the change in groundwater storage which has occurred during the lapse in time involved.

4.10 ASSESSMENT OF PERMEABILITY

4.10.1 Assessment of permeability in the laboratory

Permeability is assessed in the laboratory by using either a constant head or falling head permeameter. A constant head permeameter (Fig. 4.10a) is used to measure the permeability of granular materials such as gravels and sands. A sample is placed in a cylinder of known cross-sectional area, A, and water is allowed to move through it under a constant head. The amount of water discharged, Q, in a given period of time, t, together with the difference in head, h, measured by means of manometer tubes, over a given length of sample, l, is obtained. The results are substituted in the Darcy expression and the coefficient of permeability, k, is derived from:

$$Q/t = (Ak)h/l = Aki \qquad (4.40)$$

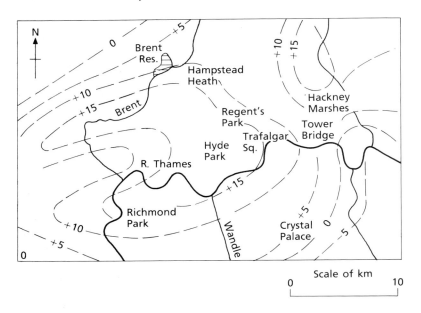

Fig. 4.9 Changes in groundwater levels in the Chalk below London, 1965–80 (all contours in metres). (After Marsh and Davies, 1984.)

where i is the hydraulic gradient.

Determination of the permeability of fine sands and silts, as well as many rock types, is made by using a falling head permeameter (Fig. 4.10b). The sample is placed in the apparatus which is then filled with water to a certain height, h_1, in the standpipe. Then the stopcock is opened and the water infiltrates through the sample, the height of the water in the standpipe falling to h_2. The times at the beginning, t_1, and end, t_2, of the test are re-

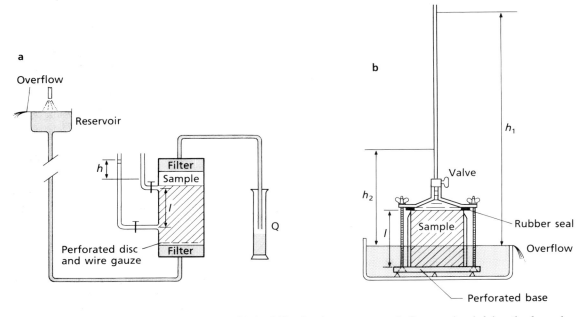

Fig. 4.10 (a) The constant head permeameter; (b) the falling head permeameter, h, Constant head; l, length of sample.

corded. These, together with the cross-sectional area, A, and length of sample, l, are then substituted in the following expression, which is derived from Darcy's law, to obtain the coefficient of permeability, k:

$$k = \frac{2.303al}{A(t_2 - t_1)} \log_{10}\left(\frac{h_1}{h_2}\right) \qquad (4.41)$$

where a is the cross-sectional area of the standpipe. The permeability of clay cannot be measured by using a permeameter, it must be determined indirectly, for example, from the consolidation test.

4.10.2 Assessment of permeability in the field

An initial assessment of the magnitude and variability of the *in situ* hydraulic conductivity can be obtained from tests carried out in boreholes as the hole is advanced. By artificially raising the level of water in the borehole (falling head test) above that in the surrounding ground, the flow rate from the borehole can be measured. However, in very permeable soils it may not be possible to raise the level of water in the borehole. Conversely, the water level in the borehole can be artificially depressed (rising head test) so allowing the rate of water flow into the borehole to be assessed. Wherever possible, a rising and a falling head test should be carried out at each required level and the results averaged.

In a rising or falling head test in which the piezometric head varies with time, the permeability is determined from the expression:

$$k = \frac{A}{F(t_2 - t_1)} \ln\left(\frac{h_1}{h_2}\right) \qquad (4.42)$$

where h_1 and h_2 are the piezometric heads at times t_1 and t_2, respectively, A is the inner cross-sectional area of the casing in the borehole, and F is an intake or shape factor. Where a borehole of diameter, D, is open at the base and cased throughout its depth, $F = 2.75D$. If the casing extends throughout the permeable bed to an impermeable contact, then $F = 2D$.

The constant head method of *in situ* permeability testing is used when the rise or fall in the level of the water is too rapid for accurate recording (i.e. occurs in less than 5 min). This test is normally conducted as an inflow test in which the flow of water into the ground is kept under a sensibly constant head (e.g. by adjusting the rate of flow into the borehole so as to maintain the water level at a mark on the inside of the casing near the top (Sutcliffe and Mostyn, 1983)). The method is only applicable to permeable ground such as gravels, sands, and broken rock, when there is a negligible or zero time for equalization. The rate of flow, Q, is measured once a steady flow into and out of the borehole has been attained over a period of some 10 min. The permeability, k, is derived from the following expression:

$$k = Q/Fh_c \qquad (4.43)$$

where F is the intake factor and h_c is the applied constant head.

The permeability of an individual bed of rock can be determined by a water-injection or packer test carried out in a drillhole (Brassington and Walthall, 1985; Bliss and Rushton, 1984). This is done by sealing off a length of uncased hole with packers and injecting water under pressure into the test section (Fig. 4.11). Usually, because it is more convenient, packer tests are carried out after the entire length of a hole has been drilled. Two packers are used to seal off selected test lengths and the tests are performed from the base of the hole upwards. The hole must be flushed to remove sediment prior to a test being performed. With double packer testing the variation in hydraulic conductivity throughout the test hole is determined. The rate of flow of water over the test length is measured under a range of constant pressures and recorded. The permeability is calculated from a flow-pressure curve. Water is generally pumped into the test section at steady pressures for periods of 15 min, readings of water absorption being taken every 15 min. The test usually consists of five cycles at successive pressures of 6, 12, 18, 12, and 6 kPa for every metre depth of packer below the surface. The evaluation of the 'permeability' from packer tests is normally based upon methods using a relationship of the form

$$k = \frac{Q}{C_s r h} \qquad (4.44)$$

where Q is the steady flow rate under an effective applied head h (corrected for friction losses), r is the radius of the drillhole and C_s is a constant depending upon the length and diameter of the test section.

Field pumping tests allow the determination of the coefficients of permeability and storage, as well as the transmissivity, of a larger mass of ground than the aforementioned tests. A pumping test involves abstracting water from a well at known discharge rate(s) and observing the resulting water levels as

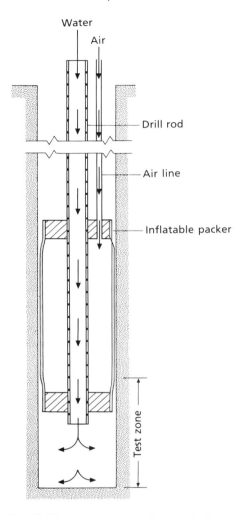

Fig. 4.11 Drillhole packer test equipment. In the double packer test the zone of rock to be tested in a drillhole is isolated by using two packers which seal off the drillhole, the water being pumped into the space between the packers. An alternative method which can be carried out only as drilling proceeds is to use a single packer for testing the zone between the bottom of the packer and the base of the drillhole. The average flow under equilibrium conditions is obtained from a metered water supply acting under a known pressure and gravity head.

drawdown occurs (Lovelock *et al.*, 1975; Anon, 1983b). At the same time the behaviour of the water table in the aquifer can be recorded in observation wells radially arranged about the abstraction well. There are two types of pumping test: (i) the constant-pumping-rate aquifer test and (ii) the step-perform-

ance test. In the former test the rate of discharge is constant whereas in a step-performance test there are a number of stages, each of equal length of time, but at different rates of discharge (Clark, 1977). The step-performance test is usually carried out before the constant-pumping-rate aquifer test. Yield draw-down graphs are plotted from the information obtained (Fig. 4.12). The hydraulic efficiency of the well is indicated by the nature of the curve(s), the more vertical and straighter they are, the more efficient the well.

4.11 ASSESSMENT OF FLOW IN THE FIELD

4.11.1 Flowmeters

A flowmeter log provides a record of the direction and velocity of groundwater movement in a drill-hole. Flowmeter logging requires the use of a velocity-sensitive instrument, a system for lowering the instrument into the hole, a depth-measuring device to determine the position of the flowmeter and a recorder located at the surface. The direction of flow of water is determined by slowly lowering and raising the flowmeter through a section of hole, 6 to 9 m in length and recording velocity measurements during both traverses. If the velocity measured is greater during the downward than the upward traverse, then the direction of flow is upwards and vice versa. A flowmeter log made while a drillhole is being pumped at a moderate rate or by allowing water to flow if there is sufficient artesian head, permits identification of the zones contributing to the discharge. It also provides information on the thickness of these zones and the relative yield at that discharge rate. Because the yield varies approximately directly with the drawdown of water level in a well, flowmeter logs made by pumping, should be pumped at least at three different rates. The drawdown of water level should be recorded for each rate.

4.11.2 Tracers

A number of different types of tracer have been used to investigate the movement of groundwater and the interconnection between surface and ground-water resources. The ideal tracer should be easy to detect quantitatively in minute concentrations; it should not change the hydraulic characteristics of, or be adsorbed by the media through which it is flowing; it should be more or less absent from, and

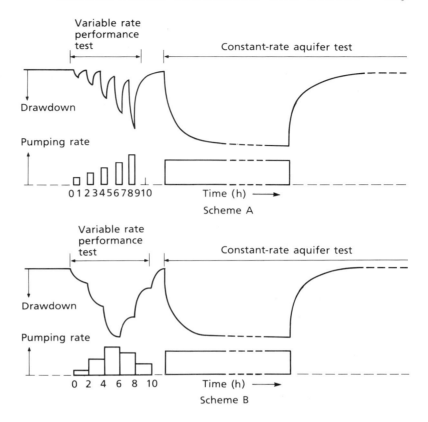

Fig. 4.12 Pumping-test procedure.

should not react with the groundwater concerned; and it should have a low toxicity. The type of tracers in use include: (i) water-soluble dyes which can be detected by colorimetry; (ii) sodium chloride or sulphate salts which can be detected chemically; (iii) strong electrolytes which can be detected by electrical conductivity; and (iv) radioactive tracers, one of the advantages of their use being that they can be detected in minute quantities in water. Radioactive tracers should have a useful half-life and should present a minimum of hazard, for example, tritium is not the best of tracers because of its relatively long half-life. In addition because it is introduced as tritiated water it is preferentially adsorbed by montmorillonite.

When a tracer is injected via a drillhole into groundwater it is subject to diffusion, dispersion, dilution, and adsorption. Dispersion is a result of very small variations in the velocity of laminar flow through porous media. Molecular diffusion is probably negligible unless the velocity of flow is unusually low. Even if these processes are not significant, flow through an aquifer may be stratified or concentrated

along discontinuities. Therefore, a tracer may remain undetected unless the observation drillholes intersect these discontinuities.

The vertical velocity of water movement in a drillhole can be assessed by using tracers. A tracer is injected at the required depth and the direction and rate of movement is monitored by a probe (Ineson and Gray, 1963).

Determination of the hydraulic conductivity in the field can be done by measuring the time it takes for a tracer to move between two test holes. Like pumping tests, this tracer technique is based on the assumption that the aquifer is homogeneous and that observations taken radially at the same distance from the well are comparable. This method of assessing hydraulic conductivity requires that injection and observation wells are close together so as to avoid excessive travel time and so that the direction of flow is known.

4.11.3 Flow nets

Flow nets provide a graphical representation of the

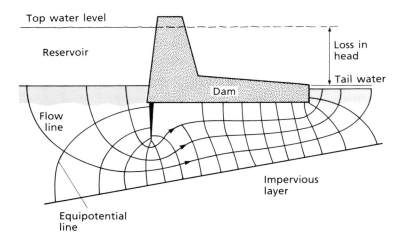

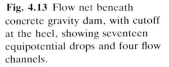

Fig. 4.13 Flow net beneath concrete gravity dam, with cutoff at the heel, showing seventeen equipotential drops and four flow channels.

flow of water through the ground and indicate the loss of head involved (Fig. 4.13). They also provide data relating to the changes in head velocity and effective pressure which occur in a foundation subjected to flowing groundwater conditions. For example, where the flow lines of a flow net move closer together this indicates that the flow increases, although their principal function is to indicate the direction of flow. The equipotential lines indicate equal losses in head or energy as the water flows through the ground, so that the closer they are, the more rapid is the loss in head. Hence, a flow net can provide quantitative data related to the flow problem in question, e.g. seepage pressures can be determined at individual points within the net.

It is possible to estimate the amount of water flowing through a soil from a flow net. If the total loss of head and the permeability of the soil are known, then the quantity of water involved can be calculated by using Darcy's law. However, it is not really as simple as that for the area through which the water flows usually varies, as does the hydraulic gradient, since the flow paths vary in length. By using the total number of flow paths, N_f, the total number of equipotential drops, N_d, and the total loss of head, h_t, together with the permeability, k, in the following expression:

$$Q = kh_t(N_f/N_d) \qquad (4.45)$$

the quantity of water flow, Q, can be estimated.

4.12 GROUNDWATER QUALITY

In any assessment of groundwater resources the quality must be considered. A study is required of changes in groundwater quality from outcrop areas to those where the aquifer is confined and also of changes which occur vertically within an aquifer. Under cover of newer strata or at some depth within an aquifer the natural groundwater flow may be negligible and mineralized water may be present. A knowledge of the variations in quality should help to prevent abstraction where this would lead to deterioration in the quality of supply.

The chemical and biological characteristics of water determine whether or not it can be used for domestic, industrial, or agricultural supply. However, the number of major dissolved constituents in groundwater is quite limited and the natural variations are not as great as might be expected. Nevertheless groundwater is a complex chemical substance which owes its composition mainly to the solution of soluble constituents in, and chemical reactions between, the water and the rocks through which it travels (Table 4.6).

The solution of carbonates, especially calcium and magnesium carbonate, is principally due to the formation of weak carbonic acid in the soil horizons where CO_2 is dissolved by soil water. Calcium in sedimentary rocks is derived from calcite, aragonite, dolomite, anhydrite, and gypsum. In igneous and metamorphic rocks calcium is supplied by the feldspars, pyroxenes, and amphiboles and the less common minerals such as apatite and wollastonite. Because of its abundance calcium is one of the most common ions in groundwater. When calcium carbonate is attacked by carbonic acid, bicarbonate is formed. Calcium carbonate and bicarbonate are the dominant constituents found in the zone of active circulation and for some distance under the cover of

Table 4.6 Chemical analysis of representative groundwaters in England

Location: Aquifer:	Gravesend Chalk	Watford Chalk	Bourne, Lincs Great Oolite	Summerfield, Worcs Sherwood Sandstone	Thornton, Northumberland Fell Sandstone
Ca	280	115	135	40	60
Mg	315	5	9	12	60
Na	2750		18		
K	98	15	4	8	
CO_3	153	147	138	56	
SO_4	700	39	150	26	38
Cl	5000	20	24	19	22
NO_3	35	ND	2	30	
Total dissolved salts	9370	384	491	213	ND
Carbonate hardness	255	245	230	93	ND
Non- carbonate hardness	1755	64	145	97	ND

ND, not determined.
Classification of hardness: soft, under 55 mg/l; slightly hard, 56–100 mg/l; moderately hard, 101–200 mg/l; hard, 201–350 mg/l; very hard, over 351 mg/l.

younger strata. The normal concentration of calcium in groundwater ranges from 10 to 100 mg/l. Such concentrations have no effect on health and it has been suggested that as much as 1000 mg/l may be harmless.

Magnesium, sodium, and potassium are less common cations, and sulphate and chloride and, to some extent, nitrate are less common anions, although in some groundwaters the latter may be present in significant concentrations. Dolomite is the common source of magnesium in sedimentary rocks. The rarer evaporite minerals such epsomite, kierserite, kainite, and carnallite are not significant contributors. Olivine, biotite, hornblende, and augite are among those minerals which make significant contributions in the igneous rocks, and serpentine, talc, diopside, and tremolite are amongst the metamorphic contributors. Despite the higher solubilities of most of its compounds (magnesium sulphate and magnesium chloride are both very soluble), magnesium usually occurs in lesser concentrations in groundwaters than calcium. In this case common concentrations of magnesium range from about 1 to 40 mg/l, concentrations above 100 mg/l are rarely encountered.

Sodium does not occur as an essential constituent of many of the principal rock forming minerals, plagioclase feldspar being the exception. Plagioclase is the primary source of most sodium in groundwaters; in areas of evaporitic deposits halite is important. Sodium salts are highly soluble and will not precipitate unless concentrations of several thousand parts per million are reached. The only common mechanism for removal of large amounts of sodium ions from water is through ion exchange, which operates if the sodium ions are in great abundance. The conversion of calcium bicarbonate to sodium bicarbonate accounts for the removal of some sodium ions from sea water which has invaded freshwater aquifers. This process is reversible. All groundwaters contain measurable amounts of sodium, up to 20 mg/l being the most common concentrations.

Common sources of potassium are the feldspars and micas of the igneous and metamorphic rocks. Potash minerals such as sylvite occur in certain evaporitic sequences but their contribution is not important. Although the abundance of potassium in the Earth's crust is similar to that of sodium, its concentration in groundwaters is usually less than one-tenth that of sodium. Most groundwaters contain less than 10 mg/l. Like sodium, potassium is highly soluble and therefore is not easily removed from water except by ion exchange.

Sedimentary rocks such as shales and clays may contain pyrite or marcasite from which sulphur can be derived. Most sulphate ions are probably derived from the solution of calcium and magnesium sulphate minerals found in evaporitic sequences, gypsum and anhydrite being the most common. The concentration of sulphate ions in water can be affected by sulphate reducing bacteria, the products of which are hydrogen sulphide and carbon dioxide. Hence, a decline in sulphate ions is frequently associated with an increase in bicarbonate ions. Concentration of sulphate in groundwaters is usually less than 100 mg/l and may be less than 1 mg/l if sulphate reducing bacteria are active.

The chloride content of groundwaters may be due to the presence of soluble chlorides from rocks, saline intrusion, connate and juvenile waters, or contamination by industrial effluent or domestic sewage. In the zone of circulation the chloride ion concentration is normally relatively small and it is a minor constituent in the Earth's crust, sodalite and apatite being the only igneous and metamorphic minerals containing chlorite as an essential constituent. Halite is one of the principal mineral sources. As with sulphate ions, the atmosphere probably makes a significant contribution to the chloride content of surface waters, these, in turn, contributing to the groundwaters. Usually the concentration of chloride in groundwater is less than 30 mg/l but concentrations of 1000 mg/l or more are common in arid regions.

Nitrate ions are generally derived from the oxidation of organic matter with a high protein content. Their presence may indicate a pollution source and their occurrence is usually associated with shallow groundwater sources. Concentrations in fresh water do not generally exceed 5 mg/l although in rural areas where nitrate fertilizer is liberally applied concentrations may exceed 600 mg/l.

Although silicon is the second most abundant element in the Earth's crust and is present in almost all the principal rock forming minerals, its low solubility means that it is not one of the most abundant constituents of groundwater. It generally contains between 5 and 40 mg/l although high values may be recorded in water from volcanic rocks.

Iron forms approximately 5% of the Earth's crust and is contained in a great many minerals in rocks as well as occurring as ore bodies. Most iron in solution is ionized.

Ion exchange affects the chemical nature of groundwater. The most common natural cation exchangers are clay minerals, humic acids, and zeolites. The replacement of Ca^{2+} and Mg^{2+} by Na^+ may occur when groundwater moves beneath argillaceous rocks into a zone of more restricted circulation. This produces soft water. Changes in temperature–pressure conditions may result in precipitation, e.g. a decrease in pressure may liberate CO_2 causing the precipitation of calcium carbonate.

Certain dissolved gases such as oxygen and carbon dioxide will alter groundwater chemistry; others will affect the use of water, for example, hydrogen sulphide in concentrations more than 1 mg/l will render water unfit for consumption because of the objectionable odour.

4.13 WELLS

The commonest way of recovering groundwater is to sink a well and lift water from it (Fig. 4.14). The most efficient well is developed so as to yield the greatest quantity of water with the least drawdown and the lowest velocity in the vicinity of the well. The specific capacity of a well is expressed in litres of yield per metre of drawdown when the well is being pumped. It is indicative of the relative permeability of the aquifer. Location of the well is obviously important if an optimum supply is to be obtained and a well site should always be selected after a careful study of the geological setting.

Completion of a well in an unconsolidated formation requires that it be cased so that the surrounding deposits are supported. Sections of the casing must be perforated to allow the penetration of water from the aquifer into the well, or screens can be used. The casing should be as permeable as, or more permeable than, the deposits it confines.

Wells which supply drinking water should be properly sealed. However, an important advantage of groundwater is its comparative freedom from bacterial pollution. For example, water which has percolated through fine-grained sands is usually cleared of bacterial pollution within about 30 m length of flow path. On the other hand flow through open-jointed limestones may transmit pollution for considerable distances. Abandoned wells should be sealed to prevent aquifers being contaminated.

The yield from a well in granular material can be increased by surging which removes the finer particles from the zone about the well. Water supply from wells in rock can be increased by driving gal-

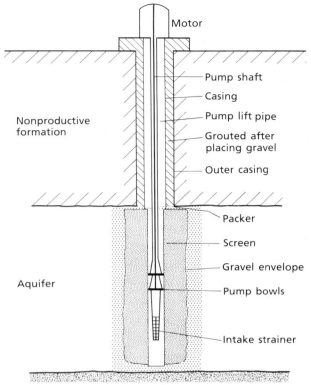

Fig. 4.14 Gravel-packed well installation.

leries or adits from the bottom of deep wells. Yields from rock formations also can be increased by fracturing the rocks with explosives, or with fluid pumped into the well under high pressure, or in the case of carbonate rocks, such as chalk, by using acid to enlarge the discontinuities. The use of explosives in sandstones has led to increases in yield of up to 40% whilst acidification of wells in carbonate rocks has increased yields by over 100%.

When water is abstracted from a well the water table in the immediate vicinity is lowered and assumes the shape of an inverted cone which is named the cone of depression (Fig. 4.15). The steepness of the cone is governed by the soil or rock type, it being flatter in highly permeable materials like openwork gravels than in less permeable chalk. The size of the cone of depression depends on the rate of pumping, equilibrium being achieved when the rate of abstraction balances the rate of recharge. But if abstraction exceeds recharge, then the cone of depression increases in size and its gradual extension from the well may mean that shallow wells within its area of influence dry up.

If these shallow wells are to continue in use, then the offending well should be rested so that the water table may regain its former level. Otherwise they must be sunk deeper, which only accentuates the problem further since this means a further depression of the water table.

4.14 SAFE YIELD AND CONTAMINATION

The abstraction of water from the ground at a greater rate than it is being recharged leads to a lowering of the water table, and upsets the equilibrium between discharge and recharge. The concept of safe yield has been used for many years to express the quantity of water that can be withdrawn from the ground without impairing the aquifer as a water source. Draft in excess of safe yield is overdraft and this can give rise to contamination, or cause serious problems due to severely increased pumping lift. Indeed this may eventually lead to exhaustion.

Estimation of the safe yield is a complex problem

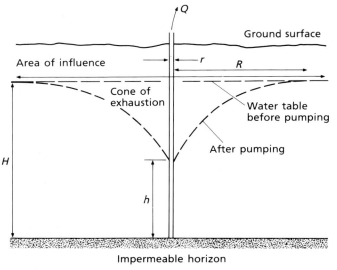

Fig. 4.15 Cone of depression or exhaustion developed around a pumped well in an unconfined aquifer. $Q = \pi k(H^2 - h^2)/(\ln R/r)$. Q, quantity; k, coefficient of permeability.

which must take into account the climatic, geological, and hydrological conditions. As such, the safe yield is likely to vary appreciably with time. Nonetheless the recharge–discharge equation, the transmissivity of the aquifer, the potential sources of contamination, and the number of wells in operation must all be given consideration if an answer is to be found. The safe yield, G, is often expressed as follows:

$$G = P - Q_s - E_T + Q_g - \Delta S_g - \Delta S_s \qquad (4.46)$$

where P is the precipitation on the area supplying the aquifer, Q_s is the surface stream flow over the same area, E_T is the evapotranspiration, Q_g is the net groundwater inflow to the area, ΔS_g is the change in groundwater storage, and ΔS_s is the change of surface storage. With the exception of precipitation all the terms of this expression can be subjected to artificial change. The equation cannot be considered an equilibrium equation or solved in terms of mean annual values. It can be solved correctly only on the basis of specified assumptions for a stated period of years.

Transmissivity of an aquifer may place a limit on the safe yield even though this equation may indicate a potentially large draft. This can only be realized if the aquifer is capable of transmitting water from the source area to the wells at a rate high enough to sustain the draft. Where contamination of the groundwater is possible, then the location of wells, their type, and the rate of abstraction must be planned in such a way that conditions permitting contamination cannot be developed.

Once an aquifer is developed as a source of water supply, then effective management becomes increasingly necessary if it is not to suffer deterioration. Management should not merely be concerned with the abstraction of water but should also consider its utilization, since different qualities of water can be put to different uses. Contamination of water supply is most likely to occur when the level of the water table has been so lowered that all the water that goes underground within a catchment area drains quickly and directly to the wells. Such lowering of the water table may cause reversals in drainage so that water drains from rivers into the groundwater system rather than the other way around. This river water may be contaminated.

4.15 ARTIFICIAL RECHARGE

Artificial recharge may be defined as an augmentation of the natural replenishment of groundwater storage by artificial means. Its main purpose is water conservation, often with improved quality as a second aim, for example, soft river water may be used to reduce the hardness of groundwater. Artificial recharge is therefore used for reducing overdraft, for conserving and improving surface runoff and for increasing available groundwater supplies.

The suitability of a particular aquifer for artificial recharge must be investigated. For example, it must have adequate storage and the bulk of the water recharged should not be lost rapidly by discharge into a nearby river. The hydrogeological and ground-water conditions must be amenable to artificial replenishment. An adequate and suitable source of water for recharge must be available. The source of water for artificial recharge may be storm runoff, river or lake water, water used for cooling purposes, industrial waste water, or sewage water. Many of these sources require some kind of pre-treatment.

Interaction between artificial recharge and groundwater may lead to precipitation, for example, of calcium carbonate and iron and magnesium salts, resulting in lower permeability. Nitrification or deni-trification, and possibly even sulphate reduction, may occur during the early stages of infiltration. Bacterial action may lead to the development of sludges, which reduce the rate of infiltration.

There are several advantages in storing water underground:
1 the cost of artificial recharge may be less than the cost of surface reservoirs and water stored in the ground is not subjected to evapotranspiration;
2 the likelihood of pollution is very much reduced;
3 an aquifer will sometimes act as a distribution system, recharge water moving from one area to another as groundwater at depth;
4 underground storage is important where suitable sites are not available at the surface, and a ground-water basin can be used more fully than surface impoundings;
5 temperature fluctuations of water stored under-ground are reduced.

Artificial recharge may be accomplished by various surface-spreading methods: utilizing basins, ditches, or flooded areas; by spray irrigation; by pumping water into the ground via vertical shafts, horizontal collector wells, pits, or trenches. The most widely practised methods are those of water spreading, which allow increased infiltration to occur over a wide area when the aquifer outcrops at or is near the surface. Therefore these methods require that the ground has a high infiltration capacity. In the basin method water is contained in a series of basins formed by a network of dykes, constructed to take maximum advantage of local topography.

4.16 SUBSIDENCE DUE TO GROUNDWATER ABSTRACTION

Subsidence due to withdrawal of groundwater has developed with most effect in those groundwater basins where there was, or is, intensive abstraction (Poland, 1981). Such subsidence is attributed to the consolidation of sedimentary deposits in which the groundwater is present, consolidation occurring as a result of increasing effective stress. The total over-burden pressure in partially saturated or saturated deposits is borne by their granular structure and the pore water. When groundwater abstraction leads to a reduction in pore water pressure by draining waters from the pores, this means that there is a gradual transfer of stress from the pore water to the granular structure. That is, the effective weight of the deposits in the dewatered zone increases since the buoyancy effect of the pore water is removed. For instance, if the water table is lowered by 1 m, this gives rise to a corresponding increase in average effective over-burden pressure of 10 kPa. By carrying this increased load the granular structure may deform in order to adjust to the new stress conditions. In particular, the porosity of the deposits concerned undergoes a reduction in volume, the surface manifestation of which is subsidence. Subsidence occurs over a longer period of time than that taken for abstraction. How-ever, in aquifers composed of sand and/or gravel the consolidation which takes place due to the increase in effective pressure is more or less immediate.

Consolidation may be elastic or non-elastic de-pending upon the character of the deposits involved and the range of stresses induced by a decline in the water level. In elastic deformation, stress and strain are proportional, and consolidation is independent of time and reversible. Non-elastic consolidation occurs when the granular structure of a deposit is rearranged to give a decrease in volume, that decrease being permanent. Generally, recoverable consolidation represents compression in the pre-consolidation stress range, while irrecoverable con-solidation represents compression due to stresses greater than the pre-consolidation pressure.

According to Lofgren (1968) the storage charac-teristics of compressible formations change sig-nificantly during the first cycle of groundwater withdrawal. Such withdrawal is responsible for permanent consolidation of any fine-grained, inter-bedded formations. The water released by con-solidation represents a one-time, and sometimes important, source of water to wells. During a second cycle of prolonged pumping overdraft, much less water is available to the wells and the water table is lowered much more rapidly. Because of the cyclic nature of groundwater abstraction, the elastic com-ponent of storage change for a given net decline in

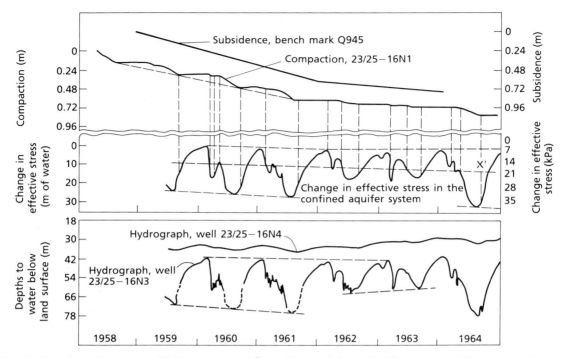

Fig. 4.16 Land subsidence, consolidation, water level fluctuations, and change in effective stress 4.8 km south of Pixley, California. (After Lofgren, 1968.)

water level may be reduced and restored many times while the inelastic component is removed but once. Figure 4.16 illustrates the relationship between fluctuations in water level in semi-confined and confined aquifers, changes in effective stress consolidation, and surface subsidence. It can be seen that each year consolidation commenced during the period of rapid decline in head, continued through the pumping season, and ceased when the head began to recover.

The amount of subsidence which occurs is governed by the increase in effective pressure, the thickness and compressibility of the deposits concerned, the length of time over which the increased loading is applied, and possibly the rate and type of stress applied. For example, the most noticeable subsidences in the Houston–Galveston region of Texas have occurred where the declines in head have been largest and where the thickness of clay in the aquifer system is greatest (Fig. 4.17). Furthermore, the ratio between maximum subsidence and groundwater-reservoir consolidation is related to the ratio between depth of burial and the lateral extent of the reservoir, small reservoirs which are deeply buried do not give

rise to noticeable subsidence, even if subjected to considerable consolidation. By contrast, extremely large underground reservoirs may develop appreciable subsidence.

The rate at which consolidation occurs depends on the rate at which the pore water can drain from the system which, in turn, is governed by its permeability. For instance, the low permeability and high specific storage of aquitards and aquicludes under virgin stress conditions means that the escape of water and resultant adjustment of pore water pressures is slow and time-dependent. Consequently, in fine-grained beds the increase in stress which accompanies the decline in head becomes effective only as rapidly as the pore water pressures are lowered towards equilibrium with the pressures in adjacent aquifers. The time required to reach this stage varies directly according to the specific storage and the square of the thickness of the zone from which drainage is occurring and inversely according to the vertical permeability of the aquitard. In fact, it may take months or years for fine-grained beds to adjust to increases in stress. Moreover, the rate of consolidation of slow-draining aquitards reduces

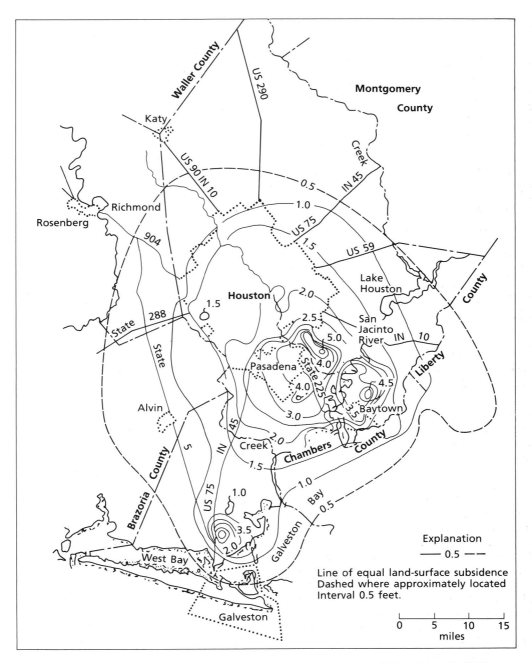

Fig. 4.17 Subsidence of the land surface in the Houston district of Texas, 1943–64. (After Gabrysch, 1976.)

with time and is usually small after a few years of loading.

In addition to being the most prominent effect in subsiding groundwater basins, surface fissuring and faulting (Fig. 4.18) may develop suddenly and therefore pose a greater potential threat to surface structures. In the USA such fissuring and faulting has occurred especially in the San Joaquin Valley, the

Fig. 4.18 Fissure developed by lowering of the water table in loess, Xian City, China. The ground to the right of the fissure is downthrown by some 500 mm. The upper storey of the building immediately behind the wall had to be dismantled due to the effects of the fissure. Eight such fissures have developed in Xian and they may be related to two important faults in the area.

Houston–Galveston region and in central Arizona. These fissures, and more particularly the faults, frequently occur along the periphery of the basin. The faults are high-angled, normal faults, with the downthrow on the side towards the centre of the basin. Generally, displacements along the faults are not great, less than a metre, but movements may continue over a period of years. Holzer *et al.* (1979) related the annual variations in the rate of faulting to annual fluctuations in groundwater levels. Such faults do not extend beneath the zone where stresses due to lowering of the groundwater level occur.

It is not only falling or low groundwater levels that cause problems. A rising or high water table can be equally troublesome. Since the mid-1960s the rate of abstraction from the Chalk and the Sherwood Sand-stone, the two principal aquifers in England, has decreased significantly so that water levels are now rising (by as much as 1 m/year in places in the Chalk, Fig. 4.9). The potential consequences of this include leaks in tunnels and deep basements, reductions in the pile capacity, the possibility of structural settlement, damage to basement floors, and the disruption of utility conduits.

The control of groundwater levels therefore has an importance which extends beyond water supply considerations. If structures are built during a period when the water table is at a particular level (possibly an artificial or atypical level), then care must be taken to ensure that changing water levels do not diminish the integrity of such structures. This obviously requires skilful long-term management of the groundwater resources to ensure that there are no large fluctuations in level. However, it should not be forgotten that aquifers are sometimes deliberately overpumped during a period of water shortage. The usual assumption is that groundwater levels will recover during the following wet season and that no harm will be done. While this may be generally true, it is important that the risks involved in this type of operation are fully realized — not only the risk of subsidence, but also the possibility that mineralized groundwater of considerable age may be drawn up into wells, or that saline intrusion or induced infiltration will occur. Consequently, overpumping should be carried out only infrequently, if at all, and only for short periods of time. Since the long-term hazards potentially far outweigh the short-term gains, the decision to overpump should be taken with caution. Routine monitoring of water quality should always be carried out in such cases and pumping should stop if the quality deteriorates significantly.

4.17 FROST ACTION IN SOIL

Frost action in a soil is influenced by the initial temperature of the soil, as well as the air temperature, the intensity and duration of the freeze period, the depth of frost penetration, the depth of water table, and the type of ground and exposure cover. If frost penetrates down to the capillary fringe in fine-grained soils, especially silts, then, under certain conditions, lenses of ice may be developed. The formation of such ice lenses may, in turn, cause frost heave and frost boil which may lead to the breakup of roads, the failure of slopes, etc. Shrinkage, which gives rise to polygonal cracking in the ground, presents another problem when soils are subjected to freezing. The formation of these cracks is attributable to thermal contraction and desiccation. Water which accumulates in the cracks is frozen and consequently helps increase their size. This water also may aid the development of lenses of ice.

4.17.1 Classification of frozen soil

Ice may occur in frozen soil as small disseminated crystals whose total mass exceeds that of the mineral grains. It may also occur as large tabular masses which range up to several metres thick, or as ice wedges. The latter may be several metres wide and may extend to 10 m or so in depth. As a consequence frozen soils need to be described and classified for engineering purposes. A method of classifying frozen soils involves the identification of the soil type and the character of the ice (Andersland and Anderson, 1978):

1 the character of the actual soil is classified according to the Unified Soil Classification System;
2 the soil characteristics consequent upon freezing are added to the description. Frozen soil characteristics are divided into two basic groups based on whether or not segregated ice can be seen with the naked eye (Table 4.7);
3 the ice present in the frozen soil is classified, this refers to inclusions of ice which exceed 25 mm in thickness.

The amount of segregated ice in a frozen mass of soil depends largely upon the intensity and rate of freezing. When freezing takes place quickly no layers of ice are visible whereas slow freezing produces visible layers of ice of various thicknesses. Ice segregation in soil also takes place under cyclic freezing and thawing conditions.

4.17.2 Mechanical properties of frozen soil

The presence of masses of ice in a soil means that, as far as engineering is concerned, the properties of both have to be taken into account. Ice has no long-term strength, that is, it flows under very small loads. If a constant load is applied to a specimen of ice, instantaneous elastic deformation occurs. This is followed by creep, which eventually develops a steady state. Instantaneous elastic recovery takes place on removal of the load, followed by recovery of the transient creep.

The relative density influences the behaviour of frozen granular soils, especially their shearing resistance, in a manner similar to that when they are unfrozen. The cohesive effects of the ice matrix are superimposed on the latter behaviour and the initial deformation of frozen sand is dominated by the ice matrix. Sand in which all the water is more or less frozen exhibits a brittle type of failure at low strains, for example, at around 2% strain. Alternatively, frozen clay, as well as often containing a lower content of ice than sand, has layers of unfrozen water (of molecular proportions) around the clay particles. These molecular layers of water contribute towards a plastic type of failure.

The water content of granular soils is almost wholly converted into ice at a very few degrees below freezing point. Hence, frozen granular soils exhibit a reasonably high compressive strength only a few degrees below freezing, and there is justification for using this parameter as a design index of their performance in the field, provided that a suitable factor of safety is incorporated. The order of increase in compressive strength with decreasing temperature is shown in Figure 4.19.

In fine-grained sediments the intimate bond between the water and the clay particles results in a significant proportion of soil moisture remaining

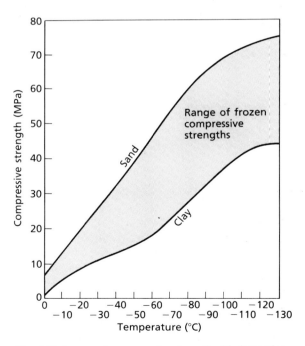

Fig. 4.19 Increase in compressive strength with decreasing temperature.

Table 4.7 Description and classification of frozen soils. (After Andersland and Anderson, 1978.)

I: *Description of soil phase (independent of frozen state)*	Classify soil phase by the Unified Soil Classification System				
	Major group		Subgroup		
	Description	Designation	Description		Designation
	Segregated ice not visible by eye	N	Poorly bonded or friable		Nf
			Well bonded	No excess ice	Nb $\begin{array}{c} n \\ \hline e \end{array}$
				Excess ice	
II: *Description of frozen soil*	Segregated ice visible by eye (ice 25 mm or less thick)	V	Individual ice crystals or inclusions		Vx
			Ice coatings on particles		Ve
			Random or irregularly oriented ice formations		Vr
			Stratified or distinctly oriented ice formations		Vs
III: *Description of substantial ice strata*	Ice greater than 25 mm thick	ICE	Ice with soil inclusions		ICE + soil type
			Ice without soil inclusions		ICE

unfrozen at temperatures as low as −25°C. The more clay material in the soil, the greater is the quantity of unfrozen moisture. As far as the unconfined compressive strength of frozen clays is concerned, there is a dramatic increase in strength with decreasing temperature. In fact, it appears to increase exponentially with the relative proportion of moisture frozen. Using silty clay as an example, the amount of moisture frozen at −18°C is only 1.25 times that frozen at −5°C, but the increase in compressive strength is more than fourfold.

When strain is plotted against time, three stages of creep are apparent under uniform load (Fig. 4.20). At first, strain increases quickly, but then settles at a uniform minimal rate of increase in its second stage. A third, plastic stage is eventually

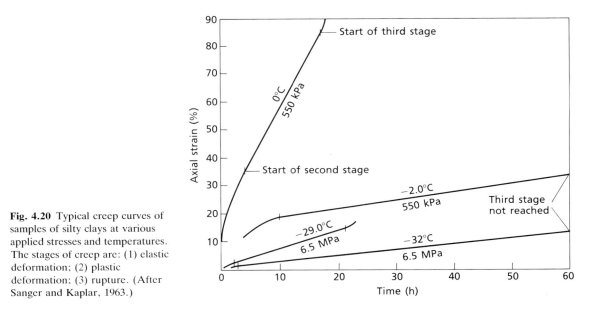

Fig. 4.20 Typical creep curves of samples of silty clays at various applied stresses and temperatures. The stages of creep are: (1) elastic deformation; (2) plastic deformation; (3) rupture. (After Sanger and Kaplar, 1963.)

reached during which complete loss of resistance occurs. This feature is well demonstrated in clay at temperatures at the freezing point of the water.

Because frozen ground is more or less impermeable this increases the problems due to thaw by impeding the removal of surface water. Also, when thaw occurs the amount of water liberated may greatly exceed that originally present in the melted-out layer of the soil (see below). As the soil thaws downwards the upper layers become saturated, and since water cannot drain through the frozen soil beneath, they may suffer a complete loss of strength. Sometimes excess water may act as a transporting agent thereby giving rise to soil flows.

Settlement is associated with thawing of frozen ground. As ice melts, settlement occurs, water being squeezed from the ground by overburden pressure or by any applied loads. Excess pore water pressures develop when the rate of ice melt is greater than the discharge capacity of the soil. Since excess pore pressures can lead to the failure of slopes and foundations, both the rate and amount of thaw settlement should be determined. Pore water pressures should also be monitored. Further consolidation, due to drainage, may occur on thawing. If the soil was previously in a relatively dense state, then the amount of consolidation is small. This situation only occurs in coarse-grained frozen soils containing very little segregated ice. On the other hand, some degree of segregation of ice is always

present in fine-grained frozen soils, for example, lenses and veins of ice may be formed when silts have access to capillary water. Under such conditions the moisture content of the frozen silts significantly exceeds the moisture content present in their unfrozen state. As a result when such ice-rich soils thaw under drained conditions they undergo large settlements under their own weight.

4.17.3 Frost heave

The following factors are necessary for the occurrence of frost heave (reuplift of the ground surface by the development of lens of ice beneath):

1 capillary saturation at the beginning and during the freezing of soil;

2 a plentiful supply of subsoil water; and

3 soil possessing fairly high capillarity together with moderate permeability.

Grain size is another important factor influencing frost heave. For example, gravels, coarse sands, and clays are not particularly susceptible to heave whilst fine sands and silts definitely are. The reason for this is that fine sands and silty soils are associated with high capillary rises but at the same time their voids are large enough to allow moisture to move quickly enough for them to become saturated rapidly. If ice lenses are present in clean gravels or coarse sands, then they simply represent small pockets of moisture which have been frozen. Casagrande (1932) suggested that the particle size critical to the development of heave was 0.02 mm. If the quantity of such particles in a soil is less than 1% no heave is to be expected, but considerable heaving may take place if the amount is over 3% in non-uniform soils and over 10% in very uniform soils.

As heaves amounting to 30% of the thickness of the frozen layer have frequently been recorded, moisture other than that initially present in the frozen layer, must be drawn from below, since water increases in volume by only 9% when frozen. In fact when a soil freezes there is an upward transfer of heat from the groundwater towards the area in which freezing is occurring. The thermal energy, in turn, initiates an upward migration of moisture within the soil. The moisture in the soil can be translocated upwards either in the vapour or liquid phase or by a combination of both. Moisture diffusion by the vapour phase occurs more readily in soils with larger void spaces than in fine-grained soils. If a soil is saturated, migration in the vapour phase cannot take place.

5   Description and behaviour of soils and rocks

5.1 SOIL CLASSIFICATION

Any system of soil classification involves grouping the different soil types into categories which possess similar properties and in so doing provides the engineer with a systematic method of soil description. Casagrande (1948) advanced one of the first comprehensive engineering classifications of soil. In the Casagrande system the coarse-grained soils are distinguished from fine-grained on the basis of particle size. Gravels and sands are the two principal types of coarse-grained soils and in this classification both are subdivided into five subgroups on a basis of grading (Table 5.1). Well graded soils are those in which the particle size distribution extends over a wide range without excess or deficiency in any particular sizes, whereas in uniformly graded soils the distribution extends over a very limited range of particle sizes. In poorly graded soils the distribution contains an excess of some particle sizes and a deficiency of others. A plasticity chart (Table 5.2) is used when classifying fine-grained soils. On this chart the plasticity index is plotted against liquid limit. The 'A' line is taken as the boundary between organic and inorganic soils, the latter lying above the line. Each of the main soil types and subgroups are given a letter, a pair of which are combined in the group symbol, the former being the prefix, the latter the suffix. Subsequently, the Unified Soil Classification (Table 5.2) was developed from the Casagrande system.

The British Soil Classification for engineering purposes (Anon, 1981c) also uses particle size as a fundamental parameter. Boulders, cobbles, gravels, sands, silts, and clays are distinguished as individual groups, each group being given the following symbol and size range:

1 boulders (B), over 200 mm,
2 cobbles (Cb), 60–200 mm,
3 gravel (G), 2–60 mm,
4 sand (S), 0.06–2 mm,
5 silt (M), 0.002–0.06 mm,
6 clay (C), less than 0.002 mm.

The gravel, sand, and silt ranges may be further divided into coarse, medium, and fine categories (Table 5.3). Mixed gravel–sand soils are categorized as shown in Table 5.4. The major soil groups are again divided into subgroups, on a basis of grading in the case of cohesionless soils, and on a basis of plasticity in the case of fine material. Granular soils are described as well graded (W) or poorly graded (P). Two further types of poorly graded granular soils are recognized: uniformly graded (Pu) and gap-graded (Pg). Silts and clays are subdivided according to their liquid limits (LL) into:

Low, under 35%
Intermediate, 35 to 50%
High, 50 to 70%
Very high, 70 to 90%
Extremely high, over 90%

Table 5.1 Symbols used in the Casagrande (1948) soil classification

Main soil type		Prefix
Coarse-grained soils	Gravel	G
	Sand	S
Fine-grained soils	Silt	M
	Clay	C
	Organic silts and clays	O
Fibrous soils	Peat	Pt
Subdivisions		*Suffix*
For coarse-grained soils	Well graded with little or no fines	W
	Well graded with suitable clay binder	C
	Uniformly graded with little or no fines	U
	Poorly graded with little or no fines	P
	Poorly graded with appreciable fines or well graded with excess fines	F
For fine-grained soils	Low compressibility (plasticity)	L
	Medium compressibility (plasticity)	I
	High compressibility (plasticity)	H

Table 5.2(a) Unified Soil Classification. Coarse-grained soils. More than half of material is larger than No. 200 sieve size†

Field identification procedures (excluding particles larger than 76 mm and basing fractions on estimated weights)			Group symbols*	Typical names
Gravels. More than half of coarse fraction is larger than No. 7 sieve size*	Clean gravels (little or no fines)	Wide range in grain size and substantial amounts of all intermediate particle sizes	GW	Well graded gravels, gravel−sand mixtures, little or no fines
		Predominantly one size or a range of sizes with some intermediate sizes missing	GP	Poorly graded gravels, gravel−sand mixtures, little or no fines
	Gravels with fines (appreciable amount of fines)	Non-plastic fines (for identification procedures see ML below)	GM	Silty gravels, poorly graded gravel−sand-silt mixtures
		Plastic fines (for identification procedures, see CL below)	GC	Clayey gravels, poorly graded gravel−sand−clay mixtures
Sands. More than half of coarse fraction is smaller than No. 7 sieve size‡	Clean sands (little or no fines)	Wide range in grain sizes and substantial amounts of all intermediate particle sizes	SW	Well graded sands, gravelly sands, little or no fines
		Predominantly one size or a range of sizes with some intermediate sizes missing	SP	Poorly graded sands, gravelly sands, little or no fines
	Sands with fines (appreciable amount of fines)	Non-plastic fines (for identification procedures, see ML below)	SM	Silty sands, poorly graded sand-silt mixtures
		Plastic fines (for identification procedures, see CL below)	SC	Clayey sands poorly graded sand−clay mixtures

* Boundary classifications: Soils possessing characteristics of two groups are designated by combinations of group symbols. For example GW–GC, well graded gravel−sand mixture with clay binder; † All sieve sizes on this chart are US standard. The No. 200 sieve size is about the smallest particle visible to the naked eye; ‡ For visual classification, the 6.3 mm size may be used as equivalent to the No. 7 sieve size.

Field identification procedure for fine-grained soils or fractions: These procedures are to be performed on the minus No. 40 sieve size particles, approximately 0.4 mm. For field classification purposes, screening is not intended, simply remove by hand the coarse particles that interfere with the tests.

Dilatancy (reacting to shaking): After removing particles larger than No. 40 sieve size, prepare a pat of moist soil with a volume of about 1 cm^3. Add enough water if necessary to make the soil soft but not sticky. Place the pat in the open palm of one hand and shake horizontally, striking vigorously against the other hand several times. A positive reaction consists of the appearance of water on the surface of the pat which changes to a livery consistency and becomes glossy. When the sample is squeezed between the fingers, the water and gloss disappear from the surface, the pat stiffens and finally it cracks and crumbles. The rapidity of appearance of water during shaking and of its disappearance during squeezing assist in identifying the character of the fines in a soil. Very fine clean sands give the quickest and most distinct reaction whereas a plastic clay has no reaction. Inorganic silts, such as a typical rock flour, show a moderately quick reaction.

Dry strength (crushing characteristics): After removing particles larger than No. 40 sieve size, mould a pat of soil to the consistency of putty, adding water if necessary. Allow the pat to dry completely by oven, sun or air drying, and then test its

Information required for describing soils		Laboratory classification criteria	
Give typical name; indicate approximate percentages of sand and gravel; maximum size; angularity, surface condition, and hardness of the coarse grains: local or geologic name and other pertinent descriptive information: and symbols in parentheses	Use grain size curve in identifying the fractions as given under field identification	Determine percentages of gravel and sand from grain-size curve. Depending on fines (fraction smaller than No. 200 sieve size) coarse-grained soils are classified as follows: Less than 5%: GW, GP, SW, SP. More than 12%: GM, GC, SM, SC. 5% to 12%: Borderline cases require use of dual symbols	$C_u = \dfrac{D_{60}}{D_{10}}$ Greater than 4 $C_e = \dfrac{(D_{30})^2}{D_{10} \times D_{60}}$ Between 1 and 3
			Not meeting all gradation requirements for GW
For undisturbed soils add information on stratification, degree of compactness, cementation, moisture conditions, and drainage characteristics			Atterberg limits below 'A' line, or PI less than 4 — Atterberg limits above 'A' line with PI greater than 7 / Above 'A' line with PI between 4 and 7 are borderline cases requiring use of dual symbols
			$C_u = \dfrac{D_{60}}{D_{10}}$ Greater than 6 $C_e = \dfrac{(D_{30})^2}{D_{10} \times D_{60}}$ Between 1 and 3
Example: Silty sand, gravelly; about 20% hard, angular gravel particles 12.5 mm maximum size; rounded and subangular sand grains coarse to fine, about 15% nonplastic fines with low dry strength; well compacted and moist in place; alluvial sand; (SM)			Not meeting all gradation requirements for SW
			Atterberg limits below 'A' line with PI less than 5 — Atterberg limits above 'A' line with PI greater than 7 / Above 'A' line with PI between 4 and 7 are borderline cases requiring use of dual symbols

strength by breaking and crumbling between the fingers. This strength is a measure of the character and quantity of the colloidal fraction contained in the soil. The dry strength increases with increasing plasticity. High dry strength is characteristic for clays of the CH group. A typical inorganic silt possesses only very slight dry strength. Silty fine sands and silts have about the same slight dry strength, but can be distinguished by the feel when powdering the dried specimen. Fine sand feels gritty whereas a typical silt has the smooth feel of flour.

Toughness (consistency near plastic limit): After removing particles larger than the No. 40 sieve size, a specimen of soil about one 1 cm^3 in size, is moulded to the consistency of putty. If too dry, water must be added and if sticky, the specimen should be spread out in a thin layer and allowed to lose some moisture by evaporation. Then the specimen is rolled out by hand on a smooth surface or between the palms into a thread about 3 mm in diameter. The thread is then folded and re-rolled repeatedly. During this manipulation the moisture content is gradually reduced and the specimen stiffens, finally loses its plasticity, and crumbles when the plastic limit is reached.

After the thread crumbles, the pieces should be lumped together and a slight kneading action continued until the lump crumbles. The tougher the thread near the plastic limit and the stiffer the lump when it finally crumbles, the more potent is the colloidal clay fraction in the soil. Weakness of the thread at the plastic limit and quick loss of coherence of the lump below the plastic limit indicate either inorganic clay of low plasticity, or materials such as kaolin-type clays and organic clays which occur below the A-line. Highly organic clays have a very weak and spongy feel at the plastic limit.

Table 5.2(b) Unified Soil Classification. Fine-grained soils. More than half of material is smaller than No. 200 sieve size[b]

Identification procedures on fraction smaller than No. 40 sieve size				Group symbols[a]	Typical Names
	Dry strength (crushing characteristics)	Dilatancy (reaction to shaking)	Toughness (consistency near plastic limit)		
Silts and clays liquid limit less than 50	None to slight	Quick to slow	None	ML	Inorganic silts and very fine sands, rock flour, silty or clayey fine sands with slight plasticity
	Medium to high	None to very slow	Medium	CL	Inorganic clays of low to medium plasticity, gravelly clays, sandy clays, silty clays, lean clays
	Slight to medium	Slow	Slight	OL	Organic silts and organic silt-clays of low plasticity
Silts and clays liquid limit greater than 50	Slight to medium	Slow to none	Slight to medium	MH	Inorganic silts micaceous or diatomaceous fine sandy or silty soils, clastic silts[b]
	High to very high	None	High	CH	Inorganic clays of high plasticity, fat clays
	Medium to high	None to very slow	Slight to medium	OH	Organic clays of medium to high plasticity
Highly organic soils	Readily identified by colour, odour, spongy feel, and frequently by fibrous texture			Pt	Peat and other highly organic soils

See footnotes to Table 5.2(a)

Each subgroup is given a combined symbol in which the letter describing the predominant size fraction is written first (e.g. GW, well graded gravels; CH, clay with high liquid limit). Any group may be referred to as organic if it contains a significant proportion of organic matter, in which case the letter O is suffixed to the group symbol (e.g. CVSO = organic clay of very high liquid limit with sand). The symbol Pt is given to peat. The British Soil Classification can be made either by rapid assessment in the field or by full laboratory procedure (Tables 5.3(a) and 5.3(b), respectively).

The proportions of boulders and cobbles are recorded separately in the British Soil Classification. Their presence should be recorded in the soil description, a plus sign being used in symbols for

Information required for
describing soils

Laboratory classification criteria

Give typical name: indicate degree
and character of plasticity, amount
and maximum size of coarse
grains, colour in wet condition,
odour if any, local or geological
name, and other pertinent
descriptive information and
symbol in parentheses

For undisturbed soils add
information on structure,
stratification, consistency in
undisturbed and remoulded states,
moisture and drainage conditions

Example:
Clayey silt, brown: slightly plastic;
small percentage of fine sand,
numerous vertical root holes; firm
and dry in place; loess; (ML)

Use grain-size curve in identifying the fractions as given under field identification

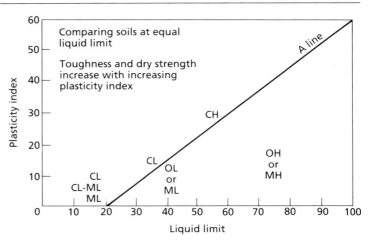

Plasticity chart for laboratory classification of fine-grained soils

soil mixtures, for example, G + Cb for gravel with cobbles. Very coarse deposits are classified as follows:

1 Boulders: over half of the very coarse material is of boulder size (over 200 mm). May be described as cobbly boulders if cobbles are an important second constituent in the very coarse fraction.

2 Cobbles: over half of the very coarse material is of cobble size (200 to 60 mm). May be described as bouldery cobbles if boulders are an important second constituent in the very coarse fraction.

Mixtures of very coarse material and soil are described as shown in Table 5.5.

5.2 COARSE-GRAINED SOILS

The microstructure of a sand or gravel refers to its particle arrangement which also includes its packing.

Table 5.3(a) Field identification and description of soils. (After Anon, 1981c.)

	Basic soil type	Particle size (mm)	Visual identification	Particle nature and plasticity	Composite soil types (mixtures of basic soil types)
Very coarse soils	Boulders		Only seen complete in pits or exposures	Particle shape:	Scale of secondary constituents with coarse soils
	Cobbles	—— 200 —— 60	Often difficult to recover from boreholes	Angular Subangular	Term — % of clay or silt
Coarse soils (over 65% sand and gravel sizes)	Gravels — Coarse / Medium / Fine	20 / 6 / 2	Easily visible to naked eye; particle shape can be described; grading can be described. Well graded: wide range of grain sizes, well distributed. Poorly graded: not well graded (May be uniform: size of most particles lies between narrow limits: or gap graded: an intermediate size of particle is markedly under-represented)	Subrounded Rounded Flat Elongate	slightly clayey / slightly silty } GRAVEL† or SAND — under 5; clayey / silty } GRAVEL or SAND — 5 to 15; very clayey / very silty } GRAVEL or SAND — 15 to 35
	Sands — Coarse / Medium / Fine	0.6 / 0.2 / 0.06	Visible to naked eye; very little or no cohesion when dry: grading can be described. Well graded: wide range of grain sizes, well distributed. Poorly graded: not well graded (May be uniform: size of most particles lies between narrow limits; or gap graded: an intermediate size of particle is markedly under-represented)	Texture: Rough Smooth Polished	Sandy GRAVEL / Gravelly SAND } Sand or gravel as important second constituent of the coarse fraction. For composite types described as: clayey: fines are plastic, cohesive: silty: fines non-plastic or of low plasticity
Fine soils (over 35% silt and clay sizes)	Silts — Coarse / Medium / Fine	0.02 / 0.006 / 0.002	Only coarse silt barely visible to naked eye; exhibits little plasticity and marked dilatancy: slightly granular or silky to the touch. Disintegrates in water; lumps dry quickly; possess cohesion but can be powdered easily between fingers	Non-plastic or low plasticity	Scale of secondary constituents with fine soils. Term — % of sand gravel. sandy / gravelly } CLAY or SILT — 35 to 65
	Clays		Dry lumps can be broken but not powdered between the fingers; they also disintegrate under water but more slowly than silt; smooth to the touch; exhibits plasticity but no dilatancy: sticks to the fingers and dries slowly: shrinks appreciably on drying usually showing cracks. Intermediate and high plasticity clays show these properties to a moderate and high degree, respectively	Intermediate plasticity (Lean clay) / High plasticity (Fat clay)	CLAY:SILT — under 35. Examples of composite types (Indicating preferred order for description). Loose, brown, subangular very sandy, fine to coarse gravel with small pockets of soft grey CLAY
Organic soils	Organic clay, silt or sand	Varies	Contains substantial amounts of organic vegetable matter		Medium dense, light brown, clayey, fine and medium SAND. Stiff, orange brown, fissured sandy CLAY
	Peats	Varies	Predominantly plant remains usually dark brown or black in colour, often with distinctive smell; low bulk density		Firm, brown, thinly laminated silt and CLAY. Plastic, brown, amorphous PEAT

† Principal soil type given in capital letters

Table 5.3(a) *Continued*

Compactness/strength		Structure			Colour
Term	**Field test**	**Term**	**Field identification**	**Interval scales**	**Colour**
Loose	By inspection of voids and particle packing	Homogeneous	Deposit consists essentially of one type	Scale of bedding spacing	Red Pink Yellow Brown Olive
Dense		Interstratified	Alternating layers of varying types or with bands or lenses of other materials. Interval scale for bedding spacing may be used	Term — Mean spacing (mm)	Green Blue White Grey Black etc.
				Very thickly bedded — Over 2000	
				Thickly bedded — 2000–600	
				Medium bedded — 600–200	
Loose	Can be excavated with a spade; 50 mm wooden peg can be easily driven	Heterogeneous	A mixture of types	Thinly bedded — 200–60	
				Very thinly bedded — 60–20	
Dense	Requires pick for excavation; 50 mm wooden peg hard to drive	Weathered	Particles may be weakened and may show concentric layering	Thickly laminated — 20–6	Supplemented as necessary with: Light
Slightly cemented	Visual examination; pick removes soil in lumps which can be abraded			Thinly laminated — under 6	Dark Mottled etc.
Soft or loose	Easily moulded or crushed in the fingers				and
Firm or dense	Can be moulded or crushed by strong pressure in the fingers	Fissured	Break into polyhedral fragments along fissures. Interval scale for spacing of discontinuities may be used		Pinkish Reddish Yellowish Brownish etc.
Very soft	Exudes between fingers when squeezed in hand	Intact	No fissures		
Soft	Moulded by light finger pressure	Homogeneous	Deposit consists essentially of one type	Scale of spacing of other discontinuities	
Firm	Can be moulded by strong finger pressure	Interstratified	Alternating layers of varying types. Interval scale for thickness of layers may be used. Usually has crumb or columnar structure	Term — Mean spacing (mm)	
Stiff	Cannot be moulded by fingers. Can be indented by thumb	Weathered		Very widely spaced — Over 2000	
Very stiff	Can be indented by thumb nail			Widely spaced — 2000–600	
				Medium spaced — 600–200	
Firm	Fibres already compressed together			Closely spaced — 200–60	
Spongy	Very compressible and open structure	Fibrous	Plant remains recognizable and retain some strength	Very closely spaced — 60–20	
Plastic	Can be moulded in hand, and smears fingers	Amorphous	Recognizable plant remains absent	Extremely closely spaced — under 20	

Table 5.3(b) British Soil Classification System for engineering purposes. (After Anon, 1981c.)

Soil groups (see note 1)		Subgroups and laboratory identification				
		Group symbol (see notes 2 & 3)	Subgroup symbol (see note 2)	Fines (% less than 0.06 mm)	Liquid limit (%)	Name
GRAVEL and SAND may be qualified sandy gravel and gravelly sand, etc., where appropriate						
Coarse soils less than 35% of the material is finer than 0.06 mm	GRAVELS More than 50% of coarse material is of gravel size (coarser than 2 mm)	G				
	Slightly silty or clayey GRAVEL	GW	GW	0 to 5		Well graded GRAVEL
		GP	GPu GPg	0 to 5		Poorly graded/uniform/gap-graded GRAVEL
	Silty GRAVEL	G–F	GM	GWM GPM	5 to 15		Well graded/poorly graded silty GRAVEL
	Clayey GRAVEL		G–C	GWC GPC	5 to 15		Well graded/poorly graded clayey GRAVEL
	Very silty GRAVEL	GF	GM	GML, etc.	15 to 35		Very silty GRAVEL; subdivide as for GC
	Very clayey GRAVEL		GC	GCL GCI GCH GCV GCE	15 to 35		Very clayey GRAVEL (clay of low, intermediate, high, very high, extremely high plasticity)
	SANDS More than 50% of coarse material is of sand size (finer than 2 mm)	S				
	Slightly silty or clayey SAND	SW	SW	0 to 5		Well graded SAND
		SP	SPu SPg	0 to 5		Poorly graded/uniform/gap-graded SAND
	Silty SAND	S–F	S–M	SWM SPM	5 to 15		Well graded/poorly graded silty SAND
	Clayey SAND		S–C	SWC SPC	5 to 15		Well graded/poorly graded clayey SAND
	Very silty SAND	SF	SM	SML, etc.	15 to 35		Very silty SAND; subdivided as for SC
	Very clayey SAND		SC	SCL SCI SCH SCV SCE	15 to 35		Very clayey SAND (clay of low, intermediate, high, very high, extremely high plasticity)

FINE SOILS more than 35% of the material is finer than 0.06 mm	Gravelly or sandy SILTS AND CLAYS 35% to 65% fines	Gravelly SILT	FG	MG	MLG, etc.	Gravelly SILT; subdivide as for CG
		Gravelly CLAY (see note 4)		CG	CLG / CIG / CHG / CVG / CEG	<35 → Gravelly CLAY of low plasticity; 35 to 50 → of intermediate plasticity; 50 to 70 → of high plasticity; 70 to 90 → of very high plasticity; >90 → of extremely high plasticity
	SILTS and CLAYS 65% to 100% fines	Sandy SILT (see note 4)	FS	MS	MLS, etc.	Sandy SILT; subdivide as for CG
		Sandy CLAY		CS	CLS, etc.	Sandy CLAY; subdivide as for CG
		SILT (M-soil)	F	M	ML, etc.	SILT; subdivide as for C
		CLAY (see notes 5 & 6)		C	CL / CI / CH / CV / CE	<35 → CLAY of low plasticity; 35 to 50 → of intermediate plasticity; 50 to 70 → of high plasticity; 70 to 90 → of very high plasticity; >90 → of extremely high plasticity
Organic soils		Descriptive letter 'O' suffixed to any group or subgroup symbol.				Organic matter suspected to be a significant constituent. Example MHO: organic SILT of high plasticity.
Peat		Pt				Peat soils consist predominantly of plant remains which may be fibrous or amorphous.

Note 1, the name of the soil group should always be given when describing soils, supplemented, if required, by the group symbol, although for some additional applications (e.g. longitudinal sections) it may be convenient to use the group symbol alone.

Note 2, the group symbol or subgroup symbol should be placed in brackets if laboratory methods have not been used for identification, e.g. (GC).

Note 3, the designation FINE SOIL, or FINES, F, may be used in place of SILT, M, or CLAY, C, when it is not possible or not required to distinguish between them.

Note 4, gravelly if more than 50% of coarse material is of gravel size. Sandy if more than 50% of coarse material is of sand size.

Note 5, SILT (M-soil), M, is material plotting below the 'A' line, and has a restricted plastic range in relation to its liquid limit, and relatively low cohesion. Fine soils of this type include clean silt-sized materials and rock flour, micaceous and diatomaceous soils, pumice, and volcanic soils, and soils containing halloysite. The alternative term 'M-soil' avoids confusion with materials of predominantly silt size, which form only a part of the group. Organic soils also usually plot below the 'A' line on the plasticity chart, when they are designated ORGANIC SILT, MO.

Note 6, CLAY, C, is material plotting above the 'A' line, and is fully plastic in relation to its liquid limit.

Table 5.4 Gravel—sand mixtures

Term	Composition of the coarse fraction
Slightly sandy GRAVEL	Up to 5% sand
Sandy GRAVEL	5—20% sand
Very sandy GRAVEL	Over 20% sand
GRAVEL—SAND	About equal proportions of gravel and sand
Very gravelly SAND	Over 20% gravel
Gravelly SAND	20—5% gravel
Slightly gravelly SAND	Up to 5% gravel

If grains approximate to spheres, then the closest type of systematic packing is rhombohedral packing, whereas the most open type is cubic packing, the porosities approximating to 26% and 48%, respectively (Table 5.6). Put another way, the void ratio of a well-sorted and perfectly cohesionless aggregate of equidimensional grains can range between values of about 0.35 and 1.00. If the void ratio is more than unity the microstructure is collapsible or metastable.

Size and sorting have a significant influence on the engineering behaviour of granular soils. Generally, the larger the particles, the higher the strength, and deposits consisting of a mixture of different sized particles are usually stronger than those which are uniformly graded. However, the behaviour of such sediments depends on their relative density:

$$\text{Relative density } (D_r) = \frac{e_{max} - e}{e_{max} - e_{min}}$$

where e_{max} is the maximum void ratio, e_{min} is the minimum void ratio, and e is the naturally occurring void ratio.

For example, if the relative density of a sand varies erratically this can give rise to differential settlement.

Table 5.5 Mixtures of very coarse materials and soil

Term	Composition
BOULDERS (or COBBLES) with a little finer material[a]	Up to 5% finer material
BOULDERS (or COBBLES) with some finer material[a]	5—20% finer material
BOULDERS (or COBBLES) with much finer material[a]	20—50% finer material
FINE MATERIAL[a] with many boulders (or cobbles)	50—20% boulders (or cobbles)
FINE MATERIAL[a] with some boulders (or cobbles)	20—5% boulders (or cobbles)
FINE MATERIAL[a] with occasional boulders (or cobbles)	Up to 5% boulders (or cobbles)

[a] Give the name of the finer material (in parentheses when it is the minor constituent), e.g. sandy GRAVEL with occasional boulders; cobbly BOULDERS with some finer material (sand with some fines).

Table 5.6 Some properties of gravels, sands, and silts

	Gravels	Sands	Silts
Specific gravity	2.5—2.8	2.6—2.7	2.64—2.66
Bulk density (Mg/m³)	1.45—2.3	1.4—2.15	1.82—2.15
Dry density (Mg/m³)	1.4—2.1	1.35—1.9	1.45—1.95
Porosity (%)	20—50	23—35	—
Void ratio	—	—	0.35—0.85
Liquid limit (%)	—	—	24—35
Plastic limit (%)	—	—	14—25
Coefficient of consolidation (m²/yr)	—	—	12.2
Effective cohesion (kPa)	—	—	75
Shear strength (kPa)	200—600	100—300	—
Angle of friction (deg)	35—45	32—42	32—36

Two basic mechanisms contribute towards the deformation of granular soil: (i) distortion of the particles and (ii) the relative motion between them. These two mechanisms are usually interdependent. At any instant during the deformation process different mechanisms may be acting in different parts of the soil and these may change as deformation continues. Interparticle sliding can occur at all stress levels, the stress required for its initiation increasing with initial stress and decreasing void ratio. Crushing and fracturing of particles begins in a minor way at small stresses, becoming increasingly important when some critical stress is reached. This critical stress is smallest when the soil is loosely packed and uniformly graded, and consists of large, angular particles with a low strength. Usually fracturing only becomes important when the stress level exceeds 3.5 MPa.

The internal shearing resistance of a granular soil is generated by friction when the grains in the zone of shearing are caused to slide, roll, and rotate against each other. The angle of shearing resistance is influenced by the grain size distribution and grain shape. The larger the grains the wider is the zone affected. The more angular the grains, the greater the frictional resistance to their relative movement since they interlock more thoroughly than do rounded ones. They therefore produce a larger angle of shearing resistance (Table 5.7).

Figure 5.1 shows that dense sand has a high peak strength and that when it is subjected to shear stress it expands up to the point of failure, after which a slight decrease in volume may occur. Conversely, loose sand compacts under shearing stress and its residual strength may be similar to that of dense sand. Both curves in Figure 5.1 exhibit strains which are approximately proportional to stress at low stress levels, suggesting a large component of elastic distortion. If the stress is reduced the unloading stress—strain curve indicates that not all the strain is recovered on unloading. The hysteresis loss represents the energy lost in crushing and repositioning

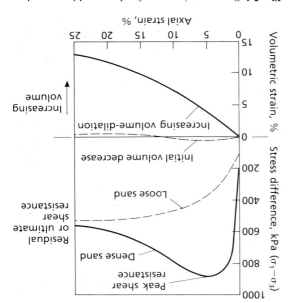

Fig. 5.1 Stress–strain curves for dense and loose sand.

of grains. At higher shear stresses the strains are proportionally greater indicating greater crushing.

The presence of water in the voids of a granular soil does not usually produce significant changes in the value of the angle of shearing resistance. However, if stresses develop in the pore water they may bring about changes in the effective stresses between the particles whereupon the shear strength and the stress—strain relationships may be radically altered.

5.3 SILTS AND LOESS

The grains in a deposit of silt are often rounded with smooth outlines. This influences their degree of packing. The latter, however, is more dependent on the grain size distribution within a silt deposit, uniformly sorted deposits not being able to achieve such close packing as those in which there is a range of grain size. This in turn influences the porosity and void ratio values, as well as bulk and dry densities (Table 5.6).

Dilatancy is characteristic of fine sands and silts. The environment is all-important for the development of dilatancy since conditions must be such that expansion can take place. What is more it has been suggested that the soil particles must be well-wetted and it appears that certain electrolytes exercise a dispersing effect, thereby aiding dilatancy. The moisture content at which a number of fine sands

Table 5.7 Effect of grain shape and grading on the peak friction angle of cohesionless soil. (After Terzaghi, 1950.)

Shape and grading	Loose	Dense
Rounded, uniform	30°	37°
Rounded, well graded	34°	40°
Angular, uniform	35°	43°
Angular, well graded	39°	45°

and silts from British formations become dilatant usually varies between 16 and 35%.

Consolidation of silt is influenced by grain size, particularly the size of the clay fraction, porosity, and natural moisture content. Primary consolidation (i.e. reduction in void space) may account for over 75% of total consolidation. In addition settlement may continue for several months after construction is completed because the rate at which water can drain from the voids under the influence of applied stress is slow.

The angle of shearing resistance decreases with increasing void ratio. It is also dependent upon the plasticity index, grain interlocking, and density. Loess (Fig. 5.2) owes its engineering characteristics largely to the way in which it was deposited since this gave it a metastable structure, in that initially the particles were loosely packed. The porosity of the structure is enhanced by the presence of fossil root holes. The latter are lined with carbonate cement, which helps bind the grains together. This

means that the initial, loosely packed structure is often preserved and the carbonate cement provides the bonding strength of loess. However, the chief binder is usually the clay matrix. On wetting, the clay bond in many loess soils becomes soft, which leads to the collapse of the structure (Lutenegger and Hallberg, 1988). The breakdown of the soil structure occurs under its own weight.

Several collapse criteria have been proposed which depend upon the void ratios at the liquid limit (e_l) and the plastic limit (e_p). According to Audric and Bouquier (1976), collapse is probable when the natural void ratio is higher than a critical void ratio (e_c) which depends on e_l and e_p. They quoted the Denisov and Feda criteria as providing fairly good estimates of the likelihood of collapse:

$$e_c = e_l \quad \text{(Denisov)}$$

$$e_c = 0.85e_l + 0.15e_p \quad \text{(Feda)}$$

Loess deposits generally consist of 50 to 90% particles of silt size. In fact sandy, silty, and clayey

Fig. 5.2 Because of the presence of fossil root holes, near-vertical slopes can form in loess. The material in the left of the foreground forms part of a large landslide, Loess Plateau, Shaanxi Province, China. The openings in the loess are caves excavated and formerly inhabited by man.

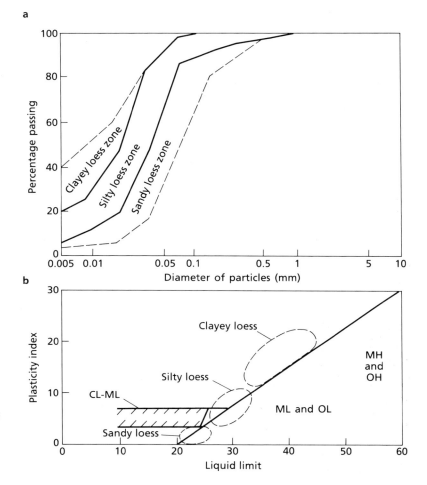

Fig. 5.3 Particle size distribution and plasticity of Missouri River Basin loess: (a) range of gradation; (b) Atterberg limits. (After Clevenger, 1958.)

loess can be distinguished (Fig. 5.3a; Table 5.8). The undisturbed densities of loess (e.g. in the Missouri basin) range from around 1.2 to 1.36 Mg/m^3. If this material is wetted or consolidated (or reworked), the density increases, sometimes to as high as 1.6 Mg/m^3 (Clevenger, 1958).

The liquid limit of loess averages about 30% (exceptionally liquid limits as high as 45% have been recorded), and their plasticity index ranges from about 4 to 9%, but averages 6% (Fig. 5.3b). As far as their angle of shearing resistance is concerned, this usually varies from 30 to 34°. In the unweathered state above the water table the unconfined compressive strength of loess may amount to several hundred kilopascals. Loess deposits are better drained (their permeability ranges from (10^{-5} to 10^{-7} m/s) than are true silts because of the fossil

root holes. As would be expected their permeability is appreciably higher in the vertical than in the horizontal direction.

Normally loess possesses a high shearing resistance and can carry high loadings without significant settlement when natural moisture contents are low. For example, moisture contents of undisturbed loess are generally around 10% and the supporting capacity of loess at this moisture content is high. However, the density of loess is the most important factor controlling its shear strength and settlement. On wetting, large settlements and low shearing resistance are encountered when the density of loess is below 1.30 Mg/m^3, whereas if the density exceeds 1.45 Mg/m^3 settlement is small and shearing resistance fairly high.

Unlike silt, loess does not appear to be frost

Table 5.8 Some geotechnical properties of loess soils

Property	Shaansi Province, China[*]		Lanzhow Province, China[†]	Czechoslovakia, near Prague[‡]	South Polish Uplands[§]
	Sandy loess	Clayey loess			
Natural moisture content (%)	9−13	13−20	11−10	21	3−26
Specific gravity					2.66−2.7
Bulk density (Mg/m³)	1.59−1.68	1.4−1.85			1.54−2.12
Dry density (Mg/m³)			1.4−1.5		1.46−1.73
Void ratio	0.8−0.92	0.76−1.11	1.05		
Porosity (%)				44−50	35−46
Grain size distribution (%)					
Sand	20.5−35.2	12−15	20−24		
Silt	54.8−69.0	64−70	57−65		
Clay	8.0−15.5	17−24	16−21		
Plastic limit (%)[a]			10−14	20	
Liquid limit (%)[b]	26−28	30−31	27−30	36	
Plasticity index (%)[c]	8−10	11−12	10−14	16	
Activity[d]				1.32	
Coefficient of collapsibility[e]	0.007−0.016	0.003−0.023		0.006−0.011	0.0002−0.06
Angle of friction					7−36

After * Lin and Wang, 1988; † Tan, 1988; ‡ Feda, 1988; § Grabowska-Olszewska, 1988.
(a) moisture content at which soil becomes plastic; (b) moisture content when soil begins to flow, i.e. to behave like a liquid; (c) difference between liquid limit and plastic limit; (d) plasticity index divided by clay fraction.
(e) the coefficient of collapsibility, i_c, can be defined as follows:

$$i_c = \frac{\Delta e}{1 + e_1}$$

where Δe is the decrease in void ratio brought about by wetting and e_1 is the initial void ratio prior to wetting. It is derived from an oedometer test.

susceptible, this being due to its more permeable character, but like silt it can exhibit quick conditions and it can be difficult, if not impossible, to compact. Because of its porous structure a 'shrinkage' factor must be taken into account when estimating earthwork.

5.4 CLAY DEPOSITS

The principal minerals in a deposit of clay tend to influence its engineering behaviour. For example, the plasticity of a clay soil is influenced by the amount of its clay fraction and the type of clay

minerals present since clay minerals greatly influence the amount of attracted water held in a soil. Variations in the plasticity of the London Clay, for instance, appear to be mainly due to changes in the total clay mineral content whilst the Weald Clay has a low plasticity because it has a small clay fraction (Table 5.9). On the other hand it would appear that there is only a general correlation between the clay mineral composition of a deposit and its activity, that is, kaolinitic and illitic clays are usually inactive, whilst montmorillonitic clays range from inactive to active. Usually active clays have a relatively high water-holding capacity and a high cation exchange capacity. They are also highly thixotropic and have a low permeability.

The undrained shear strength is related to the amount and type of clay minerals present in a clay deposit together with the presence of cementing agents. Strength is reduced with increasing content of mixed-layer clay and montmorillonite in the clay fraction. The increasing presence of cementing agents, especially calcite, enhances the strength of the clay.

Geological age also has an influence on the engineering behaviour of a clay deposit. In particular the porosity, water content, and plasticity normally decrease in value with increasing depth, whereas the strength and elastic modulus increase (Fig. 5.4).

The engineering performance of clay deposits is also very much affected by the total moisture content and by the energy with which this moisture is held. For instance, the moisture content influences their consistency and strength, and the energy with which moisture is held influences their volume change characteristics. One of the most notable characteristics of clays from the engineering point of view is their susceptibility to slow volume changes which can occur independently of loading due to swelling or shrinkage. Differences in the period and magnitude of precipitation and evapotransportation are the major factors influencing the swell–shrink response of an active clay beneath a structure.

Grim (1952) distinguished two modes of swelling in clay soils: intercrystalline swelling, and intracrystalline swelling. Intercrystalline swelling takes place when the uptake of moisture is restricted to the external crystal surfaces and the void spaces between the crystals. Intracrystalline swelling is characteristic of the smectite family of clay minerals, especially montmorillonite. The individual molecular layers which make up a crystal of montmorillonite are weakly bonded so that on wetting water enters not only between the crystals but also between the unit layers which comprise the crystals. Generally, kaolinite has the smallest swelling capacity of the clay minerals and nearly all of its swelling is of the intercrystalline type. Illite may swell by up to 15% but intermixed illite and montmorillonite may swell some 60 to 100%. Swelling in Ca montmorillonite is very much less than in the Na variety; it ranges from about 50 to 100%.

The maximum movement due to swelling beneath a building founded on expansive clay can be obtained from the following expression:

$$\text{Swell } (\%) = \frac{(PI - 10)}{10} \log_{10} S/p \qquad (5.3)$$

where PI is the plasticity index, S is the soil suction at the time of construction (in kPa), and p is the overburden plus foundation pressure acting on each layer of soil (in kPa). However, one of the most

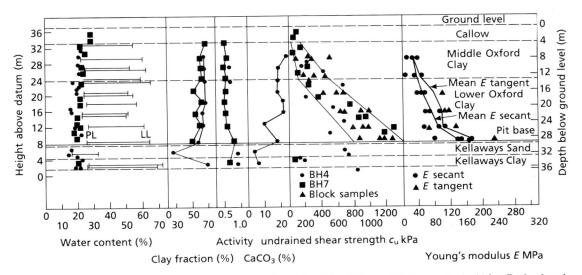

Fig. 5.4 Variation with depth of the geotechnical properties of the Oxford Clay and Kellaways Beds. (After Burland *et al.*, 1977.)

Table 5.9 Engineering properties of some British clays. (After Cripps and Taylor, 1981.)

Formation	Water content (%)		Liquid limit (%)		Plasticity index (%)	Clay fraction (%)	Porosity (%)	Undrained shear strength (kPa) (range)		Undrained shear strength (kPa) (average)	
	Weathered	Unweathered	Weathered	Unweathered				Weathered	Unweathered	Weathered	Unweathered
Palaeogene											
Barton Clay	21–32	—	45–82	—	21–55	25–70	—	20–210	50–350	40	112–150
Bracklesham Beds	—	γ 19–26	—	γ 52–68	γ 41	—	—	—	γ 143	—	—
London Clay	δ 23–49	19–28	66–100	50–105	40–65	40–72	ψ 37–59	40–190	80–800	100–175	100–400
Woolwich and Reading Beds	—	15–27	—	42–67	20–37	—	—	—	34–814	—	400
Cretaceous											
Gault Clay	32–42	18–30	70–92	μ 60–120 90–110	27–80	38–62	ψ 31–48	17–76	56–1280	λ 60	300–550
Weald Clay	25–34	5	42–82	55	28–32	20–74	ψ 15–68	—	—	—	—
Fullers Earth	—	46	—	—	—	69	57	—	—	—	—
Speeton Clay	—	—	50	—	28	—	—	—	—	—	—
Jurassic											
Kimmeridge Clay	—	18–22	—	70–81	24–59	57	35	—	70–500	—	130–470
Ampthill Clay	23–88	—	79	—	49	—	—	—	—	—	—
Middle Oxford Clay	20–33	20–28	—	58–76	31–40	35–70	—	—	45–490	—	110–360
Lower Oxford Clay	—	15–25	—	45–75	28–50	30–70	ω 30–54	52–93	96–1300	—	360–1100
Fullers Earth	26–41	33	41–77	100	20–39	38–68	48	10–120	—	50	—
Upper Lias Clay	20–38	11–23	56–68	53–70	20–39	55–65	ψ 32–48	20–180	40–1200	30–150	110–240
Lower Lias Clay	29	16–22	o 56–62	53–63	32–37	o 50–56	ω 37–44	o 28–57	12.5–45	—	—

Table 5.9 *Continued*

Formation	Effective cohesion kPa		Effective angle of friction		Residual shear strength	Coefficient of volume compression (m²/MN)		Equivalent modulus of elasticity (MPa)		Modulus of elasticity E (MPa)		Coefficient of consolidation (m²/yr)	
	Weathered	Unweathered	Weathered	Unweathered		Weathered	Unweathered	Weathered	Unweathered	Weathered	Unweathered	Weathered	Unweathered
Palaeogene													
Barton Clay	7–11	8–24	18–24	27–39	15	—	—	—	—	—	—	—	—
Bracklesham Beds	γ 0–55	34	γ 18–32	25	—	0.065–0.5	—	γ 2–15	—	—	—	—	—
London Clay	12–18	31–252	17–23	20–29	10·5–22	0.5–0.18	0.02–0.12	0.5–20	8–50	10–35	η 25–141	0.2–2.0	0.3–60
Woolwich and Reading Beds	—	—	—	—	—	—	0.02–0.06	—	17–50	—	θ 370	—	0.75–95
Cretaceous													
Gault Clay	—	25–124	—	19–53	12.19	0.12	0.01–0.08	8	12–100	10	20–162	1.0–8.9	0.09–1.4
Weald Clay	—	—	—	—	11.20	—	0.002	—	500	—	—	—	25.1
Fullers Earth	—	—	—	—	—	—	0.007	—	143	—	—	—	0.02
Speeton Clay	—	—	—	—	14	—	—	—	—	—	—	—	—
Jurassic													
Kimmeridge Clay	14–67	—	14–23	—	10–18	0.22	π 0.002	4.5	π 500			—	0.25–2.7
Ampthill Clay	23–48	—	17–32	—	10–14	0.09	—	11	—			—	—
Middle Oxford	—	10	—	31	15	—	0.077	11	56		30–40	—	0.076
Lower Oxford Clay	0.20	10–216	21.5–28	23–40	13–17	0.017–0.023	π 0.003–0.03	43–59	π 33–333	—	40–140	>40	0.49–0.93
Fullers Earth	0	—	17.4	—	12	—	π 0.005	—	π 200	—	—	—	π 0.02
Upper Lias	10–17	—	18–25	—	9–13.5	0.2	π 0.002–0.003	5	π 333–500	—	—	—	π 0.42–0.67
Lower Lias Clay	5	—	27	—	13–16	—	—	—	—	10 GN/m²	35 GN/m² 22–52	—	—

γ State of weathering not known; δ high value of water content for clay; η higher value at 46 m depth; θ from back analysis; λ 100 mm diameter samples; μ Upper Gault w_l = 60–120%, Lower Gault w_l = 90–110%; o undivided Lias shales; π deep borehole sample; ψ higher value determined from published values of bulk density, water content, and specific gravity; o both values determined from published values of bulk density, water content, and specific gravity.

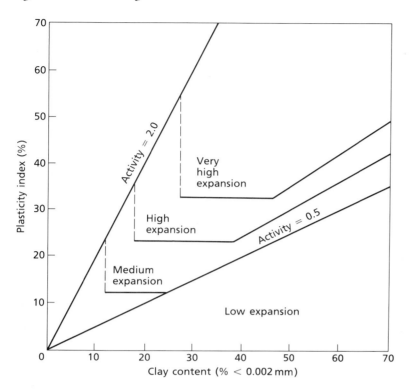

Fig. 5.5 Estimation of the degree of expansiveness of a clay soil. (After Williams and Donaldson, 1980.)

widely used soil properties to predict swell potential is the activity of a clay (Fig. 5.5).

The United States Army Engineers Waterways Experimental Station (USAEWES) classification of potential swell (Table 5.10) is based on the liquid limit (LL), plasticity index (PI), and initial (*in situ*) suction (S_i). The latter is measured in the field by a psychrometer.

The volume change which occurs due to evapotranspiration from a clay soil can be conservatively predicted by assuming the lower limit of the soil moisture content to be the shrinkage limit. Desiccation beyond this value cannot bring about further volume change. Transpiration from vegetative cover is a major cause of water loss from soils in semi-arid regions. Indeed the distribution of soil suction in the soil is primarily controlled by transpiration from vegetation. The maximum soil suction which can be developed is governed by the ability of vegetation to extract moisture from the soil. The point at which moisture is no longer available to plants is termed the permanent wilting point and this corresponds to a soil suction (pF) value of about 4.2. The complete depth of active clay profiles usually does not become fully saturated during the wet season in semi-arid regions. Changes in soil suction may be expected over a depth of some 2.0 m between the wet and dry seasons.

The moisture characteristic (moisture content versus soil suction) of a soil provides valuable data concerning the moisture contents corresponding to the field capacity (defined in terms of soil suction

Table 5.10 USAEWES classification of swell potential. (After O'Neill and Poormoayed, 1980.)

Liquid limit (%)	Plastic limit (%)	Initial (*in situ*) suction (kPa)	Potential swell (%)	Classification
Less than 50	Less than 25	Less than 145	Less than 0.5	Low
50−60	25−35	145−385	0.5−1.5	Marginal
Over 60	Over 35	Over 385	Over 1.5	High

this is a pF value of about 2.0) and the permanent wilting point, as well as the rate at which changes in soil suction take place with variations in moisture content. This enables an assessment to be made of the range of soil suction and moisture content which is likely to occur in the zone affected by seasonal changes in climate.

The extent to which the vegetation is able to increase the suction to the level associated with the shrinkage limit is obviously important. In fact, the moisture content at the wilting point exceeds that of the shrinkage limit in soils with a high content of clay and is less in those possessing low clay contents. This explains why settlement resulting from the desiccating effects of trees is more notable in low to moderately expansive soils than in expansive ones.

When vegetation is cleared from a site, its desiccating effect is also removed. Hence the subsequent regain of moisture by clay soils leads to them swelling. Swelling movements of over 350 mm have been reported for expansive clays in South Africa (Williams and Pidgeon, 1983) and similar movements in similar soils have occurred in Texas.

Desiccation cracks may extend to depths of 2 m in expansive clays, and gape up to 150 mm. The suction pressure associated with the onset of cracking is approximately pF 4.6. The presence of desiccation cracks enhances evaporation from the soil. Such cracks lead to a variable development of suction pressure, the highest suction occurring nearest the cracks. This, in turn, influences the preconsolidation pressure (the maximum effective vertical stress which has acted on a soil in the past) as well as the shear strength. It has been claimed that the effect of desiccation on clay soils is similar to that of heavy overconsolidation.

Sridharan and Allam (1982), with reference to arid and semi-arid regions, found that repeated wetting and drying of clay soils can bring about aggregation of soil particles and cementation by compounds of Ca, Mg, Al, and Fe. This enhances the permeability of the clays and increases their resistance to compression. Furthermore, interparticle desiccation bonding increases the shear strength, the aggregations offering higher resistance to stress and depending on the degree of bonding, the expansiveness of an expansive clay soil may be reduced or it may even behave as a non-expansive soil.

Volume changes in clays also occur as a result of loading and unloading which bring about consolidation and heave respectively. When a load is applied to a clay soil its volume is reduced, this principally being due to a reduction in the void ratio (Burland, 1990). If such a soil is saturated, then the load is initially carried by the pore water which causes a pressure, the excess pore water pressure, to develop. The excess of the pore water is dissipated at a rate which depends upon the permeability of the soil mass and the load is eventually transferred to the soil structure. The change in volume during consolidation is equal to the volume of the pore water expelled and corresponds to the change in void ratio of the soil. In clay soils, because of their low permeability, the rate of consolidation is slow. Primary consolidation is brought about by a reduction in the void ratio. Further consolidation may occur due to a rearrangement of the soil particles. This secondary compression is usually much less significant. The compressibility of a clay is related to its geological history, that is, to whether it is normally consolidated or overconsolidated. A normally consolidated clay is that which at no time in its geological history has been subject to vertical pressures greater than its existing overburden pressure, whereas an overconsolidated clay is one which has.

When an excavation is made in a clay with weak diagenetic bonds elastic rebound causes immediate dissipation of some stored strain energy in the soil. However, part of the strain energy is retained due to the restriction on lateral straining in the plane parallel to the ground surface. The lateral effective stresses either remain constant or decrease as a result of plastic deformation of the clay as time passes. This plastic deformation can result in significant time-dependent vertical heaving. However, creep of weakly bonded soils is not a common cause of heaving in excavations.

The value of strength derived by testing a sample of clay in the laboratory depends upon the type of test used, the time which has elapsed between sampling in the field and performing the test, the size of the specimen tested, and its orientation in the testing apparatus. It also depends on the type of sample used, for example, block samples may yield higher values than borehole samples.

An overconsolidated clay is considerably stronger at a given pressure than a normally consolidated clay (Burland, 1990) and it tends to dilate during shear whereas a normally consolidated clay consolidates. Hence, when an overconsolidated clay is sheared under undrained conditions negative pore water pressures are induced, the effective strength is increased, and the undrained strength is much higher than the drained strength (the exact opposite to a normally consolidated clay). When the negative pore water pressure gradually dissipates the strength falls as much as 60 or 80% to the drained strength.

Skempton (1964) observed that when clay is strained it develops an increasing resistance (strength), but that under a given effective pressure the resistance offered is limited, the maximum value corresponding to the peak strength. If testing is continued beyond the peak strength, then as displacement increases, the resistance decreases, again to a limiting value which is termed the residual strength. Skempton noted that in moving from peak to residual strength, cohesion falls to almost, or actually, zero and the angle of shearing resistance is reduced to a few degrees (it may be as much as 10° in some clays). He explained the drop in strength which occurred in overconsolidated clays as due to their expansion on passing peak strength and associated increasing water content, and maintained that platy clay minerals became orientated in the direction of shear and thereby offered less resistance. Failure occurs once the stress on a clay exceeds its peak strength and as failure progresses the strength of the clay along the shear surface is reduced to the residual value.

It was suggested that under a given effective pressure, the residual strength of a clay is the same whether it is normally or overconsolidated (Fig. 5.6). In other words, the residual shear strength of a clay is independent of its post-depositional history, unlike the peak undrained shear strength which is controlled by the history of consolidation as well as diagenesis. Furthermore, the value of residual shear strength (ϕ_r) decreases as the amount of clay fraction increases in a deposit. In this context, not only is the proportion of detrital minerals important but so is that of the diagenetic minerals. The latter influence the degree of induration of a deposit of clay and the

value of ϕ_r can fall significantly as the ratio of clay minerals to detrital and diagenetic minerals increases.

The shear strength of an undisturbed clay is frequently found to be greater than that obtained when it is remoulded and tested under the same conditions and at the same water content. The ratio of the undisturbed to the remoulded strength at the same moisture content is termed the sensitivity of a clay. Skempton and Northey (1952) proposed six grades of sensitivity:
1 insensitive clays (under 1);
2 low-sensitive clays (1 to 2);
3 medium-sensitive clays (2 to 4);
4 sensitive clays (4 to 8);
5 extrasensitive clays (8 to 16); and
6 quick clay (over 16).
Clays with high sensitivity values have little or no strength after being disturbed. Indeed if they suffer slight disturbance this may cause an initially fairly strong material to behave as a viscous fluid. Any subsequent regain in strength due to thixotropic hardening does not exceed a small fraction of its original value. Sensitive clays generally possess high moisture contents, frequently with liquidity indices well in excess of unity (the liquidity index is the moisture content in excess of the plastic limit expressed as a percentage of the plasticity index). A sharp increase in moisture content may cause a great increase in sensitivity, sometimes with disastrous results. Heavily overconsolidated clays are insensitive.

Fissures play an extremely important role in the failure mechanism of overconsolidated clays, for example, the strength along fissures is only slightly higher than the residual strength of the intact clay. Hence, the upper limit of the strength of fissured clay is represented by its intact strength whilst the lower limit corresponds to the strength along the fissures. The operational strength, which is somewhere between the two, however, is often signifi-

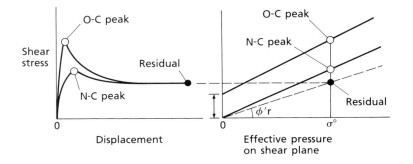

Fig. 5.6 Peak strength and residual strength of normally consolidated (N-C) and overconsolidated (O-C) clays. (After Skempton, 1964.)

Table 5.11 Strength of weathered (brown) and unweathered (blue) London Clay

Parameter	Brown	Blue
s_u (kPa)	100–175	120–250
c' (kPa)	0–31	35–252
ϕ^0	20–23	25–90

cantly higher than the fissure strength. In addition to allowing clay to soften, fissures allow concentrations of shear stress which locally exceed the peak strength of clay thereby giving rise to progressive failure. Under stress the fissures in clay seem to propagate and coalesce in a complex manner. The ingress of water into fissures means that the pore water pressure in the clay concerned increases, which in turn means that its strength is reduced. Fissures in normally consolidated clays have no significant practical consequences.

The greatest variation in the engineering properties of clays can be attributed to the degree of weathering which they have undergone (Fig. 5.7). For example, consolidation of a clay deposit gives rise to an anisotropic texture due to the rotation of the platy minerals. Second, diagenesis bonds particles together either by the development of cement, the intergrowth of adjacent grains or the action of van der Waals charges which are operative at very small grain separations. Weathering reverses these processes altering the anisotropic structure and destroying or weakening interparticle bonds (Coulthard and Bell, 1992). Ultimately, weathering, through the destruction of interparticle bonds, leads to a clay deposit reverting to a normally consolidated sensibly remoulded condition. Higher moisture contents are found in more weathered clay. This progressive degrading and softening is also accompanied by reductions in strength and deformation moduli with a general increase in plasticity. The reduction in strength has been illustrated by Cripps and Taylor (1981) who quoted strength parameters from brown (weathered) and blue (unweathered) London Clay (Table 5.11). These values indicate that the undrained shear strength (s_u) is reduced by approximately half and that the effective cohesion (c') can suffer significant reduction on weathering. The effective angle of shearing resistance is also reduced and at $\phi = 20°$ the value corresponds to a fully softened condition.

5.5 TROPICAL SOILS

Ferruginous and aluminous clay soils are frequent products of weathering in tropical latitudes (Mitchell and Sitar, 1982; Anon, 1990). They are characterized by the presence of iron and aluminium oxides and hydroxides. These compounds, especially those of iron, are responsible for the red, brown, and yellow colours of the soils. The soils may be fine-grained or they may contain nodules or concretions. Concretions occur in the matrix when there are higher concentrations of oxides in the soil. More extensive accumulations of oxides give rise to laterite.

Laterite is a residual ferruginous claylike deposit (Fig. 5.8) which generally occurs below a hardened ferruginous crust or hardpan. The ratios of silica (SiO_2) to sesquioxides (Fe_2O_3, Al_2O_3) in laterites usually are less than 1.33; those betwen 1.33 and 2.0 are indicative of lateritic soils; and those greater than 2.0 are indicative of non-lateritic types. Laterites tend to occur in areas of gentle topography which are not subject to significant erosion. During drier periods the water table is lowered. The small amount of iron which has been mobilized in the ferrous state by the groundwater is then oxidized, forming haematite, or if hydrated — goethite. The movement of the water table leads to the gradual accumulation of iron oxides at a given horizon in the soil profile. A cemented layer of laterite is formed which may be a continuous or honeycombed mass, or nodules may be formed, as in laterite gravel. Concretionary layers are often developed near the surface in lowland areas because of the high water table.

Laterite hardens on exposure to air. Hardening may be due to a change in the hydration of iron and aluminium oxides.

Laterite commonly contains all size fractions from clay to gravel and sometimes even larger material (Fig. 5.9a). Usually at or near the surface the liquid limits of laterites do not exceed 60% and the plasticity indices are less than 30% (Fig. 5.9b). Consequently, laterites are of low to medium plasticity (Vaughan *et al.*, 1988; Vaughan, 1990). The activity of laterites may vary between 0.5 and 1.75. Values of common properties of laterite are given in Table 5.12.

Lateritic soils, particularly where they are mature, furnish a good bearing stratum (Brand, 1985; Blight, 1990). The hardened crust has a low compressibility and therefore settlement is likely to be negligible. In

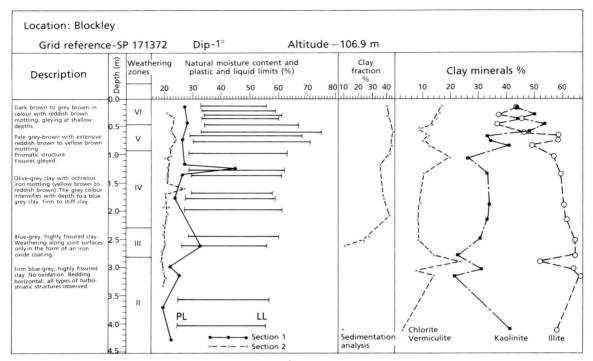

Fig. 5.7 Geotechnical profile of the Lower Lias Clay, Blockley, Gloucestershire. A measure of the average orientation of clay particles seen in a thin section beneath the petrological microscope is afforded by the birefringence ratio (the ratio between the minimum and maximum light transmitted under crossed polars). This varies from 0 with perfect parallel orientation to 1 with perfect random orientation. With increasing weathering the birefringence ratio increases indicating

such instances, however, the strength of laterite may decrease with increasing depth.

Red earths or latosols are residual ferruginous soils in which oxidation readily occurs (Morin, 1982). Such soils tend to develop in undulating country and most of them appear to have been derived from the first cycle of weathering of the parent material. They differ from laterite in that they behave as a clay and do not possess strong concretions. They do, however, grade into laterite.

Black clays are typically developed on poorly drained plains in regions with well defined wet and dry seasons, where the annual rainfall is not less than 1250 mm. Generally the clay fraction in these soils exceeds 50%, silty material varying between 20 and 40%, and sand forming the remainder. The organic content is usually less than 2%. The liquid limits of black clays may range between 50 and 100% with plasticity indices of between 25 and 70% (Ola, 1978). The shrinkage limit is frequently around 10 to 12%. Montmorillonite is commonly present in the clay fraction and is the chief factor determining the behaviour of these clays, for example, they undergo appreciable volume changes on wetting and drying due to the montmorillonite content. These volume changes, however, tend to be confined to an upper critical zone of the soil, which is frequently less than 1.5 m thick (Horn, 1982). Below this the moisture content remains more or less the same, at around 25%.

Table 5.12 Some common properties of laterites

Moisture content (%)	10–49
Liquid limit (%)	33–90
Plastic limit (%)	13–31
Clay fraction	15–45
Dry unit weight (kN/m^3)	15.2–17.3
Cohesion, c_u (kPa)	466–782
Angle of internal friction, ϕ_u	28–35
Unconfined compressive strength (kPa)	220–825
Compression index	0.0186
Coefficient of consolidation (m^2/year)	262
Young's modulus (kPa)	5.63×10^4

Cont.

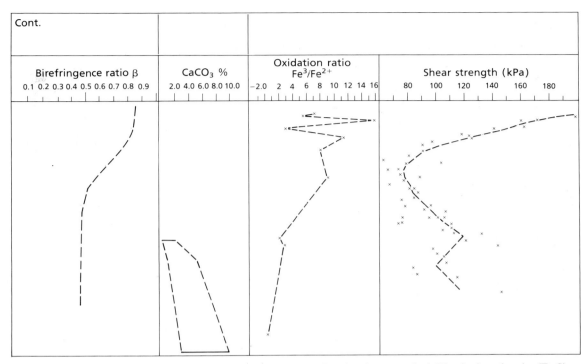

| Birefringence ratio β | CaCO₃ % | Oxidation ratio Fe³/Fe²⁺ | Shear strength (kPa) |

that the fabric of the clay becomes more disordered as the ground surface is approached. Weathering of pyrite (Fe_2S) produces sulphate and sulphuric acid. The latter reacts with calcium carbonate. The oxidation of iron compounds tends to increase as the surface is approached which leads to increasing shear strength. PL, plastic limit; LL, liquid limit. (After Coulthard and Bell, 1992.)

In arid and semi-arid regions the evaporation of moisture from the surface of the soil may lead to the precipitation of salts in the upper layers. The most commonly precipitated material is calcium carbonate. These caliche deposits are referred to as calcrete (Fig. 5.10). The development of calcrete is inhibited beyond a certain aridity since the low precipitation is unable to dissolve and drain calcium carbonate towards the water table. Consequently, in arid climates gypcrete may take the place of calcrete. As the carbonate content increases it first occurs as scattered concentrations of flaky habit, then as hard concretions. Once it exceeds 60% the concentration becomes continuous. The calcium carbonate in calcrete profiles decreases from top to base, as generally does the hardness.

Dispersive clay soils deflocculate in the presence of relatively pure water forming colloidal suspensions (colloidal clay particles going into suspension even in quiet water). Consequently, such soils are highly susceptible to erosion (Fig. 5.11) and piping. They occur in areas where the annual rainfall is less than 860 mm (Bell *et al.*, 1991). Such soils contain a higher content of dissolved sodium in their pore water than ordinary soils (they may contain up to 12% sodium). There are no significant differences in the clay contents of dispersive and non-dispersive soils, except that soils with less than 10% clay particles may not have enough colloids to support dispersive piping. Usually dispersive soils contain a moderate to high content of clay.

Dispersive erosion depends on the mineralogy and chemistry of the clay on the one hand, and the dissolved salts in the pore and eroding water on the other. The presence of exchangeable sodium is the main chemical factor contributing towards dispersive clay behaviour. This is expressed in terms of the exchangeable sodium percentage (ESP):

$$ESP = \frac{exchangeable\ sodium}{cation\ exchange\ capacity} \times 100 \qquad (5.4)$$

where the units are given in me/100 g of dry soil. Above a threshold value of ESP of 10 soils have their free salts leached by seepage of relatively pure water and are prone to dispersion. Soils with ESP

Fig. 5.8 Excavation in laterite, Singapore.

values above 15% are highly dispersive, according to Gerber and Harmse (1987). On the other hand those soils with low cation exchange values (15 me/100 g of clay) are non-dispersive at ESP values of 6% or below. Similarly, soils with high cation-exchange capacity values and a plasticity index greater than 35% swell to such an extent that dispersion is not significant. High ESP values and piping potential generally exist in soils in which the clay fraction is composed largely of smectitic clays. Some illitic soils are highly dispersive. High values of ESP and high dispersibility are rare in clays composed largely of kaolinites.

Damage due to internal erosion of dispersive clay leads to the formation of pipes and internal cavities on slopes. Piping is initiated by dispersion of clay particles along desiccation cracks, fissures, and root holes. Indications of piping take the form of small leakages of mud-coloured water. Also severe erosion damage forming deep gullies occurs on slopes after rainfall.

5.6 TILLS AND OTHER GLACIALLY ASSOCIATED DEPOSITS

Till is usually regarded as being synonymous with boulder clay. It is deposited directly by ice whilst stratified drift is deposited in meltwater associated with glaciers.

The character of till deposits varies appreciably and depends on the lithology of the material from which it was derived, on the position in which it was transported in the glacier, and on the mode of deposition. The underlying bedrock material usually constitutes up to about 80% of basal or lodgement tills, depending on its resistance to abrasion and plucking. Argillaceous rocks, such as shales and mudstones, are more easily abraded and so produce fine-grained tills which are richer in clay minerals and therefore are more plastic than other tills. Mineral composition also influences the natural moisture content which is slightly higher in tills containing appreciable quantities of clay minerals or mica.

Lodgement till is plastered on to the ground beneath a moving glacier in small increments as the basal ice melts. Because of the overlying weight of ice such deposits are overconsolidated. Due to abrasion and grinding the proportion of silt and clay-size material is relatively high in lodgement till (e.g. the clay fraction varies from 15 to 40%). Lodgement till is commonly stiff, dense, and relatively incompressible (Sladen and Wrigley, 1983). Hence it is practically impermeable. Fissures are frequently

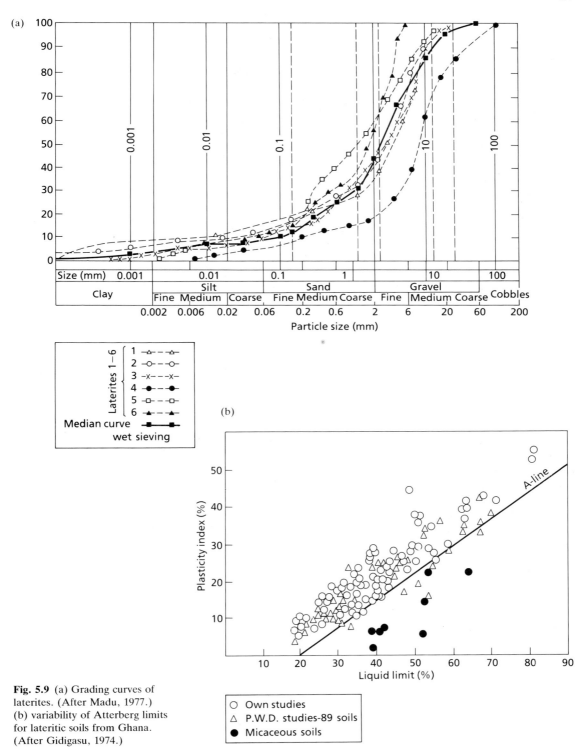

Fig. 5.9 (a) Grading curves of laterites. (After Madu, 1977.) (b) variability of Atterberg limits for lateritic soils from Ghana. (After Gidigasu, 1974.)

Fig. 5.10 Calcrete in the north of Namib-Naukluft Park, Namibia (about 50 km east of Walvis Bay).

Fig. 5.11 Gully erosion in dispersive soils developed above Karoo mudstones in northern Natal, South Africa.

present in lodgement till, especially if it is clay matrix dominated.

Ablation till accumulates on the surface of the ice when englacial debris melts out and as the glacier decays the ablation till is slowly lowered to the ground. It is therefore normally consolidated. Ablation till has a high proportion of far-travelled material and may not contain any of the local bedrock. Because it has not been subjected to much abrasion, ablation till is characterized by abundant large stones that are angular, and the proportion of sand and gravel is high whereas clay is present only in small amounts (usually less than 10%). Because the texture is loose, ablation till can have an extremely low *in situ* density. Since ablation till consists of the load carried at the time of ablation, it usually forms a thinner deposit than lodgement till.

The particle size distribution and fabric (stone orientation, layering, fissuring, and jointing) are among the most significant features as far as the engineering behaviour of a till is concerned. McGown and Derbyshire (1977) therefore used the percentage of fines to distinguish granular, well-graded, and matrix dominated tills, the boundaries being placed at 15 and 45% respectively. The fabric of tills includes features of primary and secondary origins such as folds, thrusts, fissures (macrofabric), disposition of clasts (macro- and mesofabric), and the organization of the matrix.

Tills are frequently gap-graded, the gap generally occurring in the sand fraction (Fig. 5.12). Large, often very local, variations can occur in the gradings of till which reflect local variations in the formation processes, particularly the comminution processes. The range in the proportions of coarse and fine fractions in tills dictates the degree to which the properties of the fine fraction influence the properties of the composite soil. The variation in the engineering properties of the fine soil fraction is greater than that of the coarse fraction, and this often tends to dominate the engineering behaviour of the till.

The specific gravity of till deposits is often remarkably uniform, varying from 2.77 to 2.78. These values suggest the presence of fresh minerals in the fine fraction, that is, rock flour rather than clay minerals. Rock flour behaves more like granular material than cohesive material and has a low plasticity. The consistency limits of tills are dependent

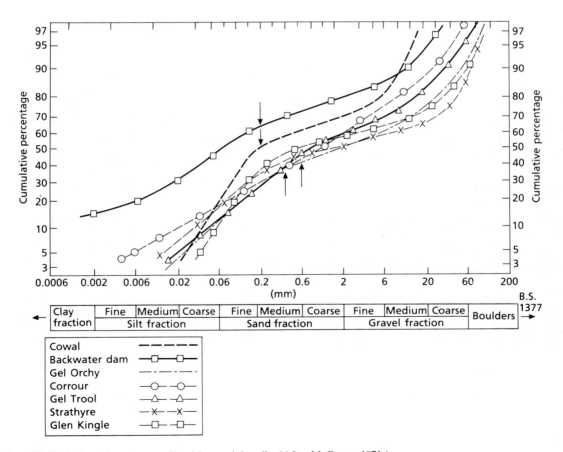

Fig. 5.12 Typical gradings of some Scottish morainic soils. (After McGown, 1971.)

Table 5.13 Strength of tills of north Norfolk and Holderness. (After Bell, 1991; Bell and Forster, 1991.)

	Unconfined compressive (kPa)			Triaxial			
	Intact	Remoulded	Sensitivity	c_u	ϕ_u^0	c'	ϕ'
A *North Norfolk*							
1 Hunstanton Till							
(Holkham)							
Max	184	164	1.22 (L)	43	9	18	34
Min	152	128	1.18 (L)	22	3	8	26
Mean	158	134	1.19 (L)	29	5	12	29
2 Marly Drift							
(Weybourne)							
Max	120	94	1.49 (L)	49	3	16	28
Min	104	70	1.28 (L)	16	0	7	21
Mean	110	81	1.34 (L)	27	1	11	24
3 Contorted Drift							
(Trimingham)							
Max	180	168	1.67 (L)	46	10	20	33
Min	124	76	1.08 (L)	20	3	6	27
Mean	160	136	1.23 (L)	26	6	11	30
4 Cromer Till							
(Happisburgh)							
Max	224	188	1.19 (L)	48	6	19	32
Min	154	140	1.10 (L)	26	2	12	26
Mean	176	156	1.13 (L)	35	4	14	29
B *Holderness*							
1 Hessle Till							
(Dimlington, Hornsea)							
Max	138	116	1.31 (L)	98	8	80	24
Min	96	74	1.10 (L)	22	5	10	13
Mean	106	96	1.19 (L)	35	7	26	25
2 Withernsea Till							
(Dimlington)							
Max	172	148	1.18 (L)	62	19	42	34
Min	140	122	1.15 (L)	17	5	17	16
Mean	160	136	1.16 (L)	30	9	23	25
3 Skipsea Till							
(Dimlington)							
Max	194	168	1.15 (L)	50	21	35	36
Min	182	154	1.08 (L)	17	10	22	24
Mean	186	164	1.13 (L)	29	12	28	30
4 Basement Till							
(Dimlington)							
Max	212	168	1.27 (L)	59	17	42	36
Min	163	140	1.19 (L)	22	6	19	20
Mean	186	156	1.21 (L)	38	9	34	29

c = Cohesion (kPa); ϕ = angle of friction; L = low sensitivity.

upon water content, grain size distribution, and the properties of the fine-grained fraction (Bell, 1991; Bell and Forster, 1991). Generally, however, the plasticity index is small and the liquid limit of tills decreases with increasing grain size (Sladen and Wrigley, 1983).

The compressibility and consolidation of tills are principally determined by the clay content, as is the shear strength (Table 5.13). For example, the value of compressibility index tends to increase linearly with increasing clay content whilst for moraines of very low clay content (less than 2%), this index remains about constant (C_c = 0.01). The shear strength of till can range from 150 kPa to over 1.5 MPa, the latter being in heavily overconsolidated tills.

Fissures in till can have a very definite preferred orientation. They tend to be variable in character, spacing, orientation, and areal extent. Opening up and softening along these fissures gives rise to a rapid reduction of undrained shear strength along the fissures and along fissures in till it may be as little as one-sixth that of intact soil. The distinction between the nature of the various fissure coatings (sand, silt, or clay-size material) is of importance in determining the shear strength behaviour of fissured tills. Deformation and permeability are also con-trolled by the nature of the fissure surfaces and coatings.

Eyles and Sladen (1981) recognized four zones of weathering within the soil profile of lodgement till in the coastal area of Northumberland (Table 5.14). As the degree of weathering of the till increases, so does the clay fraction and moisture content. This, in turn, leads to changes in the liquid and plastic limits and in the shear strength (Table 5.15 and Fig. 5.13).

Deposits of stratified drift are often subdivided into two categories:
1 those which develop in contact with the ice — the ice contact deposits; and
2 those which accumulate beyond the limits of the ice, forming in streams, lakes or seas — the proglacial deposits.
Outwash fans range in particle size from coarse sands to boulders. When they are first deposited their porosity may be anything from 25 to 50% and they tend to be very permeable. The finer silt–clay fraction is transported further downstream. Kames, kame terraces, and eskers usually consist of sands and gravels (section 3.4.3, p. 80).

The most familiar proglacial deposits are varved clays. The thickness of the individual varve is fre-quently less than 2 mm although much thicker layers have been noted in a few deposits. Generally, the

Table 5.14 A weathering scheme for Northumberland lodgement tills. (After Eyles and Sladen, 1981.)

Weathering state	Zone	Description
Highly weathered	IV	Oxidized till and surficial material Strong oxidation colours High rotten boulder content Leached of most primary carbonate Prismatic gleyed jointing Pedological profile usually leached brown earth
Moderately weathered	III	Oxidized till Increased clay content Low rotten boulder content Little leaching of primary carbonate Usually dark brown or dark red-brown Base commonly defined by fluvioglacial sediments
Slightly weathered	II	Selective oxidation along fissure surfaces where present, otherwise as Zone 1
Unweathered	I	Unweathered till No post-depositionally rotted boulders No oxidation No leaching of primary carbonate Usually dark grey

Table 5.15 Typical geotechnical properties for Northumberland lodgement tills. (After Eyles and Sladen, 1981.)

Property	Weathering zones	
	I	III & IV
Bulk density (kg/m^3)	2150–2300	1900–2200
Natural moisture content (%)	10–15	12–25
Liquid limit (%)	25–40	35–60
Plastic limit (%)	12–20	15–25
Plasticity index	0–20	15–40
Liquidity index	−0.20 to −0.05	III −0.15 to +0.05
		IV 0 to +30
Grading of fine (<2 mm) fraction		
% clay	20–35	30–50
% silt	30–40	30–50
% sand	30–50	10–25
Average activity	0.64	0.68
c' (kPa)	0–15	0–25
ϕ' (degrees)	32–37	27–35
ϕ'_r (degrees)	30–32	15–32

Table 5.16 Some properties of varved and laminated clays

	Varved clay* (Elk Valley)	Laminated clay[†] (Teesside)
Moisture content (%)	35	22–35 (30)
Plastic limit (%)	22	18–31 (26)
Liquid limit (%)	34	29–78 (56)
Plasticity index[1] (%)	15.5	19–49 (33)
Liquidity index	0.36	−0.12–0.35 (0.15)
Activity[2]	0.36	0.47–0.65 (0.54)
Linear shrinkage (%)		9–14 (11)
Compression index	0.405–0.587 (0.496)	0.55
Undrained shear strength (kPa)[3]		20–102 (62)

* After George, 1986; [†] After Bell and Coulthard, 1991. Range with average value in brackets.
[1] Low plasticity, <35%; intermediate plasticity; 35–50%.
[2] Inactive, <0.75.
[3] Soft, <40; firm, 40–75; stiff, 75–150 kPa.

coarser layer is of silt size and the finer of clay size. Varved clays tend to be normally consolidated or lightly overconsolidated, although it is usually difficult to make the distinction. In many cases the precompression may have been due to ice loading. The range of liquid limits for varved clays tends to vary between 30 and 80% whilst that of plastic limits often varies between 15 and 30% (Table 5.16; Fig. 5.14a). These limits allow the material to be classified as inorganic silty clay of low to high plasticity or compressibility. In some varved clays the natural moisture content is near the liquid limit. They are consequently soft and have sensitivities generally of the order of 4. Their activity tends to range between active and normal, and some of these clays may be expansive (Fig. 5.14b). The average strength of some varved clays from Ontario, for example, is about 42 kPa, with a range of 24 to 49 kPa. The effective stress parameters of apparent cohesion and angle of shearing resistance range from 0.7 to 19.5 kPa, and 22 to 25°, respectively (Metcalf and Townsend, 1961).

The material comprising quick clays is predominantly smaller than 0.002 mm. Many deposits, however, seem to be very poor in clay minerals, containing a high proportion of ground-down, fine quartz. The fabric of these soils contains aggre-

gations. Granular particles, whether aggregations or primary minerals, are rarely in direct contact, being linked generally by bridges of fine particles. Clay minerals usually are non-oriented and clay coatings on primary minerals tend to be uncommon, as are cemented junctions. Networks of platelets occur in some soils.

Quick clays generally exhibit little plasticity, their plasticity index often varying from 8 to 15%. Their liquidity index normally exceeds 1, and their liquid limit is often less than 40%. Quick clays are usually inactive, their activity frequently being less than 0.5 (Table 5.17). The most extraordinary property possessed by quick clays is their very high sensitivity (Locat et al., 1984), that is, a large proportion of their undisturbed strength is permanently lost following shear. The small fraction of the original strength regained after remoulding may be attributable to the development of some different form of interparticle bonding. The reason why only a small fraction of the

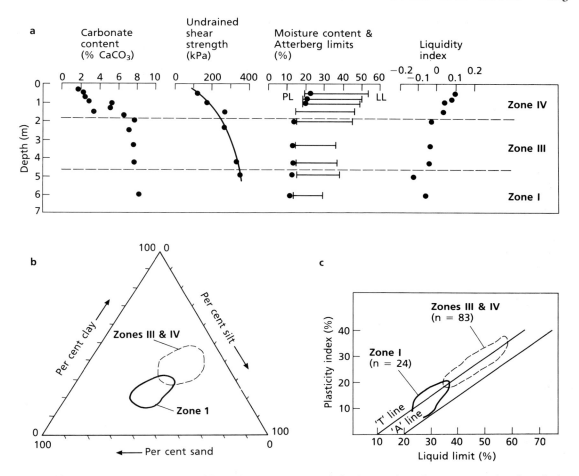

Fig. 5.13 (a) Northumberland lodgement tills: carbonate content, undrained strength, moisture content, Atterberg limits and liquidity index versus depth at a single representative site (Sandy Bay); (b) particle size distribution envelopes for weathered and unweathered Northumberland lodgement tills; (c) a plasticity chart showing envelopes for weathered and unweathered Northumberland lodgement tills. *n*, number of determinations. (After Eyles and Sladen, 1981.)

original strength can ever be recovered is that the rate at which it develops is so slow.

5.7 ORGANIC SOILS: PEAT

Peat is an accumulation of partially decomposed and disintegrated plant remains which have been fossilized under conditions of incomplete aeration and high water content (Hobbs, 1986). Physicochemical and biochemical processes cause this organic material to remain in a state of preservation over a long period of time.

Macroscopically, peaty material can be divided into three basic groups: (i) amorphous granular,

(ii) coarse fibrous, and (iii) fine fibrous. The amorphous granular peats have a high colloidal fraction, holding most of their water in an adsorbed rather than free state. In the other two types the peat is composed of fibres, these usually being woody. In the coarse variety a mesh of second-order size exists within the interstices of the first-order network, whilst in fine fibrous peat the interstices are very small and contain colloidal matter.

The ash percentage of peat consists of the mineral residue remaining after its ignition, which is expressed as a fraction of the total dry weight. Ash contents may be as low as 2% in some highly organic peats, or may be as high as 50%. The mineral

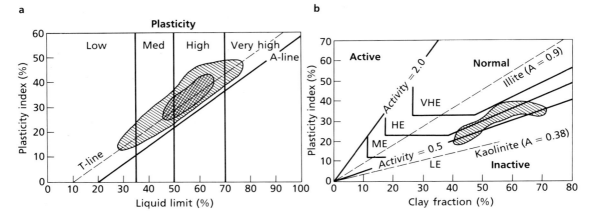

Fig. 5.14 (a) Plasticity chart showing distribution of over 100 specimens of Tees Laminated Clay. Inner shaded area contains the results from over two-thirds of the tests; (b) activity chart showing distribution of Tees Laminated Clays (shaded area). VHE, very high expansion; HE, high expansion; ME, medium expansion; LE, low expansion. (After Bell and Coulthard, 1991.)

material is usually quartz sand and silt. In many deposits the mineral content increases with depth. The mineral content influences the engineering properties of peat.

The void ratio of peat ranges between 9, for dense, amorphous granular peat, and 25, for fibrous types with high contents of sphagnum. It usually tends to decrease with depth within a peat deposit. Such high void ratios give rise to phenomenally high water contents. The latter is the most distinctive characteristic of peat, and most of the peculiarities in the physical characteristics of peat are attributable to the amount of moisture present. This varies according to the type of peat; it may be as low as 500% in some amorphous granular varieties whilst, by contrast, values exceeding 3000% have been recorded from coarse fibrous varieties.

The volumetric shrinkage of peat increases up to a maximum and then remains constant, the volume being reduced almost to the point of complete dehydration. The amount of shrinkage which can occur generally ranges between 10 and 75% of the original volume of the peat and it can involve reductions in void ratio from over 12 down to about 2.

Table 5.17 Engineering properties of sensitive soils. (After Gillott, 1979.)

Location*	Depth (m)	Natural moisture content (%)	Preconsolidation pressure (kPa)	Undrained strength (kPa)	Sensitivity	Liquid limit (%)	Plastic limit (%)	Liquidity index	Activity
O	13.7	60	450	160	—	49	23	1.4	0.35
Q	5.2	75	150	50	—	70	26	1.1	0.64
Q	14.3	81	150	50	—	65	28	1.4	0.45
O	2.6	65	60	20	100	55	22	1.3	0.73
Q	12.2	28	590	230	—	23	16	1.7	0.18
O	5.2	78	320	120	—	65	28	1.3	0.44
BC	20.1	38	—	20	30	28	22	2.7	0.22
BC	14.0	29	—	—	4	23	16	1.9	0.33
BC	35.4	37	—	60	5	28	23	2.8	0.17
A	61.3	17	—	—	—	26	21	0.8	—
A	60.7	—	—	—	—	23	20	—	—

* O, Ontario; Q, Quebec; BC, British Columbia; A, Alaska.

Amorphous granular peat has a higher bulk density than the fibrous types, that is, in the former it can range up to $1.2\,Mg/m^3$ whilst in woody fibrous peats it may be half this figure. However, the dry density is a more important engineering property of peat, influencing its behaviour under load. Dry densities of drained peat fall within the range of 65 to $120\,kg/m^3$. The dry density is influenced by the mineral content and higher values than those quoted can be obtained when peats possess high mineral residues. The specific gravity of peat ranges from as low as 1.1 up to about 1.8, again being influenced by the content of mineral matter. Due to its extremely low submerged density, which may be between 15 and $35\,kg/m^3$, peat is especially prone to rotational failure or failure by spreading, particularly under the action of horizontal seepage forces.

In an undrained bog the unconfined compressive strength is negligible, the peat possessing a consistency approximating to that of a liquid. The strength is increased by drainage to values between 20 and $30\,kPa$ and the modulus of elasticity to between 100 and $140\,kPa$ (Hanrahan, 1954). When loaded, peat deposits undergo high deformations but their modulus of deformation tends to increase with increasing load.

If the organic content of a soil exceeds 20% by weight, consolidation becomes increasingly dominated by the behaviour of the organic material (Berry and Poskitt, 1972), for example, on loading, peat undergoes a decrease in permeability of several orders of magnitude. Moreover, residual pore water pressure affects primary consolidation, and considerable secondary consolidation further complicates settlement prediction.

Differential and excessive settlement is the principal problem confronting the engineer working on peaty soil (Lefebvre et al., 1984; Berry et al., 1985). When a load is applied to peat, settlement occurs because of the low lateral resistance offered by the adjacent unloaded peat. Serious shearing stresses are induced even by moderate loads. Worse still, should the loads exceed a given minimum, then settlement may be accompanied by creep, lateral spreading or, in extreme cases, by rotational slip and upheaval of adjacent ground. At any given time the total settlement in peat due to loading involves settlement with and without volume change. Settlement without volume change is the more serious for it can give rise to the types of failure mentioned. What is more it does not enhance the strength of peat.

5.8 DESCRIPTION OF ROCKS AND ROCK MASSES

Description is the initial step in an engineering assessment of rocks and rock masses. It should therefore be both uniform and consistent in order to gain acceptance. The data collected regarding rocks and rock masses should be recorded on data sheets for subsequent processing. A data sheet for the description of rock masses and another for discontinuity surveys are shown in Figures 5.15 and 2.17 respectively.

Intact rock may be described from a geological or engineering point of view. In the first case the origin and mineral content of a rock are of prime importance, as is its texture and any change which has occurred since its formation. In this respect the name of a rock provides an indication of its origin, mineralogical composition and texture (Table 5.18). Only a basic petrographical description of the rock is required when describing a rock mass. The micropetrographic description of rocks for engineering purposes includes the determination of all parameters which cannot be obtained from a macroscopic examination of a rock sample, such as mineral content, grain size, and texture, and which have a bearing on the mechanical behaviour of the rock or rock mass. In particular a microscopic examination should include a modal analysis, determination of microfractures and secondary alteration, determination of grain size, and, where necessary, fabric analysis. The ISRM (Hallbauer et al., 1978) recommends that the report of a petrographic examination should be confined to a short statement on the origin, classification, and details relevant to the mechanical properties of the rock concerned. Wherever possible this should be combined with a report on the mechanical parameters.

The texture of a rock, in particular its grain size (Table 5.19), relative grain size, grain shape, and fabric have been referred to in Chapter 1. Grain size exerts some influence on the physical properties of a rock, for example, finer-grained rocks are usually stronger than coarser-grained varieties.

The overall colour of a rock should be assessed by reference to a colour chart (e.g. the rock colour chart of the Geological Society of America). This is because it is difficult to make a quantitative assessment with the eye alone.

Rock material tends to deteriorate in quality as a result of weathering and/or alteration. Qualitative classifications based on the estimation and descrip-

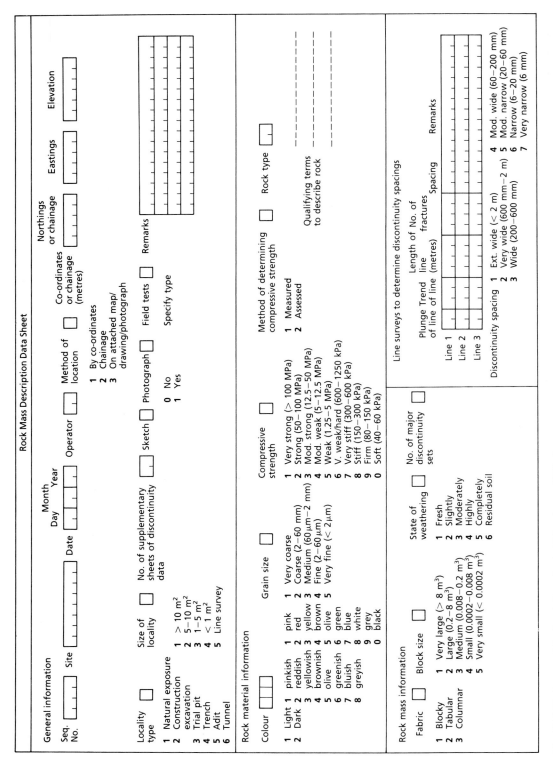

Fig. 5.15 Rock mass data description sheet. (After Anon, 1977a.)

Table 5.18 Rock type classification. (After Anon, 1979.)

Grain size (mm)	Size category / Group	Detrital sedimentary — Grains of rock, quartz, feldspar, and clay minerals (Bedded)	At least 50% of grains are of carbonate	Pyroclastic — At least 50% of grains are of fine-grained igneous rock	Chemical/organic
60	Very coarse-grained (Boulders, Cobbles) — Rudaceous	Grains are of rock fragments. Rounded grains: conglomerate	Carbonate gravel — Calci-rudite	Rounded grains: agglomerate; Angular grains: Volcanic Breccia, Lapilli tuff	Saline rocks, Halite, Anhydrite
2	Coarse-grained (Gravel) — Rudaceous	Angular grains: breccia			Gypsum
	Medium-grained (Sand) — Arenaceous	Grains are mainly mineral fragments. Sandstone: Grains are mainly mineral fragments. Quartz arenite: 95% quartz, voids empty or cemented. Arkose: 75% quartz, up to 25% feldspar: voids empty or cemented. Greywacke: 75% quartz, 15% fine detrital material: rock and feldspar fragments	Carbonate sand — Calci-arenite	Tuff	Limestone, Dolomite
			Limestone and Dolomite (undifferentiated)	*Volcanic ash*	Chert, Flint
0.06	Fine-grained (Silt) — Argillaceous	Siltstone: 50% fine-grained particles	Carbonate silt — Calci-siltite, chalk	Fine-grained tuff	
0.002	Very fine-grained (Clay) — Argillaceous	Claystone: 50% very fine grained particles	Carbonate mud — Calci-lutite	Very fine-grained tuff	Peat, Lignite, Coal
	Glassy, Amorphous				

(Argillaceous fine fraction: Mudstone, shale: fissile mudstone, Marlstone)

Continued on p. 168

Table 5.18 *Continued*

Genetic/group	Metamorphic		Igneous			
Usual structure	Foliated		Massive			
Composition	Quartz, feldspars, micas, acicular dark minerals		Light coloured minerals are quartz, feldspar, mica		Dark and light minerals	Dark minerals
			Acid rocks	Intermediate	Basic rocks	Ultrabasic
Grain size (mm)						
60 Very coarse-grained			Pegmatite			Pyroxenite and peridotite
2 Coarse-grained	Gneiss (ortho-, para-. Alternate layers of granular and flaky minerals)	Marble / Granulite	Granite	Diorite	Gabbro	Serpentinite
Medium-grained	Migmatite Schist	Quartzite Hornfels Amphibolite	Microgranite	Microdiorite	Dolerite	
0.06 Fine-grained	Phyllite		Rhyolite	Andesite	Basalt	
0.002 Very fine grained	Slate Mylonite					
Glassy Amorphous			Obsidian and pitchstone		Tachylyte	
			Volcanic glasses			

Table 5.19 Description of grain size

Term	Particle size	Equivalent soil grade
Very coarse-grained	Over 60 mm	Boulders and cobbles
Coarse-grained	2–60 mm	Gravel
Medium-grained	0.06–2 mm	Sand
Fine-grained	0.002–0.06 mm	Silt
Very fine-grained	Less than 0.002 mm	Clay

tion of physical disintegration and chemical decomposition of originally sound rock are given in Chapter 3.

As far as the engineering properties of rocks are concerned the IAEG (Anon, 1979) grouped the dry density and porosity of rocks into five classes as shown in Table 5.20. Determination of the strength and deformability of intact rock is obtained with the aid of laboratory tests. There are several scales of unconfined compressive strength; three are given in Table 5.21. If the strength of rock is not measured, then it can be estimated as shown in Table 5.22. Obviously such estimates can only be very approximate. The point-load test provides an indication of the tensile strength of intact rock and Franklin and Broch (1972) suggested the scale for the point-load strength shown in Table 5.23. As far as deformability is concerned, five classes have been proposed by the IAEG (Anon, 1979), and are shown in Table 5.24.

The seismic velocity refers to the velocity of propagation of shock waves through a rock mass. Its value is governed by the mineral composition, density, porosity, elasticity, and degree of fracturing within a rock mass. The IAEG (Anon, 1979) recognized five classes of sonic velocity (Table 5.25).

The durability of rocks is referred to in Chapter 3, the description of discontinuities in Chapter 2, and permeability in Chapter 4.

Classifications of intact rock are based upon some selected mechanical properties. The specific purpose for which a classification is developed obviously plays an important role in determining which mechanical properties of the intact rock are chosen. The object of the classification is to provide a reliable basis for assessing rock quality. In fact a classification of intact rock for engineering purposes should be

Table 5.20 Description of dry density and porosity. (After Anon, 1979.)

Class	Dry density (Mg/m^3)	Description	Porosity (%)	Description
1	Less than 1.8	Very low	Over 30	Very high
2	1.8–2.2	Low	30–15	High
3	2.2–2.55	Moderate	15–5	Medium
4	2.55–2.75	High	5–1	Low
5	Over 2.75	Very high	Less than 1	Very low

Table 5.21 Description of unconfined compressive strength

Geological Society (Anon, 1970)		IAEG (Anon, 1979)		ISRM (Anon, 1981b)	
Strength (MPa)	Description	Strength (MPa)	Description	Strength (MPa)	Description
Less than 1.25	Very weak	1.5–15	Weak	Under 6	Very low
1.25–5.00	Weak	15–50	Moderately strong	6–20	Low
5.00–12.50	Moderately weak	50–120	Strong	20–60	Moderate
12.50–50	Moderately strong	120–230	Very strong	60–200	High
50–100	Strong	Over 230	Extremely strong	Over 200	Very high
100–200	Very strong				
Over 200	Extremely strong				

Table 5.22 Estimation of the strength of intact rock. (After Anon, 1977a.)

Description	Approximate unconfined compressive strength (MPa)	Field estimation
Very strong	Over 100	Very hard rock, more than one blow of geological hammer required to break specimen
Strong	50−100	Hard rock, hand-held specimen can be broken with a single blow of hammer
Moderately strong	12.5−50	Soft rock, 5 mm indentations with sharp end of pick
Moderately weak	5.0−12.5	Too hard to cut by hand
Weak	1.25−5.0	Very soft rock — material crumbles under firm blows with the sharp end of a geological hammer

Table 5.23 Description of point load strength. (After Franklin and Broch, 1972.)

	Point load strength index (MPa)	Approximate uniaxial compressive strength (MPa)
Extremely high strength	Over 10	Over 160
Very high strength	3−10	50−160
High strength	1−3	15−60
Medium strength	0.3−1	5−16
Low strength	0.1−0.3	1.6−5
Very low strength	0.03−0.1	0.5−1.6
Extremely low strength	Less than 0.03	Less than 0.5

Table 5.24 Description of deformability. (After Anon, 1979.)

Class	Deformability (MPa × 10³)	Description
1	Less than 5	Very high
2	5−15	High
3	15−30	Moderate
4	30−60	Low
5	Over 60	Very low

Table 5.25 Description of sonic velocity. (After Anon, 1979.)

Class	Sonic velocity (m/s)	Description
1	Less than 2500	Very low
2	2500−3500	Low
3	3500−4000	Moderate
4	4000−5000	High
5	Over 5000	Very high

relatively simple, being based on significant mechanical properties so that it has a wide application, for example, Deere and Miller (1966) based their engineering classification of intact rock on the unconfined compressive strength and Young's modulus (Fig. 5.16).

5.9 ENGINEERING ASPECTS OF IGNEOUS AND METAMORPHIC ROCKS

The plutonic igneous rocks are characterized by

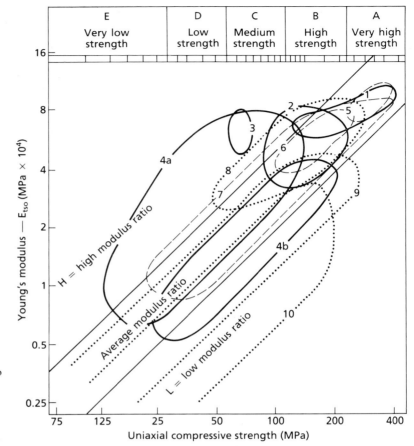

Fig. 5.16 Engineering classification of intact rock based on uniaxial compressive strength and modulus ratio. Fields are shown for igneous, sedimentary, and metamorphic rocks. Metamorphic: 1, quartzite; 2, gneiss; 3, marble; 4a, schist, steep foliation; 4b, schist, flat foliation. Igneous: 5, diabase; 6, granite; 7, basalt and other flow rocks. Sedimentary: 8, limestone and dolomite; 9, sandstone; 10, shale.

Table 5.26 Some physical properties of igneous and metamorphic rocks

	Specific gravity	Unconfined compressive strength (MPa)	Point load strength (MPa)	Shore scleroscope hardness	Schmidt hammer hardness	Young's modulus (GPa)
Mount Sorrel Granite	2.68	176.4	11.3	77	54	60.6
Eskdale Granite	2.65	198.3	12.0	80	50	56.6
Dalbeattie Granite	2.67	147.8	10.3	74	69	41.1
Markfieldite	2.68	185.2	11.3	78	66	56.2
Granophyre (Cumbria)	2.65	204.7	14.0	85	52	84.3
Andesite (Somerset)	2.79	204.3	14.8	82	67	77.0
Basalt (Derbyshire)	2.91	321.0	16.9	86	61	93.6
Slate* (North Wales)	2.67	96.4	7.9	41	42	31.2
Slate† (North Wales)		72.3	4.2			
Schist* (Aberdeenshire)	2.66	82.7	7.2	47	31	35.5
Schist† (Aberdeenshire)		71.9	5.7			
Gneiss	2.66	162.0	12.7	68	49	46.0
Hornfels (Cumbria)	2.68	303.1	20.8	79	61	109.3

* Tested normal to cleavage or schistosity; † Tested parallel to cleavage or schistosity.

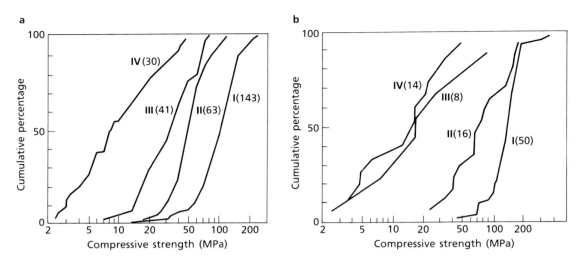

Fig. 5.17 Compressive strength: (a) granite; (b) volcanics, from Hong Kong. Roman numerals indicate grade of weathering; figures in brackets represent the number of samples tested. (After Lumb, 1983.)

granular texture, massive structure, and relatively homogeneous composition. In their unaltered state they are essentially sound and durable with adequate strength for any engineering requirement (Table 5.26). In some instances, however, intrusive rocks may be highly altered by weathering (Chapter 3; Fig. 5.17a) or hydrothermal attack; furthermore fissure zones are by no means uncommon in granites. The rock mass may be very much fragmented along such zones, indeed it may be reduced to sand-size material, and it may have undergone varying degrees of kaolinization. Generally, the weathered product of plutonic rocks has a large clay content although that of granitic rocks is sometimes porous with a permeability comparable to that of medium-grained sand.

Older volcanic deposits do not prove a problem in foundation engineering, ancient lavas having unconfined compressive strengths frequently in excess of 200 MPa (Fig. 5.17b). But volcanic deposits of geologically recent age at times prove treacherous, particularly if they have to carry heavy loads. This is because they often represent markedly anisotropic sequences in which lavas, pyroclasts, and mudflows are interbedded. In addition, weathering during periods of volcanic inactivity may have produced fossil soils, these being of much lower strength. The individual lava flows may be thin and transected by a polygonal pattern of cooling joints. They also may be vesicular or contain pipes, cavities, or even tunnels.

Pyroclasts usually give rise to extremely variable ground conditions due to wide variations in strength, durability, and permeability. Their behaviour very much depends upon their degree of induration, for example, many agglomerates have enough strength to support heavy loads and also have a low permeability. By contrast, ashes are invariably weak and often highly permeable. One particular hazard concerns ashes, not previously wetted, which are metastable and exhibit a significant decrease in their void ratio on saturation. Ashes are frequently prone to sliding. Montmorillonite is not an uncommon constituent in the weathered products of basic ashes.

Slates, phyllites, and schists are characterized by textures which have marked preferred orientation. This preferred alignment of platy minerals accounts for cleavage and schistosity which typify these metamorphic rocks and means that slate is notably fissile. Such rocks are appreciably stronger across, than along, the lineation (Table 5.26). Not only does cleavage and schistosity adversely affect the strength of metamorphic rocks, it also makes them more susceptible to decay. However, slates, phyllites, and schists weather slowly but the areas of regional metamorphism in which they occur have suffered extensive folding so that in places rocks may be fractured and deformed. They are variable in quality, some providing excellent foundations for heavy structures; others, regardless of the degree of their deformation or weathering, are so poor as to be wholly unsuitable. For example, talc, chlorite, and sericite schist are weak rocks containing planes of schistosity only a millimetre or so apart. Some schists become slippery upon weathering and therefore fail under a moderately light load.

The engineering performance of gneiss is usually similar to that of granite. However, some gneisses are strongly foliated which means that they possess a texture with a preferred orientation. Generally, this will not significantly affect their engineering behaviour. They may, however, be fissured in places and this can mean trouble.

Fresh, thermally metamorphosed rocks such as quartzite and hornfels are very strong and afford good ground conditions. Marble has the same advantages and disadvantages as other carbonate rocks.

5.10 ENGINEERING BEHAVIOUR OF SEDIMENTARY ROCKS

Sandstones may vary from thinly laminated micaceous types to very thickly bedded varieties. Moreover they may be cross-bedded and are invariably jointed. With the exception of shaly sandstone, sandstone is not subject to rapid surface deterioration on exposure.

The dry density and especially the porosity of a sandstone are influenced by the amount of cement and/or matrix material occupying the pores. Usually the density of a sandstone tends to increase with increasing depth below the surface.

The compressive strength and deformability of a sandstone is influenced by its porosity, the amount and type of cement, and/or matrix material, grain contact, as well as the composition of the individual grains (Fig. 5.18; Bell, 1978b; Bell and Culshaw, 1992; Dobereiner and De Freitas, 1986; Hawkins and McConnell, 1991). If cement binds the grains together a stronger rock is produced than one in which a similar amount of detrital matrix performs the same function. However, the amount of cementing material is more important than the type of cement, although if two sandstones are equally well cemented, one having a siliceous, the other a calcareous cement, then the former is the stronger.

Pore water plays a very significant role as far as the compressive strength and deformation characteristics of a sandstone are concerned. As can be seen from Table 5.27 it can reduce the unconfined compressive strengths by 30 to 60%.

The mineral content of shales influences their geotechnical properties, the most important factor in this respect being the quartz–clay minerals ratio. For example, the liquid limit of clay shales increases with increasing clay-mineral content, the amount of montmorillonite, if present, being especially important. Mineralogy also affects the activity of an

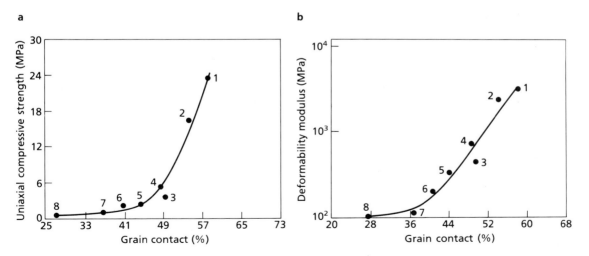

Fig. 5.18 Dependence of strength and deformability on grain contact in weak sandstones: 1, Keuper Waterstones (perpendicular to bedding); 2, Keuper Waterstones (parallel to bedding); 3, Lahti Sandstone; 4, Bauru Sandstone; 5, Bauru Sandstone; 6, Kidderminster Sandstone (perpendicular to bedding); 7, Bauru Sandstone; 8, Kidderminster Sandstone (parallel to bedding). Grain contact equals length of grain contact divided by total length of grain surface. (After Dobereiner and De Freitas, 1986.)

Table 5.27 Some physical properties of arenaceous sedimentary rocks

	Fell Sandstone (Rothbury)	Chatsworth Grit (Stanton in the Peak)	Sherwood Sandstone (Edwinstowe)	Keuper Waterstones (Edwinstowe)	Horton Flags* (Helwith Bridge)	Bronllwyn Grit* (Llanberis)
Specific gravity	2.69	2.69	2.68	2.73	2.70	2.71
Dry density (Mg/m^3)	2.25	2.11	1.87	2.26	2.62	2.63
Porosity	9.8	14.6	25.7	10.1	2.9	1.8
Dry unconfined compressive strength (MPa)	74.1	39.2	11.6	42.0	194.8	197.5
Saturated unconfined compressive strength (MPa)	52.8	24.3	4.8	28.6	179.6	190.7
Point load strength (MPa)	4.4	2.2	0.7	2.3	10.1	7.4
Scleroscope hardness	42.0	34.0	18.0	28.0	67.0	88.0
Schmidt hardness	37.0	28.0	10.0	21.0	62.0	54.0
Young's modulus (GPa)	32.7	25.8	6.4	21.3	67.4	51.1
Permeability ($\times 10^{-9}$ m/s)	1740.0	1960.0	3500.0	22.4	—	—

* Greywackes.

argillaceous material; again this increases with clay mineral content, particularly with increasing content of montmorillonite. Activity influences the slaking characteristics of a shale.

Consolidation with concomitant recrystallization on the one hand and the parallel orientation of platy minerals, notably micas, on the other give rise to the fissility of shales. An increasing content of siliceous or calcareous material gives a less fissile shale whilst carbonaceous or organic shales are exceptionally fissile. Moderate weathering increases the fissility of shale by partially removing the cementing agents along the laminations or by expansion due to the hydration of clay particles. Intense weathering produces a soft claylike soil.

The natural moisture content of shales varies from less than 5%, increasing to as high as 35% for some clayey shales. When the natural moisture content of shales exceeds 20% they frequently are suspect as they tend to develop potentially high pore water pressures. Usually the moisture content in the weathered zone is higher than in the unweathered shale beneath. Depending upon the relative humidity, many shales slake almost immediately when exposed to air. Desiccation of shale following exposure, leads to the creation of negative pore water pressures and consequent tensile failure of weak intercrystalline bonds. This leads to the production of shale particles

of coarse sand, fine gravel size. Alternate wetting and drying causes breakdown of shales (Taylor, 1988; Varley, 1990; Bell, 1992a). Low-grade compaction shales, in particular, undergo complete disintegration after several cycles of drying and wetting. Mudstones tend to break down along irregular fracture patterns which, when well developed, can mean that these rocks disintegrate within one or two cycles of wetting and drying.

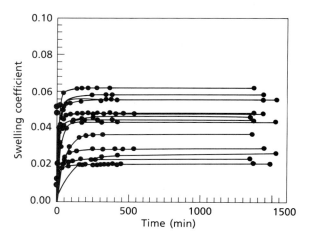

Fig. 5.19 Free swell of samples of carbonaceous mudrock. (After Jermy and Bell, 1991.)

The swelling properties of certain shales (Fig. 5.19) have proven extremely detrimental to the integrity of many civil engineering structures. Swelling is attributable to the absorption of free water by certain clay minerals, notably montmorillonite, in the clay fraction of a shale. Highly fissured overconsolidated shales have greater swelling tendencies than poorly fissured clayey shales, the fissures providing access for water.

The porosity of shale may range from slightly under 5% to just over 50% with natural moisture contents of 3 to 35%. Argillaceous materials are capable of undergoing appreciable suction before pore water is removed, drainage commencing when the necessary air-entry suction is achieved (about pF 2). Under increasing suction pressure the incoming air drives out water from a shale and some shrinkage takes place in the fabric before air can offer support. Generally, as the natural moisture content and liquid limit increase so the effectiveness of soil suction declines.

Cemented shales are invariably stronger and more durable than compacted shales. In weak compaction shales cohesion may be lower than 15 kPa and the angle of friction as low as 5°. By contrast, Underwood (1967) quoted values of cohesion and angle of friction of 750 kPa and 56°, respectively for dolomitic shales of Ordovician age, and 8 to 23 MPa and 45 to 64°, respectively for calcareous and quartzose shales from the Cambrian period. Generally, shales with a cohesion less than 20 kPa and an apparent angle of friction of less than 20° are likely to present problems. The elastic moduli of compaction shales range between 140 and 1400 MPa; well-cemented shales have elastic moduli in excess of 14 000 MPa (Table 5.28). Values of Young's modulus can be up to five times greater when shale is tested normal as opposed to parallel to the direction of lamination.

Unconfined compressive strength tests on Accra Shales carried out by De Graft-Johnson *et al.* (1973) indicated that the samples usually failed at strains between 1.5 and 3.5%. The compressive strengths varied from 200 kPa to 20 MPa.

Severe settlements may take place in low-grade compaction shales. Conversely, uplift frequently occurs in excavations in shales and is attributable to swelling and heave. Rebound on unloading of shales during excavation is attributed to heave due to the release of stored strain energy. The greatest amount of rebound occurs in heavily overconsolidated compaction shales.

Sulphur compounds are frequently present in shales, clays, and mudstones. An expansion in volume large enough to cause structural damage can occur when sulphide minerals such as pyrite and marcasite suffer oxidation to give sulphates (Fasiska *et al.*, 1974).

Clay shales usually have permeabilities of the order 1×10^{-8} m/s to 10^{-12} m/s. However, sandy and silty shales and closely jointed cemented shales may have permeabilities as high as 1×10^{-6} m/s.

The engineering properties of carbonate sediments are influenced by grain size and those postdepositional changes which bring about induration. Because induration can take place at the same time as deposition is occurring, this means that carbonate sediments can sustain high overburden pressures which, in turn, means that they can retain high porosities to considerable depths. Indeed a layer of cemented grains may overlie one that is poorly cemented. Eventually, however, high overburden pressures, creep, and recrystallization produce crystalline limestone with very low porosity.

Representative values of some physical properties of carbonate rocks are listed in Table 5.29. It can be seen that generally the density of these rocks increases with age, whilst the porosity is reduced (Bell, 1981a). Diagenetic processes mainly account for the lower porosities of the older limestones. The high porosity values of the Upper Chalk may be due to the presence of hollow tests and the complex shapes of the constituent particles. This chalk is also poorly cemented.

Limestone when dolomitized undergoes an increase in porosity of a few per cent and therefore tends to possess a lower compressive strength than limestone which has not been dolomitized (Attewell, 1971; Al-Jassar and Hawkins, 1979). For example, the Great Limestone of the north of England has a compressive strength ranging from 110 to 210 MPa with an average porosity of 4%. When dolomitized its average porosity is 7.5% with a compressive strength of between 70 and 165 MPa.

Joints in limestone have generally been subjected to various degrees of dissolution so that some may gape. Sinkholes may develop where joints intersect and these may lead to an integrated system of subterranean galleries and caverns. The latter are characteristic of thick massive limestones. The progressive opening of discontinuities by dissolution leads to an increase in mass permeability.

An important effect of solution of limestone is enlargement of the pores which enhances water circulation thereby encouraging further solution.

Table 5.28 An engineering evaluation of shale. (After Underwood, 1967.)

Laboratory tests and in situ observations	Physical properties Average range of values		Probable in situ behaviour*						
	Unfavourable	Favourable	High pore pressure	Low bearing capacity	Tendency to rebound	Slope stability problems	Rapid sinking	Rapid erosion	Tunnel support problems
Compressive strength (kPa)	350–2070	2070–3500	√	√					√
Modulus of elasticity (MPa)	140–1400	140–14000		√					√
Cohesive strength (kPa)	35–700	700–>10500			√	√			
Angle of internal friction (deg)	10–20	20–65			√	√			√
Dry density (Mg/m³)	1.12–1.78	1.78–2.56	√		√	√		√(?)	
Potential swell (%)	3–15	1–3			√	√			
Natural moisture content (%)	20–35	5–15	√			√			
Coefficient of permeability (m/s)	10^{-7}–10^{-12}	>10^{-7}	↓				↓		
Predominant clay minerals	Montmorillonite or illite	Kaolinite and chlorite	√			√			
Activity	0.75–>2.0	0.35–0.75		√		√			
Wetting and drying cycles	Reduces to grain sizes	Reduces to flakes		√		√	√	√(?)	√
Spacing of rock defects	Closely spaced	Widely spaced				√			√
Orientation of rock defects	Adversely oriented	Favourably oriented			√	√			√
State of stress	>Existing over-burden load	≈Over-burden load			√	√			√

Note: According to S. Irmay (*Israel Journal of Technology*, 1968, **6**, No. 4, 165–72), the maximum possible $\phi = 47.5$.
* The ticks relate to the unfavourable range of values.

Table 5.29 Some physical properties of carbonate rocks

	Carboniferous Limestone (Buxton)	Magnesium Limestone (Anston)	Ancaster Freestone (Ancaster)	Bath Stone (Corsham)	Middle Chalk (Hillington)	Upper Chalk (Northfleet)
Specific gravity	2.71	2.83	3.70	2.71	2.70	2.69
Dry density (Mg/m^3)	2.58	2.51	2.27	2.30	2.16	1.49
Porosity	2.9	10.4	14.1	15.6	19.8	41.7
Dry unconfined compressive strength (MPa)	106.2	54.6	28.4	15.6	27.2	5.5
Saturated unconfined compressive strength (MPa)	83.9	36.6	16.8	9.3	12.3	1.7
Point load strength (MPa)	3.5	2.7	1.9	0.9	0.4	—
Scleroscope hardness	53.0	43.0	38.0	23.0	17.0	6.0
Schmidt hardness	51.0	35.0	30.0	15.0	20.0	9.0
Young's modulus (GPa)	66.9	41.3	19.5	16.1	30.0	4.4
Permeability ($\times 10^{-9}$ m/s)	0.3	40.9	125.4	160.5	1.4	13.9

This brings about an increase in stress within the remaining rock framework which reduces the strength of the rock mass and leads to increasing stress corrosion. On loading the volume of the voids is reduced by fracture of the weakened cement between the particles and by the reorientation of the intact aggregations of rock that become separated by loss of bonding. Most of the resultant settlement takes place rapidly within a few days of the application of load.

The dry density of chalk has a notable range, for example, low values have been recorded from the Upper Chalk of Kent (1.35–1.64 Mg/m^3) whilst those from the Middle Chalk of Norfolk and the Lower Chalk of Yorkshire frequently exceed 2.0 Mg/m^3. The porosity of chalk tends to range between 30 and 50%.

Chalk compresses elastically up to a critical pressure, the apparent preconsolidation pressure. Marked breakdown and substantial consolidation occurs at higher pressures. The coefficients of consolidation (c_v) and compressibility (m_v) are around 1135 m^2/year and 0.019 m^2/MN, respectively. The unconfined compressive strength of chalk ranges from moderately weak (much of the Upper Chalk) to moderately strong (much of the Lower Chalk of Yorkshire and the Middle Chalk of Norfolk). However, the unconfined compressive strength of chalk undergoes a marked reduction when it is saturated (Bell *et al.*, 1990), for example, the Upper Chalk from Kent may suffer a loss on saturation amounting to approximately 70%. The Upper Chalk from Kent is particularly deformable, a typical value of Young's modulus being 5×10^3 MPa. In fact it exhibits elastic–plastic deformation, with perhaps incipient creep prior to failure. The deformation properties of chalk in the field depend upon its hardness, and the spacing, tightness, and orientation of its discontinuities. The values of Young's modulus are also influenced by the amount of weathering chalk has undergone (Table 5.30).

The discontinuities are the fundamental factors governing the mass permeability of chalk. Chalk is also subject to dissolution along discontinuities. However, subterranean solution features tend not to develop in chalk since it is usually softer than limestone and so collapses as solution occurs. Nevertheless, solution pipes and swallow holes are present in chalk.

Representative specific gravities and dry densities of gypsum, anhydrite, rock salt and potash are given in Table 5.31, as are porosity values. Anhydrite is a strong rock, gypsum and potash are moderately strong, whilst rock salt is moderately weak (Bell, 1981b). Evaporitic rocks exhibit varying degrees of plastic deformation before failing, for example, in rock salt the yield strength may be as little as one-tenth the ultimate compressive strength, whereas anhydrite undergoes comparatively little plastic de-

Table 5.30 Correlation between grades and the mechanical properties of Middle Chalk at Mundford. (After Ward *et al.*, 1969*.)

Grade	Description	Approx. range of E (MPa)	Approx. value of E_{dv}^* (MPa) (after Abbiss, 1979)†	Range of compression wave velocities (km/s) (after Grainger *et al.*, 1973)‡	Bearing pressure causing yield (kPa)	Creep properties	SPT N value (after Wakeling, 1970)§‖	Rock mass factor (after Burland and Lord, 1969)¶
V	Structureless melange. Unweathered and partly weathered angular chalk blocks and fragments set in a matrix of deeply weathered remoulded chalk. Bedding and jointing are absent	Below 500	Below 500	0.65–0.75	Below 200	Exhibits significant creep	Below 15	0.1
IV	Friable to rubbly chalk. Unweathered or partially weathered chalk with bedding and jointing present. Joints and small fractures closely spaced, ranging from 10 to 60 mm apart	500–1000	800	1.0–1.2	200–400	Exhibits significant creep	15–20	0.1–0.2
III	Rubbly to blocky chalk. Unweathered medium to hard chalk with joints 60–200 mm apart. Joints open up to 8 mm, sometimes with secondary staining and fragmentary infillings	1000–2000	4000	1.6–1.8	400–600	For pressures not exceeding 400 kPa creep is small and terminates in a few months	20–25	0.2–0.4
II	Medium hard chalk with widely spaced, closed joints. Joints more than 200 mm apart. Fractures irregularly when excavated, does not break along joints. Unweathered	2000–5000	7000	2.2–2.3	Over 1000	Negligible creep for pressure of at least 400 kPa	25–35	0.6–0.8
I	Hard, brittle chalk with widely spaced, closed joints. Unweathered	Over 5000	Over 2.3	Over 10000	Over 1000	Negligible creep for pressure of at least 400 kPa	Over 35	Over 0.8

For footnotes see opposite

Table 5.31 Some physical properties of evaporitic rocks

	Gypsum (Sherburn in Elmet)	Anhydrite (Sandwith)	Rock salt (Winsford)	Potash (Loftus)
Specific gravity	2.36	2.93	2.2	2.05
Dry density (Mg/m^3)	2.19	2.82	2.09	1.98
Porosity (%)	4.6	2.9	4.8	5.1
Unconfined compressive strength (MPa)	27.5	97.5	11.7	25.8
Point load strength (MPa)	2.1	3.7	0.3	0.6
Schleroscope hardness	27.0	38.0	12.0	9.0
Schmidt hardness	25.0	40.0	8.0	11.0
Young's modulus (GPa)	24.8	63.9	3.8	7.9
Permeability ($\times 10^{-10}$ m/s)	6.2	0.3	—	—

formation prior to rupture. Creep may account for anything between 20 and 60% of the strain at failure when these evaporitic rocks are subjected to incremental creep tests. Rock salt is most prone to creep.

Gypsum is more readily soluble than limestone, 2100 mg/l can be dissolved in non-saline waters as compared with 400 mg/l (Table 3.4). Sinkholes and caverns can therefore develop in thick beds of gypsum more rapidly than they can in limestone. Cavern collapse has led to extensive cracking and subsidence at the ground surface. However, gypsum is weaker than limestone and therefore collapses more readily. Massive anhydrite can be dissolved to produce uncontrollable runaway situations in which seepage flow rates increase in a rapidly accelerating manner.

Heave is another problem associated with anhydrite. This takes place when anhydrite is hydrated to form gypsum, in so doing there is a volume increase of between 30 and 58% which exerts pressures that have been variously estimated between 2 and 69 MPa. It is thought that no great length of time is required to bring about such hydration. When it occurs at shallow depths it causes expansion but the process is gradual and is usually accompanied by the removal of gypsum in solution. At greater depths anhydrite is effectively confined during the process. This results in a gradual build-up of pressure and the stress is finally liberated in an explosive manner.

Salt is even more soluble than gypsum (Table 3.4) and the evidence of slumping, brecciation and collapse structures in rocks which overlie saliferous strata bears witness to the fact that salt has gone into solution in past geological times. It is generally believed, however, that in areas underlain by saliferous beds, measurable surface subsidence is unlikely to occur except where salt is being abstracted as brine (Bell, 1993c). Perhaps this is because equilibrium has been attained between the supply of unsaturated groundwater and the salt available for solution.

Footnotes to Table 5.30

* Ward *et al.* (1969) emphasized that their classification was specifically developed for the site at Mundford and hence its application elsewhere should be made with caution.

† Abbiss, C.P. (1979) A comparison of the stiffness of the Chalk at Mundford from a seismic survey and large-scale tank test. *Geotechnique*, **29**, 461–8.

‡ Grainger, P., McCann, D.M. and Gallois, R.W. (1973) The application of seismic refraction to the study of fracturing in the Middle Chalk at Mundford, Norfolk. *Geotechnique*, **23**, 219–32.

§ Wakeling, T.R.M. (1970) A comparison of the results of standard site investigation methods against the results of a detailed geotechnical investigation in Middle Chalk at Mundford, Norfolk. In *In Situ Investigations in Soils and Rocks*. British Geotechnical Society, London, pp. 17–22.

‖ The correlation between SPT N value and grade may be different in the Upper Chalk (see Dennehy, J.P. (1976) Correlation of the SPT value with chalk grade for some zones of the Upper Chalk. *Geotechnique*, **26**, 610–14).

¶ Burland, J.B. and Lord, J.A. (1970). The deformation behaviour of Middle chalk at Mundford, Norfolk: a comparison between full-scale performance and *in situ* and laboratory measurements. In *In Situ Investigations in Soils and Rocks*. British Geotechnical Society, London, 3–16.

6 Geological materials used in construction

6.1 BUILDING OR DIMENSION STONE

Stone has been used as a construction material for thousands of years. One of the reasons for this is its ready availability locally. Furthermore, stone requires little energy for extraction and processing. Indeed stone is used more or less as it is found except for the seasoning, shaping, and dressing that is necessary before it is used for building purposes. Yet other factors, as many ancient buildings testify, are its attractiveness, durability, and permanence.

A number of factors determine whether a rock will be worked as a building stone:
1 the volume of material that can be quarried;
2 the ease with which it can be quarried;
3 the wastage consequent upon quarrying;
4 the cost of transportation;
5 its appearance and physical properties.

As far as volume is concerned, the life of a quarry should be at least 20 years. The amount of overburden that has to be removed affects the economics of quarrying and there comes a point when its removal makes operations uneconomic. Weathered rock represents waste; therefore, the ratio of fresh to weathered rock is another factor of economic importance. The ease with which a rock can be quarried depends mainly upon geological structures, notably the geometry of joints and bedding planes, where present. Ideally, rock for building stone should be massive, and free from closely spaced joints or other discontinuities as these control block size. The stone should be free of fractures and other flaws. In the case of sedimentary rocks, where beds dip steeply, quarrying has to take place along the strike. Steeply dipping rocks also can give rise to problems of slope stability when excavated. However, if beds of rock dip gently it is advantageous to develop the quarry floor along the bedding planes. The massive nature of igneous rocks such as granite, means that a quarry can be developed in any direction, within the constraints of planning permission.

A uniform appearance is generally desirable in building stone. Its appearance largely depends upon its colour, which in turn is determined by its mineral composition. Texture also affects the appearance of a stone, as does the way in which it weathers, e.g. the weathering of certain minerals, such as pyrite, may produce ghastly stains. Generally speaking, rocks of light colour are used as building stone.

For usual building purposes a compressive strength of 35 MPa is satisfactory and the strength of most rocks used for building stone is well in excess of this figure (Table 6.1). In certain instances tensile strength is important, for example, tensile stresses may be generated in a stone subjected to ground movements. However, the tensile strength of a rock, or more particularly its resistance to bending, is a fraction of its compressive strength (Chapter 5). Hardness is a factor of small consequence, except where a stone is subjected to continual wear such as in steps or pavings.

The texture and porosity (Table 6.1) of a rock affect its ease of dressing, and the amount of expansion, freezing, and dissolution it may undergo, fine-grained rocks are more easily dressed than coarse varieties. Water retention in a rock with small pores is greater than in one with large pores, and so the former are more prone to frost attack.

The durability of a stone is a measure of its ability to resist weathering and so to retain its original size, shape, strength, and appearance over an extensive period of time (Sims, 1991; Bell, 1993a) and one of the most important factors which determines whether or not a rock will be worked for building stone. The amount of weathering undergone by a rock in field exposures or quarries affords some indication of its qualities of resistance. However, there is no guarantee that the durability is the same throughout a rock mass and if it changes it is far more difficult to detect, for example, than a change in colour.

Stone can be damaged by alternate wetting and drying. Water in the pores of a stone of low tensile strength can expand enough when warmed to cause its disruption, for example, when the temperature of water is raised from 0°C to 60°C it expands some 1.5% and this can exert a pressure of up to 52 MPa in rock pores. Water can cause expansion within granite ranging from 0.004 to 0.009%, in marble from 0.001 to 0.0025%, and in quartz arenites from

0.01 to 0.044%. The stresses imposed upon masonry by expansion and contraction brought about by changes in temperature and moisture content can result in masonry between abutments spalling at the joints. Blocks may even be shattered and fall out of place.

Frost damage is one of the major factors causing deterioration in a building stone. Sometimes small fragments are separated by frost action from the surface of a stone but the major effect is gross fracture. Frost damage is most likely to occur on steps, copings, sills, and cornices where rain and snow can collect. Damage to susceptible stone may be reduced if it is placed in a sheltered location. Most igneous rocks, and the better quality sandstones and limestones, are immune. As far as frost susceptibility is concerned the porosity, tortuosity, pore size, and degree of saturation all play an important role. As water turns to ice it increases in volume resulting in pressure increase within the pores. This action is further enhanced by the displacement of pore water away from the developing ice front. Once ice has formed the ice pressures rapidly increase with decreasing temperature, so that at approximately $-22°C$ ice can exert a pressure of 200 MPa (Winkler, 1973). Usually coarse-grained rocks withstand freezing better than fine-grained ones. The critical pore size for freeze−thaw durability appears to be about 0.005 mm, that is, rocks with larger mean pore diameters allow outward drainage and escape of fluid from the frontal advance of the ice line and are therefore less frost susceptible. Fine-grained rocks which have over 5% sorped water are often very susceptible to frost damage whilst those containing less than 1% are very durable. Freezing tests have proved an unsatisfactory method of assessing frost resistance. Capillary tests have been used in France and Belgium to assess the frost susceptibility of building stone whilst in Britain a crystallization test (Table 6.1) has been used (Anon, 1983a).

Deleterious salts, when present in a building stone, generally are derived from the ground or the atmosphere, although soluble salts may occur in the pores of the parental rock. Their presence in a stone gives rise to different effects. They may cause efflorescence by crystallizing on the surface of a stone. In subflorescence crystallization takes place just below the surface and may be responsible for surface scabbing. The pressures produced by crystallization of salts in small pores are appreciable, for example, halite (NaCl) exerts a pressure of 200 MPa;

Fig. 6.1 Honeycomb weathering developed in Coal Measures sandstone, former almshouse, Palace Green, University of Durham, Durham, England.

gypsum ($CaSO_4.H_2O$), 100 MPa; anhydrite ($CaSO_4$), 120 MPa; kieserite ($MgSO_4.H_2O$), 100 MPa; and are often sufficient to cause disruption. Crystallization caused by freely soluble salts such as sodium chloride, sodium sulphate, or sodium hydroxide can lead to a stone surface crumbling or powdering. Deep cavities may be formed in magnesian limestone when it is attacked by magnesium sulphate. Salt action can give rise to honeycomb weathering in some sandstones (Fig. 6.1) and porous limestones. Conversely, surface induration of a stone by the precipitation of salts may give rise to a protective hard crust, that is, case hardening. If the stone is the sole supplier of these salts, then the interior is correspondingly weaker.

The rate of weathering of silicate rocks is usually slow although once weathering penetrates the rock, the rate accelerates. Even so building stones cut from igneous rocks generally suffer negligible decay in climates like that of Britain. By contrast, some basalts used in Germany have proved exceptional in this respect in that they have deteriorated rapidly, crumbling after about 5 years of exposure. On petrological examination these basalts were found to contain analcite, the development within which of microcracks is presumed to have produced the deterioration. Certain igneous stones change colour when weathered, for example, within several

Table 6.1 Some physical properties of British building stones. (After Bell, 1992b, 1993b.)
(a) *Limestones*

Stone Age Location	Orton Scar Lr. Carboniferous Orton	Anstone Magnesian Limestone Kiveton Park	Doulting Inferior Oolite Shepton Mallet	Ancaster Lincolnshire Limestone Ancaster	Bath Great Oolite Monks Park	Portland Portland Isle of Portland	Purbeck Purbeck Swanage
Property							
Specific gravity	2.72	2.83	2.7	2.7	2.71	2.7	2.7
Dry density (Mg/m^3)	2.59	2.51	2.34	2.27	2.3	2.25	2.21
Porosity (%)	4.4	10.4	12.8	19.3	18.2	22.4	9.6
Microporosity (% saturation)	54	23	30	60	77	43	6.2
Saturation coefficient	0.68	0.64	0.69	0.84	0.94	0.58	0.62
Unconfined compressive strength (MPa)	96.4	54.6	35.6	28.4	15.6	20.2	24.1
Young's modulus (GPa)	60.9	41.3	24.1	19.5	16.1	17.0	17.4
Velocity of sound (m/s)	4800	3600	2900	2900	2800	3000	3700
Crystallization test (% wt loss)	1	5	8	20	52	13	3
Durability classification*	A	B	C	D	E	C	B

(b) *Sandstones*

Property	Carstone Cretaceous Snettisham	Hollington Trias Nr Uttoxeter	Lazonby Permian Nr Penrith	Ladycross Coal Measures Nr Hexham	Stancliffe Namurian Nr Matlock	Blaxter Lr Carboniferous Elsdon	Monmouth Old Red Sandstone Nr Monmouth
Colour	Deep yellow to brown	Pink to red, mottled buff	Dark pink to red	Light grey to buff	Buff	Buff	Red to pinkish brown
Grain Size	Medium to coarse grained	Fine to medium grained	Fine to medium grained	Fine to medium grained	Fine to medium grained	Fine to medium grained	Fine to medium grained
Specific gravity	2.75	2.71	2.68	2.69	2.67	2.67	2.69
Dry density (Mg/m³)	1.83	2.04	2.38	2.36	2.38	2.24	2.43
Porosity (%)	32.8	23.5	9.3	11.6	11.5	16.6	8.8
Unconfined compressive strength (MPa)	4	29	40	82	72	50	22
Young's modulus (GPa)	1.7	13.6	21.8	41.2	41.5	35.4	17.4
Acid immersion test	Failed	Passed	Passed	Passed	Passed	Passed	Failed
Crystallization test	—	79	37	9	20	56	—
Saturation coefficient	0.76	0.71	0.47	0.62	0.63	0.59	0.59
Durability classification*	F	D	B,C	A	A,B	B,C	E,F

* A, excellent — E, performs best in sheltered positions in inland locations where pollution is low and frost activity infrequent, F, generally unsatisfactory.

weeks some light grey granites may alter to various shades of pink, red, brown, or yellow, caused by the hydration of the iron oxides.

Building stones derived from sedimentary rocks may undergo a varying amount of decay in urban atmospheres, where weathering is accelerated due to the presence of aggressive impurities such as SO_2, SO_3, NO_3, Cl_2, and CO_2 in the air, which produce corrosive acids. Limestones are most suspect (Fig. 6.2). For instance, weak sulphuric acid reacts with the calcium carbonate of limestones to produce calcium sulphate. The latter often forms just below the surface of a stone and the expansion which takes place upon crystallization causes slight disruption. If this reaction continues, then the outer surface of the limestone begins to flake off.

The degree of resistance that a sandstone offers to

Fig. 6.2 Black crust (a mixture of gypsum, derived from the reaction of limestone with sulphur dioxide dissolved in rainwater, and soot particles, which has an open crystalline structure that permits penetration of moisture) developed on a column of limestone (Lincolnshire Limestone, Jurassic), Lincoln Cathedral, Lincoln, England.

weathering depends upon its mineralogical composition, texture, porosity, amount and type of cement/matrix, and the presence of any planes of weakness. Accordingly, the best type of sandstone for external building purposes is a quartz arenite which is well bonded with siliceous cement, has a low porosity, and is free from visible laminations. The tougher the stone, however, the more expensive it is to dress. Sandstones are chiefly composed of quartz grains which are highly resistant to weathering but other minerals present in lesser amounts may be suspect, for example, feldspars may be kaolinized. Calcareous cements react with weak acids in urban atmospheres, as do iron oxides which produce rusty surface stains. The reactions caused by acid attack may occasionally lead to the surface of a stone flaking off irregularly or, in extreme cases, to it crumbling. Laminated sandstone usually weathers badly when it is used in the exposed parts of buildings, and decays in patches.

Exposure of a stone to intense heating causes expansion of its component minerals with subsequent exfoliation at its surface. The most suspect rocks appear to be those which contain high proportions of quartz and alkali feldspars, such as granites and sandstones. Indeed, quartz is one of the most expansive minerals, expanding by 3.76% between normal temperatures and 570°C. When limestones and marbles are heated to 900°C or dolostones to 800°C, superficial calcination begins to produce surface scars. Finely textured rocks offer a higher degree of heat resistance than do coarse-grained varieties.

Stone preservation involves the use of chemical treatments which prolong the life of a stone, either by preventing or retarding the progress of stone decay or by restoring the physical integrity of the decayed stone (Bell and Coulthard, 1990). A stone preservative therefore will avert or compensate for the harmful effects of time and the environment but must not change the natural appearance or architectural value of the stone to any appreciable extent.

Stone may be 'got' (a quarryman's term):
1 by splitting along bedding and/or joint surfaces by using a wedge and feathers;
2 by drilling a series of closely spaced holes (often with as little as 25 mm spacing between them) in line in order to split a large block from the face;
3 by cutting it from the quarry face with a wire saw;
4 by using a flame cutter (primarily for winning granitic rocks).
It is claimed that this technique is the only way of cutting stone in areas of high stress relief.

If explosives are used to work building stone, then the blast should only weaken the rock along joint and/or bedding planes, and not fracture the material. The object is to obtain large blocks which can be sawn to size. Hence the blasting pattern and amount of charge (black powder) are very important and every effort should be made to keep rock wastage and hair cracking to a minimum.

When stone is won from a quarry, it contains a certain amount of pore water termed quarry sap. As this dries out, the stone hardens. Consequently, it is wise to shape the material as soon as possible after it has been got from the quarry. Blocks are first sawn to the required size (Fig. 6.3), after which they may be planed or turned, before final finishing. Careless operation of dressing machines or tooling of the stone may produce bruising. Subsequently, scaling may develop at points where the stone was bruised, so spoiling its appearance.

Granite is ideally suited for building, engineering,

and monumental purposes. Its crushing strength varies between 160 and 240 MPa. It has exceptional weathering properties and most granites are virtually indestructible under normal climatic conditions. There are examples of granite polished over 100 years ago on which the polish has not deteriorated to any significant extent. Indeed, it is accepted that the polish on granite is such that it is only after exposure to very heavily polluted atmospheres for a considerable length of time that any sign of deterioration is apparent. The maintenance cost of granite compared with other materials is therefore very much less and in most cases there is no maintenance cost at all for a considerable number of years.

Limestones show a variation in their colour, texture, and porosity and those which are fossiliferous are highly attractive when cut and polished. However, carbonate stone can undergo dissolution by acidified water. This results in dulling of polish, surface discoloration, and structural weakening.

Fig. 6.3 Cutting a block of sandstone (Chatsworth Grit, Namurian), Stanton-on-the-Peak, Derbyshire, England. (Courtesy of Ann Twyfords, Ltd.)

Fig. 6.4 Weathered statuary, Wells Cathedral, Wells, England.

Carvings and decoration are subdued and eventually may disappear (Fig. 6.4) and natural features such as grain, fossils, etc., are emboldened.

The colour and strength of a sandstone are largely attributable to the type and amount of cement binding the constituent grains. The cement content also influences the porosity and therefore water absorption. Sandstones which are of use for building purposes are found in most of the geological systems, the exception being those of the Cainozoic era. The sandstones of this age generally are too soft and friable to be of value.

6.2 ROOFING AND FACING MATERIALS

Rocks used for roofing purposes must possess a sufficient degree of fissility to allow them to split into thin slabs, as well as being durable and impermeable. Consequently, slate is one of the best roofing materials available and has been used extensively. Today, however, more and more tiles are being used for roofing, these being cheaper than stone, which has to be got and cut to size.

Slates are derived from argillaceous rocks which, because they were involved in major earth movements, were metamorphosed. They are characterized by their cleavage, which allows the rock to break into thin slabs. Some slates, however, may possess a grain which runs at an angle to the cleavage planes and may tend to fracture along it. Thus in slate used for roofing purposes the grain should run along its length. Welsh slates are differently coloured, they may be grey, blue, purple, red, or mottled. The green coloured slates of the Lake District, England, are obtained from the Borrowdale Volcanic Series and in fact are cleaved tuffs. They are somewhat coarser grained than Welsh slates but more attractive. The colour of slate varies: red slates contain more than twice as much ferric as ferrous oxide; a slate may be greenish coloured if the reverse is the case; manganese is responsible for the purplish colour of some slates; blue and grey slates contain little ferric oxide.

The specific gravity of a slate is about 2.7 to 2.9 with an approximate density of $2.59\,Mg/m^3$. The maximum permissible water absorption of a slate is 0.37%. Calcium carbonate may be present in some slates of inferior quality which may result in them flaking and eventually crumbling upon weathering. Accordingly, a sulphuric acid test is used to test their quality. Top quality slates, which can be used under moderate to severe atmospheric pollution con-

Fig. 6.5 Coarse-grained 'green-slate' being split by hand for facing stone and so producing a riven finish. (Courtesy of Broughton Moor Stone, Ltd, Cumbria, England.)

ditions, reveal no signs of flaking, lamination, or swelling after the test.

There is a large amount of wastage when explosives are used to quarry slate. Accordingly slates are sometimes quarried by using a wire saw. The slate, once won, is then sawn into blocks, and then into slabs about 75 mm thick. These slabs are split into slate tiles by hand.

An increasingly frequent method of using stone at the present day is as relatively thin slabs, applied as a facing to a building to enhance its appearance. Facing stone also provides a protective covering. Various thicknesses are used from 40 mm in the case of granite, marble, and slate in certain positions at ground floor level, to 20 mm at first-floor level or above. If granite or syenite is used as a facing stone, then it should not be overdried, but should retain some quarry sap, otherwise it becomes too tough and hard to fabricate. Limestone and sandstone slabs are somewhat thicker, varying between 50 and

Fig. 6.6 Polishing facing stone. The polishers in the centre of the photograph are fitted with heads impregnated with carborundum. These are usually in three grades, the coarser-graded heads being used first. (Courtesy of Pisani, Ltd, Cromford, Derbyshire, England.)

100 mm. Because of their comparative thinness facing stones should not be too rigidly fixed otherwise differential expansion, due to changing temperatures, can produce cracking.

When fissile stones are used as facing stone and are given a riven or honed finish they are extremely attractive (Fig. 6.5). Facing stones usually have a polished finish (Fig. 6.6), then they are even more attractive. Such stones are almost self-cleansing.

Rocks used for facing stones have a high tensile strength in order to resist cracking. The high tensile strength also means that thermal expansion is not a great problem when slabs are spread over large faces.

6.3 CRUSHED ROCK: CONCRETE AGGREGATE

Crushed rock is produced for a number of purposes, the chief of which are for concrete and road aggregate. Approximately 75% of the volume of concrete consists of aggregate, therefore its properties have a significant influence on the engineering behaviour of concrete. Aggregate is divided into coarse and fine varieties, the former usually consists of rock material that passes the 37 mm sieve and is retained on the 4.8 mm sieve. The latter passes through this sieve and is caught on the 100 mesh sieve. Fines passing through the 200 mesh sieve should not exceed 10% by weight of the aggregate.

High explosives such as gelignite, dynamite, or trimonite are used in drillholes when quarrying crushed rocks. ANFO, a mixture of diesel oil and ammonium nitrate, also is used frequently. The holes are drilled at an angle of about 10° to 20° from vertical for safety reasons, and are usually located 3 to 6 m from the working face and a similar distance apart. Usually one (sometimes two), row of holes is drilled. The explosive does not occupy the whole length of a drillhole, lengths of explosive alternating with zones of stemming, which is commonly quarry dust or sand. Stemming occupies the top six or so metres of a drillhole. A single detonation fires a cordex instantaneous fuse, which has been fed into each hole. It is common practice to have millisecond delay intervals between firing individual holes. In this way the explosions are complementary. The object of blasting is to produce a stone of workable size. Large stone must be further reduced by using a drop-ball or by secondary blasting. The height of the face largely depends upon the stability of the rock mass concerned. When the height of a working face begins to exceed 20 to 30 m it may be worked in tiers (Fig. 6.7). Increasing overburden often means that quarrying operations become uneconomic. In certain instances, however, rock may be mined.

After quarrying the rock is fed into a crusher and then screened to separate the broken rock material into different grade sizes. Because crushed rock can be produced on a large scale by comparatively inexpensive operations it is relatively cheap. Fortunately, the raw materials which can be used for crushed rock are widespread, and indeed in large construction operations if the rock excavated is suit-

Fig. 6.7 Quarrying granite for aggregate in Hong Kong.

able, then it can be crushed and used as aggregate.

The crushing strength of rock used for aggregate generally ranges between 70 and 300 MPa. Aggregates that are physically unsound lead to the deterioration of concrete, inducing cracking, popping, or spalling. On drying, cement shrinks. If the aggregate is strong the amount of shrinkage is minimized and the cement−aggregate bond is good.

The shape of aggregate particles is an important property and is mainly governed by the fracture pattern within a rock mass. Rocks like basalts, dolerites, andesites, granites, quartzites, and limestones tend to produce angular fragments when crushed. However, argillaceous limestones when crushed produce an excessive amount of fines. The crushing characteristics of a sandstone depend upon the closeness of its texture and the amount and type of cement. Angular fragments may produce a mix which is difficult to work, that is, it can be placed less easily and offers less resistance to segregation. Nevertheless, angular particles are said to produce a denser concrete. Rounded, smooth fragments produce workable mixes. The less workable the mix, the more sand, water, and cement must be added to produce a satisfactory concrete. Fissile rocks such as those which are strongly cleaved, schistose, foliated, or laminated have a tendency to split and, unless crushed to a fine size, give rise to tabular or planar shaped particles. Planar and tabular fragments not only make concrete more difficult to work but they also pack poorly and so reduce its compressive strength and bulk weight. Furthermore, they tend to lie horizontally in the cement so allowing water to collect beneath them, which inhibits the development of a strong bond on their under surfaces.

The surface texture of aggregate particles largely determines the strength of the bond between the cement and themselves (French, 1991). A rough surface creates a good bond whereas a smooth surface does not.

As concrete sets hydration takes place and alkalis (Na_2O and K_2O) are released. These react with siliceous material. Table 6.2 lists some of the reactive rock types. If any of these types of rock are used as aggregate in concrete made with high-alkali cement the concrete is liable to expand and crack (Fig. 6.8), thereby losing strength. Expansion due to alkali−aggregate reaction also has occurred when greywacke has been used as aggregate. When concrete is wet the alkalis that are released are dissolved by its water content and as the water is used up during hydration so the alkalis are concentrated in the remaining liquid. This caustic solution attacks reactive aggregates to produce alkali−silica gels. The osmotic pressures developed by these gels as they absorb more water eventually may rupture the cement around reacting aggregate particles. The gels gradually occupy the cracks thereby produced and these eventually extend to the surface of the concrete. If alkali reaction is severe a polygonal pattern of cracking developes on the surface. These troubles can be avoided if a preliminary petrological examination is made of the aggregate. That is, material that contains over 0.25% opal, over 5% chalcedony, or over 3% glass or cryptocrystalline acidic-to-intermediate volcanic rock, by weight, will

Table 6.2 Rocks which react with high-alkali cements. (After McConnell *et al.*, 1950.)

Reactive rocks	Reactive component
Siliceous rocks	
Opaline cherts	Opal
Chalcedonic cherts	Chalcedony
Siliceous limestones	Chalcedony and/or opal
Volcanic rocks	
Rhyolites and rhyolitic tuffs	
Dacites and dacitic tuffs	Glass, devitrified glass, and
Andesites	tridymite
Metamorphic rocks	
Phyllites	Hydromica (illite)
Miscellaneous rocks	
Any rocks containing veinlets, inclusions, coatings, or detrital grains of opal, chalcedony, or tridymite. Quartz highly fractured by natural processes	

be sufficient to produce an alkali reaction in concrete unless low-alkali cement is used. This contains less than 0.6% of Na_2O and K_2O. If aggregate contains reactive material surrounded by or mixed with inert matter, a deleterious reaction may be avoided. The deleterious effect of alkali–aggregate reaction can also be avoided if a pozzolan is added to the mix to react with the alkalis.

Reactivity may be related not just to composition but also to the percentage of strained quartz that a rock contains. For instance, Gogte (1973) maintained that rock aggregates containing 40% or more of strongly undulatory or highly granulated quartz were highly reactive whilst those with between 30 and 35% were moderately reactive. He also showed that basaltic rocks with 5% or more secondary chalcedony or opal, or about 15% palagonite, showed deleterious reactions with high-alkali cements. Sandstones and quartzites containing 5% or more chert behaved in a similar manner.

Certain argillaceous dolostones have been found to expand when used as aggregates in high-alkali cement, thereby causing failure in concrete. This phenomenon has been referred to as alkali–

Fig. 6.8 Alkali–aggregate reaction in concrete of Charles Cross multistorey car park, Plymouth, England.

carbonate rock reaction and its explanation has been attempted by Gillott and Swenson (1969). They proposed that the expansion of such argillaceous dolostones in high-alkali cements was due to the uptake of moisture by the clay minerals. This was made possible by dedolomitization which provided access for moisture. Moreover, they noted that expansion only occurred when the dolomite crystals were less than 75 μm.

It is usually assumed that shrinkage in concrete will not exceed 0.045%, this taking place in the cement. However, basalt, gabbro, dolerite, mudstone, and greywacke have been shown to be shrinkable, that is, they have large wetting and drying movements of their own, so much so that they affect the total shrinkage of concrete. Clay and shale absorb water and are likely to expand if they are incorporated in concrete, and on drying they shrink causing injury to the cement. Consequently, the proportion of clay material in a fine aggregate should not exceed 3%. Granite, limestone, quartzite, and felsite remain unaffected.

6.4 ROAD AGGREGATE

Aggregate constitutes the basic material for road construction and forms the greater part of a road surface. Thus, it has to bear the main stresses imposed by traffic and has to resist wear. The rock material used should therefore be fresh and have a high strength (Fookes, 1991).

Aggregate used as road metal must, as well as having a high strength, have a high resistance to impact and abrasion, polishing and skidding, and frost action. It must also be impermeable, chemically inert and possess a low coefficient of expansion. The principal tests used to assess the value of a roadstone are:
1 the aggregate crushing test;
2 the aggregate impact test;
3 the aggregate abrasion test;
4 the test for the assessment of the polished-stone value.

Other tests are those for water absorption, specific gravity, and density, and the aggregate shape tests (Anon, 1975a). Some typical values are given in Table 6.3.

The properties of an aggregate are obviously related to the texture and mineralogical composition of the rock from which it was derived. Most igneous and contact metamorphic rocks meet the requirements demanded of good roadstone, but many regional metamorphic rocks are either cleaved or schistose and are therefore unsuitable for roadstone because they tend to produce flaky particles when crushed. Such particles do not achieve a good interlock and therefore impair the development of dense mixtures for surface dressing. The amount and type of cement and/or matrix material which bind grains together in a sedimentary rock influence roadstone performance.

The way in which alteration develops can strongly influence roadstone durability. Weathering may reduce the bonding strength between grains to such an extent that they are easily plucked out. Chemical alteration is not always detrimental to the mechanical properties. Indeed, a small amount of alteration may improve the resistance of a rock to polishing (see below). On the other hand, resistance to abrasion decreases progressively with increasing content of altered minerals, as does the crushing strength. The combined hardness of the minerals in

Table 6.3 Some representative values of roadstone properties of some common aggregates

Rock type	Water absorption	Specific gravity	Aggregate crushing value	Aggregate impact value	Aggregate abrasion value	Polished stone value
Basalt	0.9	2.91	14	13	14	58
Dolerite	0.4	2.95	10	9	6	55
Granite	0.8	2.64	17	20	15	56
Micro-granite	0.5	2.65	12	14	13	57
Hornfels	0.5	2.81	13	11	4	59
Quartzite	1.8	2.63	20	18	15	63
Limestone	0.5	2.69	14	20	16	54
Greywacke	0.5	2.72	10	12	7	62

a rock together with the degree to which they are cleaved, as well as the texture of the rock, also influence its rate of abrasion. The crushing strength is related to porosity and grain size — the higher the porosity and the larger the grain size, the lower the crushing strength.

One of the most important parameters of road aggregate is the polished-stone value, which influences skid resistance. A skid-resistant surface is one which is able to retain a high degree of roughness whilst in service. The rate of polish is initially proportional to the volume of the traffic and straight stretches of road are less subject to polishing than bends. The latter may polish up to seven times more rapidly. Stones are polished when fine detrital powder is introduced between tyre and surface. Investigations have shown that detrital powder on a road surface tends to be coarser during wet periods than in dry ones. This suggests that polishing is more significant when the road surface is dry than wet, the coarser detritus more readily roughening the surface of stone chippings. An improvement in skid resistance can be brought about by blending aggregates.

Rocks within the same major petrological group may differ appreciably in their polished-stone characteristics. The best resistance to polish occurs in rocks containing a small proportion of softer alteration materials. Coarser grain size and the presence of cracks in individual grains also tends to improve resistance to polishing. In the case of sedimentary rocks the presence of hard grains set in a softer matrix produces a good resistance to polish. Sandstones, greywackes, and gritty limestones offer a good resistance to polishing, but unfortunately not all of them possess sufficient resistance to crushing and abrasion to render them useful in the wearing course of a road. Purer limestones show a significant tendency to polish. In igneous and contact-metamorphic rocks a good resistance to polish is due to variation in hardness between the minerals present.

The petrology of an aggregate determines the nature of the surfaces to be coated, the adhesion attainable depending on the affinity between the individual minerals and the binder, as well as the surface texture of the individual aggregate particles. If the adhesion between the aggregate and binder is less than the cohesion of the binder, then stripping may occur. Insufficient drying and the non-removal of dust before coating are, however, the principal causes of stripping. Acid igneous rocks generally do not mix well with bitumen as they have a poor ability to absorb it. Basic igneous rocks such as basalts possess a high affinity for bitumen, as does limestone.

Igneous rocks are commonly used for roadstone. Dolerite, which has been used extensively, has a high strength and resists abrasion and impact, but its polished-stone value usually does not meet motorway specification in Britain, although it is suitable for trunk roads. Felsite, basalt, and andesite are also much sought after. The coarse-grained igneous rocks such as granite are not generally so suitable as the fine-grained types, as they crush more easily. On the other hand the very fine-grained and glassy volcanics are often unsuitable since when crushed they produce chips with sharp edges, and they tend to develop a high polish.

Igneous rocks with a high silica content resist abrasion better than those in which the proportion of ferromagnesian minerals is high, that is, acid rocks like rhyolites are harder than basic rocks such as basalts. Hornfels and quartzite, which are the products of thermal metamorphism, have high strength and resistance to wear and make good roadstones, but many rocks of regional metamorphic origin, because of their cleavage and schistosity, are unsuitable. Coarse-grained gneisses are similar to granites. The sedimentary rocks, limestone and greywacke are frequently used as roadstone. Greywacke, in particular, has a high strength, resists wear, and develops a good skid resistance. Some quartz arenites are used, as are gravels. In fact, the use of gravel aggregates is increasing.

6.5 GRAVELS AND SANDS

6.5.1 Gravel

Gravel deposits usually represent local accumulations, e.g. channel fillings. In such instances they are restricted in width and thickness but may have considerable length. Fan-shaped deposits of gravels or aprons may accumulate at the snouts of ice masses, or blanket deposits may develop on transgressive beaches. The latter deposits are usually thin and patchy whilst the former are frequently wedge-shaped.

A gravel deposit consists of a framework of pebbles between which are voids. The voids are rarely empty, being occupied by sand, silt, or clay material. River and fluvioglacial gravels are notably bimodal, the principal mode being in the gravel grade, the secondary in the sand grade. Marine gravels, however, are often unimodal and tend to be more uniformly sorted than fluvial types of similar grade size.

The shape and surface texture of the pebbles in a gravel deposit are influenced by the agent responsible for its transportation and the length of time taken in transport, although shape is also dependent on the initial shape of the fragment, which in turn is controlled by the fracture pattern within the parental rock. The shape of gravel particles is classified in *BS 812* (Anon, 1975a) as rounded, irregular, angular, flaky, and elongated (Fig. 6.9). It also defines a flakiness index, an elongation index, and an angu-

larity number. The flakiness index of an aggregate is the percentage of particles, by weight, whose least dimension (thickness) is less than 0.6 times their mean dimension. The elongation index of an aggregate is the percentage, by weight, of particles whose greatest dimension (length) is greater than 1.8 times their mean dimension. The angularity number is a measure of relative angularity based on the percentage of voids in the aggregate. The least angular aggregates are found to have about 33% voids and the angularity number is defined as the amount by which the percentage of voids exceeds 33. The angularity number ranges from 0 to about 12. The same British Standard recognizes the following types

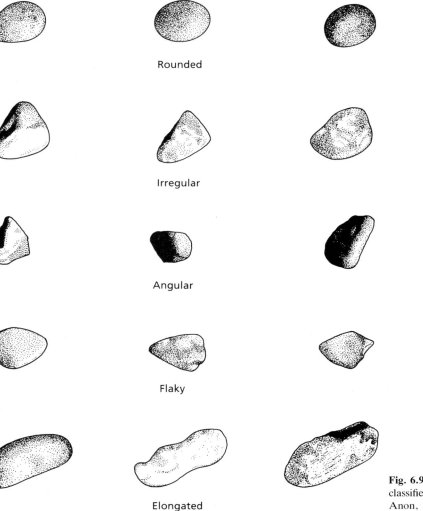

Rounded

Irregular

Angular

Flaky

Elongated

Fig. 6.9 Particle shapes as classified *BS 812*:1975. (After Anon, 1975a.)

of surface texture: glassy, smooth, granular, rough, crystalline, and honeycombed.

The composition of a gravel deposit not only reflects the type of rocks in the source area but it is also influenced by the agent(s) responsible for its formation and the climatic regime in which it was or is being deposited. Furthermore, relief influences the character of a gravel deposit, for example, under low-relief, gravel production is small and the pebbles tend to be chemically inert residues such as vein quartz, quartzite, chert, and flint. By contrast, high relief and rapid erosion yield coarse, immature gravels. All the same, a gravel achieves maturity much more rapidly than does a sand under the same conditions. Gravels which consist of only one type of rock fragment are called oligomictic. Such deposits are usually thin and well-sorted. Polymictic gravels usually consist of a varied assortment of rock fragments and occur as thick, poorly sorted deposits.

Gravel particles generally possess surface coatings that may be the result of weathering or may represent mineral precipitates derived from circulating groundwaters. The latter type of coating may be calcareous, ferruginous, siliceous, or occasionally, gypsiferous. Clay may also form a coating about pebbles. Surface coatings generally reduce the value of gravels for use as concrete aggregate, thick and/or soft and loosely adhering surface coatings are particularly suspect. Clay and gypsum coatings, however, can often be removed by screening and washing. Siliceous coatings tend to react with the alkalis in high-alkali cements and are therefore detrimental to the concrete.

In a typical gravel pit the material is dug from the face by a mechanical excavator. This loads the material into trucks or onto a conveyor which transports it to the primary screening and crushing plant. After crushing the material is further screened and washed. This sorts the gravel into various grades and separates it from the sand fraction. The latter is usually sorted into coarser and finer grades, the coarser is used for concrete and the finer is preferred for mortar. Because gravel deposits are highly permeable, if the water table is high, then the gravel pit will flood. The gravels then have to be worked by dredging. Sea-dredged aggregates are becoming increasingly important.

6.5.2 Sand

The textural maturity of sand varies appreciably. A high degree of sorting coupled with a high degree of rounding characterizes a mature sand. The shape of sand grains, however, is not greatly influenced by length of transport. Maturity is also reflected in their chemical or mineralogical composition and it has been argued that the ultimate sand is a concentration of pure quartz. This is because the less-stable minerals disappear due to mechanical or chemical breakdown during erosion and transportation or even after the sand has been deposited.

Sands are used for building purposes to give bulk to concrete, mortars, plasters, and renderings. For example, sand is used in concrete to lessen the void space created by the coarse aggregate. A sand consisting of a range of grade sizes gives a lower proportion of voids than one in which the grains are of uniform size. Indeed grading is probably the most important property as far as the suitability of a sand for concrete is concerned (Anon, 1981a). In any concrete mix, consideration should be given to the total specific surface of the coarse and fine aggregates, since this represents the surface which has to be lubricated by the cement paste to produce a workable mix. Poorly graded sands can be improved by adding the missing grade sizes to them, so that a high-quality material can be produced with correct blending. Generally, a sand with rounded particles produces a slightly more workable concrete than one consisting of irregularly shaped particles.

Sands used for building purposes are usually siliceous in composition and should be as free from impurities as possible. They should contain no significant quantity of silt or clay (less than 3% by weight), since these need a high water content to produce a workable concrete mix. This in turn leads to shrinkage and cracking on drying. Furthermore, clay and shaly material tend to retard setting and hardening, or they may spoil the appearance. If sand particles are coated with clay they form a poor bond with cement and produce a weaker and less durable concrete. The presence of feldspars in sands used in concrete has sometimes given rise to hair cracking, and mica and particles of shale adversely affect the strength of concrete. Organic impurities may adversely affect the setting and hardening properties of cement by retarding hydration and thereby reduce its strength and durability. Organic and coaly matter also cause popping, pitting, and blowing. If iron pyrites occurs in sand, then it gives rise to unsightly rust stains when used in concrete. The salt content of marine sands is unlikely to produce any serious adverse effects in good-quality concrete although it will probably give rise to efflorescence. Salt can be removed by washing sand.

High-grade quartz sands are also used for making silica bricks used for refractory purposes. Glass sands must have a silica content of over 95% (over 96% for plate glass). The amount of iron oxides present in glass sands must be very low, in the case of clear glass under 0.05%. Uniformity of grain size is another important property as this means that the individual grains melt in the furnace at approximately the same temperature.

6.5.3 Gravel and sand deposits

Scree material or talus accumulates along mountain slopes as a result of freeze–thaw action. Talus is frequently composed of one rock type. The rock debris has a wide range of size distribution and the particles are angular. Because scree simply represents broken rock material, then it is suitable for use as aggregate, if the parent rock is suitable. Such scree deposits, if large enough, only need crushing and screening and therefore are generally more economical to work than the parent rock.

The composition of a river gravel deposit reflects the rocks of its drainage basin. Sorting takes place with increasing length of river transportation, the coarsest deposits being deposited first, although during flood periods large fragments can be carried great distances. Thus, river deposits possess some degree of uniformity as far as sorting is concerned. Naturally, differences in gradation occur in different deposits within the same river channel but the gradation requirements for aggregate are generally met with or they can be made satisfactory by a small amount of processing. Moreover, as the length of river transportation increases so softer material is progressively eliminated, although in a complicated river system with many tributaries new sediment is being constantly added. Deposits found in river beds are usually characterized by rounded particles. This is particularly true of gravels. River transportation also roughens the surfaces of pebbles.

River terrace deposits are similar in character to those found in river channels. The pebbles of terrace deposits may possess secondary coatings, due to leaching and precipitation. These are frequently of calcium carbonate which does not impair the value of the deposit but if they are siliceous, then this could react with alkalis in high-alkali cement and therefore could be detrimental to concrete. The longer the period of post-depositional weathering to which a terrace deposit is subjected, the greater is the likelihood of its quality being impaired.

Alluvial cones are found along valleys located at the foot of mountains. They are poorly stratified and contain rock debris with a predominantly angular shape and great variety in size.

Gravels and sands of marine origin are increasingly used as natural aggregate. The winnowing action of the sea leads to marine deposits being relatively clean and uniformly sorted. Thus, they may require some blending. The particles are generally well rounded with roughened surfaces. Gravels and sands which occur on beaches generally contain deleterious salts and therefore require vigorous washing. By contrast, much of the salt may have been leached out of the deposits found on raised beaches.

Wind blown sands are uniformly sorted. They are composed predominantly of well-rounded quartz grains which have frosted surfaces.

Glacial deposits are poorly graded, commonly containing an admixture of boulders and rock flour. Furthermore, glacial deposits generally contain a wide variety of rock types and the individual rock fragments are normally subangular. The selective action of physical and chemical breakdown processes is retarded when material is entombed in ice and therefore glacial deposits often contain rock material that is unsuitable for use as aggregate and are usually of limited value.

Conversely, fluvioglacial deposits are frequently worked for this purpose. These deposits were laid down by meltwaters which issued from bodies of ice. They take the form of eskers, kames, and outwash fans. The influence of water on these sediments means that they have undergone varying degrees of sorting. They may be composed of gravels or, more frequently, of sands. The latter are often well sorted and may be sharp, thus forming ideal building material.

6.6 LIME, CEMENT, AND PLASTER

Lime is made by heating limestone, including chalk, to a temperature of between 1100 and 1200°C in a current of air, at which point CO_2 is driven off to produce quicklime (CaO). Approximately 56 kg of lime can be obtained from 100 kg of pure limestone. Slaking and hydration of quicklime take place when water is added, giving calcium hydroxide. Carbonate rocks vary from place to place both in chemical composition and physical properties so that the lime produced in different districts varies somewhat in its behaviour. Dolostones also produce lime, however, the resultant product slakes more slowly than does that derived from limestones.

Portland cement is manufactured by burning pure

limestone or chalk with suitable argillaceous material (clay, mud, or shale) in the proportions 3 to 1. The raw materials are first crushed and ground to a powder, and then blended. They are then fed into a rotary kiln and heated to a temperature of over 1800°C. Carbon dioxide and water vapour are driven off and the lime fuses with the aluminium silicate in the clay to form a clinker. This is ground to a fine powder and less than 3% gypsum is added to retard setting. Lime is the principal constituent of Portland cement but too much lime produces a weak cement. Silica constitutes approximately 20% and alumina 5%. Both are responsible for the strength of the cement, but a high silica content produces a slow setting, whilst a high alumina content gives a quick-setting cement. The percentage of iron oxides is low and in white Portland cement it is kept to a minimum. The proportion of magnesium oxide (MgO) should not exceed 4% otherwise the cement is unsound. Similarly, the sulphate content must not exceed 2.75%. Sulphate resisting cement is made by the addition of a very small quantity of tricalcium aluminate to normal Portland cement.

When gypsum ($CaSO_4.nH_2O$) is heated to a temperature of 170°C it loses three-quarters of its water of crystallization, becoming calcium sulphate hemihydrate or plaster of Paris. Anhydrous calcium sulphate forms at higher temperatures. These two substances are the chief materials used in plasters. Gypsum plasters have now more or less replaced lime plasters.

6.7 CLAY DEPOSITS AND REFRACTORY MATERIALS

The principal clay minerals belong to the kandite, illite, smectite, vermiculite, and palygorskite families. The kandites, of which kaolinite is the chief member, are the most abundant clay minerals. Deposits of kaolin or china clay are associated with granite masses which have undergone kaolinization. The soft china clay is excavated by strong jets of water under high pressure, the material being washed to the base of the quarry. This process helps separate the lighter kaolin fraction from the quartz. The lighter material is pumped to the surface of the quarry where it is fed into a series of settling tanks. These separate mica, which is itself removed for commercial use, from china clay. Washed china clay has a comparatively coarse size, approximately 20% of the constituent particles being below 0.01 mm in size, accordingly the material is non-plastic. Kaolin is used for the manufacture of white earthenwares

and stonewares, in white Portland cement and for special refractories.

Ball clays are composed almost entirely of kaolinite and as between 70 and 90% of the individual particles are below 0.01 mm in size these clays have a high plasticity. Their plasticity at times is enhanced by the presence of montmorillonite. They contain a low percentage of iron oxide and consequently when burnt give a light cream colour. They are used for the manufacture of sanitary ware and refractories.

If a clay or shale can be used to manufacture refractory bricks it is termed a fireclay. Such material should not fuse below 1600°C and should be capable of taking a glaze. Ball clays and china clays are in fact fireclays, fusing at 1650°C and 1750°C respectively. However, they generally are too valuable except for making special refractories. Most fireclays are highly plastic and contain kaolinite as their predominant material. Some of the best fireclays are found beneath coal seams, and in the United Kingdom fireclays are almost entirely restricted to strata of Coal Measures age. The material in a bed of fireclay which lies immediately beneath a coal seam is often of better quality than that found at the base of the bed. Since fireclays represent fossil soils which have undergone severe leaching, they consist chiefly of silica and alumina and contain only minor amounts of alkalis, lime, and iron compounds. This accounts for their refractoriness (alkalis, lime, magnesium oxide, and iron oxides in a clay deposit tend to lower its temperature of fusion and act as fluxes). Very occasionally a deposit contains an excess of alumina and in such cases it possesses a very high refractoriness. After the fireclay has been quarried or mined, it is usually left to weather for an appreciable period of time to allow it to break down before it is crushed. The crushed fireclay is mixed with water and moulded. Bricks, tiles, and sanitary ware are made from fireclay.

Bentonite is formed by the alteration of volcanic ash, the principal clay mineral being either montmorillonite or beidellite. When water is added to bentonite it swells to many times its original volume to produce a soft gel. Bentonite is markedly thixotropic and this, together with its plastic properties, has given the clay a wide range of uses. For example, it is added to poorly plastic clays to make them more workable and to cement mortars for the same purpose. In the construction industry it is used as a material for clay grouting, for drilling mud, slurry trenches, and diaphragm walls.

6.7.1 Evaluation of mudrocks for brickmaking

The suitability of a raw material for brickmaking is determined by its physical, chemical, and mineralogical character and the changes which occur when it is fired. The unfired properties such as plasticity, workability (i.e. the ability of the clay to be moulded into shape without fracturing and to maintain its shape when the moulding action ceases), dry strength, dry shrinkage, and vitrification range are dependent upon the source material but the fired properties such as colour, strength, total shrinkage on firing, porosity, water absorption, bulk density, and tendency to bloat are controlled by the nature of the firing process. The price that can be charged for a brick depends largely upon its attractiveness, that is, its colour and surface appearance. The ideal raw material should possess moderate plasticity, good workability, high dry strength, total shrinkage on firing of less than 10%, and a long vitrification range. However, the suitability of a mudrock for brick manufacture can be determined only by running it through a production line or by pilot plant firing tests.

The mineralogy of the raw material influences its behaviour during the brickmaking process and hence the properties of the finished product (Bell, 1992a). Mudrocks consist of clay minerals and non-clay minerals, mainly quartz. The clay mineralogy varies from one deposit to another. Although bricks can be made from most mudrocks, the varying proportions of the different clay minerals have a profound effect on the processing and on the character of the fired brick. Those clays which contain a single predominant clay mineral have a shorter temperature interval between the onset of vitrification and complete fusion, than those consisting of a mixture of clay minerals. This is more true of montmorillonitic and illitic clays than those composed chiefly of kaolinite. Also those clays which consist of a mixture of clay minerals do not shrink as much when fired as those composed predominantly of one type of expansive clay mineral. Mudrocks containing significant amounts of disordered kaolinite tend to have moderate to high plasticity and, therefore, are easily workable. They produce lean clays which undergo little shrinkage during brick manufacture. They also possess a long vitrification range and produce a fairly refractory product. However, mudrocks containing appreciable quantities of well-ordered kaolinite are poorly plastic and less workable. Illitic mudrocks are more plastic and less refractory than those in which disordered kaolinite is dominant, and

fire at somewhat lower temperatures. Smectites are the most plastic and least refractory of the clay minerals. They show high shrinkage on drying since they require high proportions of added water to make them workable. As far as the unfired properties of the raw materials are concerned the non-clay minerals present act mainly as a diluent, but they may be of considerable importance in relation to the fired properties. The non-clay material also may enhance the working properties, for example, colloidal silica improves the workability by increasing plasticity.

The presence of quartz, in significant amounts, gives strength and durability to a brick. This is because during the vitrification period quartz combines with the basic oxides of the fluxes released from the clay minerals on firing to form glass, which improves the strength. However, as the proportion of quartz increases, the plasticity of the raw material decreases.

The accessory minerals in mudrocks play a significant role in brickmaking. The presence of carbonates is particularly important and can influence the character of the bricks produced. When heated above 900°C carbonates break down yielding carbon dioxide and leave behind reactive basic oxides, particularly those of calcium and magnesium. The escape of carbon dioxide can cause lime popping or bursting if large pieces of carbonate (e.g. shell fragments) are present, thereby pitting the surface of a brick. To avoid lime popping the material must be finely ground to pass a 20 mesh sieve. The residual lime and magnesia form fluxes which give rise to low viscosity silicate melts. The reaction lowers the temperature of the brick and hence, unless additional heat is supplied, lowers the firing temperature, and shortens the range over which vitrification occurs. The reduction in temperature can result in inadequately fired bricks. If excess oxides remain in the brick it will hydrate on exposure to moisture, thereby destroying the brick. The expulsion of significant quantities of carbon dioxide can increase the porosity of bricks, reducing their strength. Engineering bricks must be made from a raw material which has a low carbonate content.

Sulphate minerals in mudrocks are detrimental to brickmaking. For instance, calcium sulphate does not decompose within the range of firing temperature of bricks. It is soluble and, if present in trace amounts in the fired brick, causes efflorescence when the brick is exposed to the atmosphere. Soluble sulphates dissolve in the water used to mix the clay. During drying and firing they often form a white scum on the surface of a brick. Barium carbonate may be added to render such salts insoluble and so prevent scumming.

Iron sulphides, such as pyrite and marcasite, frequently occur in mudrocks. When heated in oxidizing conditions, the sulphides decompose to produce ferric oxide and sulphur dioxide. In the presence of organic matter oxidation is incomplete yielding ferrous compounds which combine with silica and basic oxides, if present, to form black, glassy spots. This may lead to a black vitreous core being present in some bricks which can reduce strength significantly. If the vitrified material forms an envelope around the ferrous compounds and heating continues until this decomposes, then the gases liberated cannot escape, causing bricks to bloat and distort. Under such circumstances the rate of firing should be controlled in order to allow gases to be liberated prior to the onset of vitrification. Too high a percentage of pyrite or other iron bearing minerals gives rise to rapid melting which can lead to difficulties on firing.

Pyrite, and other iron-bearing minerals such as haematite and limonite, provide the iron which primarily is responsible for the colour of bricks (section 6.7.2). The presence of other constituents, notably calcium, magnesium, or aluminium oxides, tend to reduce the colouring effect of iron oxide, whereas the presence of titanium oxide enhances it. High original carbonate content tends to produce yellow bricks.

Organic matter commonly occurs in mudrock. It may be concentrated in lenses or seams, or be finely disseminated throughout the mudrock. Incomplete oxidation of the carbon upon firing may result in black coring or bloating. Even minute amounts of carbonaceous material can give black coring in dense bricks if it is not burned out. Black coring can be prevented by ensuring that all carbonaceous material is burnt out below the vitrification temperature. This means that if a raw material contains much carbonaceous material, it may be necessary to admit cool air into the firing chamber to prevent the temperature rising too quickly. On the other hand, the presence of oily material in a clay can be an advantage for it can reduce the fuel costs involved in brickmaking. For example, the Lower Oxford Clay in parts of England contains a significant proportion of oil, so that when it is heated above approximately 300°C it becomes almost self-firing.

Mineralogical and chemical information is essential for determining the brickmaking characteristics of a mudrock. Differential thermal analysis and thermogravimetric analysis can identify clay minerals in mudrocks but provide only very general data on relative abundance. X-ray diffraction methods are used to determine the relative proportions of clay and other minerals present (Fig. 6.10a). The composition of the clay minerals present also can be determined by plotting ignition loss against moisture absorption (Fig. 6.10b). The moisture absorption characterizes the type of clay mineral present while ignition loss provides some indication of the quantity present.

The presence of most of the accessory minerals can be determined by relatively simple analysis. Wet chemical methods can be used to determine total calcium, magnesium, iron, and sulphur. The organic matter can be estimated by oxidation or calorimetric methods. Carbon and sulphur are important impurities, which are critical to the brickmaking process, and the quantities present are always determined.

Physical tests can provide some indication of how the raw material will behave during brick manufacture. The particle size distribution affects the plasticity of the raw material, and hence its workability; plasticity is related to shrinkage and ignition loss during firing. Plasticity can be assessed in terms of the Atterberg limits or, less frequently, by means of the Pfefferkorn test. Shrinkage, loss on ignition, and the temperature range for firing are determined from biquettes in a laboratory furnace. Other tests which are carried out in relation to brickmaking such as water absorption, bulk density, strength, liability to efflorescence are described in *BS 3921* (Anon, 1985).

Sufficient quantities of suitable raw material must be available at a site before a brickfield can be developed. The volume of suitable mudrock must be determined along with the amount of waste, that is, the overburden and unsuitable material within the sequence that is to be extracted. The first stages of the investigation involve topographical and geological surveys, followed by a drillhole programme. This leads to a lithostratigraphic and structural evaluation of the site. It also should provide data on the position of the water table and the stability of the slopes which will be produced during excavation of the brick pit.

6.7.2 Bricks

The nature of brick production has changed in the last 40 years. Most of the small units which produced bricks for local needs have closed so that brickmaking

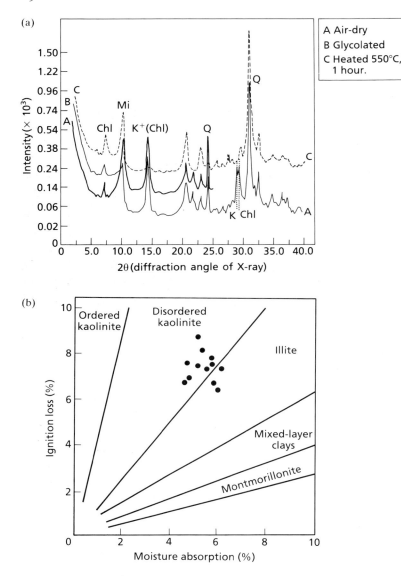

Fig. 6.10 (a) X-ray diffraction traces of a sample of Coal Measures shale used for brickmaking. Chl, chlorite; Mi, mica; K, kaolinite; Q, quartz; (b) clay−mineral determination, Keeling's method. (After Bell, 1992a.)

is now concentrated in large units which supply bricks over much wider areas and which require large sources of raw material. The technology of brickmaking has changed in response to economic pressures which favour the concentration of brickmaking in large, highly mechanized units.

Mudrocks are dug from a brick pit by a variety of mechanical excavators, including face shovels, draglines, and continuous strippers (Fig. 6.11). The material then may be stockpiled to allow weathering to aid its breakdown. It is then crushed and sieved before being moulded, dried, and fired. There are

four main methods of brick production in the United Kingdom:
1 the wire-cut process:
2 the semi-dry pressed method:
3 the stiff-plastic method; and
4 moulding by hand or machine.
One of the distinguishing factors between these methods is the moisture content of the raw material when the brick is fashioned. This varies from as little as 10% in the case of semi-dry pressed bricks to 25% or more in hand-moulded bricks. Hence, the natural moisture content of a mudrock can have a bearing

Fig. 6.11 Continuous strip working of the Lower Oxford Clay at Whittlesey, near Peterborough, England. (Courtesy London Brick Company.)

on the type of brickmaking operation, for example, many mudrocks have a natural moisture content in excess of 15% and are therefore unsuitable for the semi-dry pressed or even the stiff-plastic methods of production unless they are dried.

Three stages can be recognized in brick burning. During the water-smoking stage, which takes place up to approximately 600°C, water is given off. Pore water and the water with which the clay was mixed is driven off at about 110°C, whilst water of hydration disappears between 400°C and 600°C. The next stage is that of oxidation during which the combustion of carbonaceous matter and the decomposition of iron pyrites takes place, and carbon dioxide, sulphur dioxide, and water vapour are given off. The final stage is that of vitrification. Above 800°C the centre of the brick gradually develops into a highly viscous mass, the fluidity of which increases as the temperature rises. The components are now bonded together by the formation of glass. Bricks are fired at temperatures around 1000°C to 1100°C for about 3 days. The degree of firing depends on the fluxing oxides, principally H_2O, Na_2O, CaO, and Fe_2O. Mica is one of the chief sources of alkalis in clay deposits. Because illites are more intimately associated with micas than kaolinites, illitic clays usually contain a higher proportion of fluxes and so are less refractory than kaolinitic clays.

The strength of the brick depends largely on the degree of vitrification. Theoretically, the strength of bricks made from mudrocks containing fine-grained clay minerals such as illite should be higher than those containing the coarser-grained kaolinite. Illitic clays, however, vitrify more easily and there is a tendency to underfire, particularly if they contain fine-grained calcite or dolomite. Kaolinitic clays are much more refractory and can stand harder firing. Greater vitrification is therefore achieved.

Permeability also depends on the degree of vitrification. Mudrocks containing a high proportion of clay minerals produce less permeable products than clays with a high proportion of quartz, but the former types of clays may give a high-drying shrinkage and high moisture absorption.

The colour of the mudrock prior to burning gives no indication of the colour it will have after leaving the kiln. Indeed, a chemical analysis can only offer an approximate guide to the colour of the finished brick. The iron content of a clay, however, is important in this respect. For example, as there is less scope for iron substitution in kaolinite than in illite, this often means that kaolinitic clays give a whitish or pale yellow colour on firing whilst illitic clays generally produce red or brown bricks. More particularly a clay possessing about 1% of iron oxides when burnt tends to produce a cream or light yellow colour; 2 to 3% gives buff, and 4 to 5% red.

Other factors, however, must be taken into

account, for example, a clay containing 4% Fe_2O_3 under oxidizing conditions, burns pink below 800°C, turns brick-red at about 1000°C and at 1150°C, as vitrification approaches completion, it adopts a deep-red colour. By contrast, under reducing conditions ferrous silicate develops and the clay product has a blackish colour. Reducing conditions are produced if carbonaceous material is present in the clay or they may be brought about by the addition of coal or sawdust to the clay before it is burnt. Blue bricks are also produced under reducing conditions. The clay should contain about 5% iron together with lime and alkalis. An appreciable amount of lime in a clay tends to lighten the colour of the burnt product, for example, 10% of lime does not affect the colour at 800°C, but at higher temperatures, with the formation of calcium ferrites, a cream colour is developed. This occurs in clays with 4% of Fe_2O_3 or less. The presence of manganese in a clay may impart a purplish shade to the burnt product.

7 Site investigation and engineering geological maps

The general objective of a site investigation is to assess the suitability of a site for the proposed structure. It involves exploring the ground conditions at and below the surface (Anon, 1981c) and is a prerequisite for the successful and economic design of engineering structures and earthworks. A site investigation also should attempt to foresee and provide against difficulties that may arise during construction due to ground and/or other local conditions and should continue after construction begins. It is essential that the assessment of ground conditions which constitute the basic design assumption is checked as construction proceeds and designs should be modified accordingly if conditions differ from those predicted. An investigation of a site for an important structure requires not only the exploration but also the sampling of strata likely to be significantly affected by the structural load. Data appertaining to the groundwater conditions, the extent of weathering, and the discontinuity pattern in rock masses are also important. In certain areas special problems need investigating, for example, potential subsidence in areas of shallow, abandoned mine workings.

7.1 DESK STUDY AND PRELIMINARY RECONNAISSANCE

The effort expended in any desk study depends on the complexity and size of the proposed project and on the nature of the ground conditions. Detailed searches for information can be extremely time consuming and may not be justified for small schemes at sites where the ground conditions are relatively simple or well known. In such cases a study of the relevant topographical and geological maps and memoirs, and possibly aerial photographs may suffice. On large projects literature and map surveys may save time and thereby reduce the cost of the associated site explorations. The data obtained during such searches should help the planning of the subsequent site exploration and should prevent duplication of effort. In some parts of the world, however, little or no literature or maps are available.

Topographical, geological, and soil maps can provide valuable information which can be used during the planning stage of a construction operation. The former are particularly valuable when planning routeways. Geological maps afford a generalized picture of the geology of an area. Generally, the stratum boundaries and positions of the structural features, especially faults, are interpolated. Consequently, their accuracy cannot always be trusted. Map memoirs may accompany maps, and these provide a detailed survey of the geology of the area in question.

From the engineer's point of view one of the shortcomings of conventional geological maps is that the boundaries are stratigraphical and more than one type of rock may be included in a single mappable unit. Also, geological maps lack the quantitative information required by the engineer concerning physical properties of the rocks and soils, the nature of the discontinuities, the amount and degree of weathering, the hydrogeological conditions, etc. However, special maps can be prepared from a geological map. These maps could interpret the geology in terms of potential supplies of construction materials. Geological maps can be used to indicate those rocks which should be investigated as potential producers of water.

Remote sensing commonly represents one of the first stages of land assessment in underdeveloped areas. It involves the identification and analysis of phenomena on the Earth's surface by using devices borne by aircraft or spacecraft. Most techniques used in remote sensing depend upon recording energy from part of the electromagnetic spectrum, ranging from gamma rays, through the visible spectrum, to radar. The two principal systems of remote sensing are: (i) infrared linescan (IRLS); and (ii) side-looking airborne radar (SLAR). The scanning equipment used measures both emitted and reflected radiation and the employment of suitable detectors and filters permits the measurement of certain spectral bands. Signals from several bands of the spectrum can be recorded simultaneously by multispectral scanners. Lasers are being developed for use in remote sensing.

IRLS is dependent upon the fact that all objects emit electromagnetic radiation generated by the thermal activity of their component atoms. Identification of grey tones is the most important aspect for interpretation of thermal imagery since these provide an indication of the radiant temperatures of a surface. Warm areas give rise to light tones, and cool areas to dark ones. The data can be processed in colour as well as black and white, colours substituting for grey tones. Relatively cool areas are depicted as purple and relatively warm areas as red on a positive print. Thermal inertia is important since rocks with high thermal inertia, such as dolostone or quartzite, are relatively cool during the day and warm at night. Rocks and soils with low thermal inertia, for example, shale, gravel, or sand, are warm during the day and cool at night, that is, the variation in temperature of materials with high thermal inertia during the daily cycle is much less than those with low thermal inertia. Because clay soils possess relatively high thermal inertia they appear warm in pre-dawn imagery whereas sandy soils, because of their relatively low thermal inertia, appear cool. The moisture content of a soil influences the image produced, that is, soils which possess high moisture content appear cool irrespective of their type. Consequently, high moisture content may mask differences in soil types. Fault zones are often picked out because of their higher moisture content. Similarly, the presence of old landslides can frequently be discerned due to their moisture content differing from that of their surroundings.

In SLAR, short pulses of energy, in a selected part of the radar waveband, are transmitted sideways to the ground from antennae on both sides of an aircraft. The pulses strike the ground, are reflected back to the aircraft, and are transformed into black and white photographs. Mosaics of photographs are suitable for the identification of regional geological features and for preliminary identification of terrain units. Lateral overlap of radar cover can give a stereoscopic image, which offers a more reliable assessment of the terrain and can provide appreciable detail of landforms.

The large areas of the ground surface which satellite images cover give a regional physiographic setting and permit the distinction of various landforms. Accordingly, such imagery can provide a geomorphological framework from which a study of the component landforms is possible. The character of the landforms may afford some indication of the type of material of which they are composed and geomorphological data aid the selection of favourable sites for investigation on larger-scale aerial surveys.

The amount of useful information which can be obtained from aerial photographs varies with the nature of the terrain and the type and quality of the photographs. A study of aerial photographs allows the area concerned to be divided into topographical and geological units, and enables the engineering geologist to plan fieldwork and to select locations for sampling. This should result in a shorter, more profitable period in the field.

Examination of consecutive pairs of aerial photographs with a stereoscope allows observation of a three-dimensional image of the ground surface, which means that heights can be determined and contours can be drawn, thereby producing a topographic map. However, the relief presented in this image is exaggerated so that slopes appear steeper than they actually are.

Aerial photographs may be combined in order to cover larger regions. The simplest type of combination is the uncontrolled print laydown, which consists of photographs, laid along side each other, which have not been accurately fitted into a surveyed grid. Photomosaics represent a more elaborate type of print laydown, requiring more care in their production, and controlled photomosaics are based on a number of geodetically surveyed points. They can be regarded as having the same accuracy as topographic maps.

There are four main types of film used in normal aerial photography: (i) black and white; (ii) infrared monochrome; (iii) true colour; and (iv) false colour. Black and white film is used for topographic survey work and for normal interpretation purposes. The other types of film are used for special purposes, for example, infrared monochrome film makes use of the fact that near-infrared radiation is strongly absorbed by water. Accordingly, it is of particular value when mapping shorelines, the depth of shallow underwater features and the presence of water on land, as for instance, in channels, at shallow depths underground or beneath vegetation. Furthermore, it is more able to penetrate haze than conventional photography. True colour photography generally offers much more refined imagery and colour photographs have an advantage over black and white ones as far as photogeological interpretation is concerned. False colour is the term frequently used for infrared colour photography. False colour provides a more sensitive means of identifying exposures of bare, grey rocks than any other type of film. Lineaments, variations in water content in soils and rocks, and changes in vegetation which may not be readily apparent on black and white photographs are often

Table 7.1 Types of photogeological investigation

Structural geology	Mapping and analysis of folding. Mapping of regional fault systems and recording any evidence of recent fault movements. Determination of the number and geometry of joint systems
Rock types	Recognition of the main lithological types (crystalline and sedimentary rocks, unconsolidated deposits)
Soil surveys	Determining main soil type boundaries, relative permeabilities and cohesiveness, periglacial studies
Topography	Determination of relief and landforms
Stability	Slope instability (especially useful in detecting old failures which are difficult to appreciate on the ground) and rock fall areas, quick clays, loess, peat, mobile sand, soft ground, features associated with old mine workings
Drainage	Outlining of catchment areas, stream divides, areas of subsurface drainage such as karstic areas, areas liable to flooding. Tracing swampy ground, perennial or intermittent streams, and dry valleys. Levees and meander migration. Flood-control studies. Forecasting effect of proposed obstructions. Runoff characteristics. Shoals, shallow water, stream gradients and widths
Erosion	Areas of wind, sheet and gully erosion, excessive deforestation, stripping for opencast work, coastal erosion
Groundwater	Outcrops and structure of aquifers. Water-bearing sands and gravels. Seepages and springs, possible productive fracture zones. Sources of pollution. Possible recharge sites
Reservoirs and dam sites	Geology of reservoir site, including surface permeability classification. Likely seepage problems. Limit of flooding and rough relative values of land to be submerged. Bedrock gullys, faults, and local fracture pattern. Abutment characteristics. Possible diversion routes. Ground needing clearing. Suitable areas for irrigation
Materials	Location of sand and gravel, clay, rip-rap, borrow and quarry sites with access routes
Routes	Avoidance of major obstacles and expensive land. Best-graded alternatives and ground conditions. Sites for bridges. Pipe and power-line reconnaissance. Best routes through urban areas

clearly depicted by false colour. A summary of the types of geological information which can be obtained from aerial photographs is given in Table 7.1

The preliminary reconnaissance involves a walk over the site noting, where possible, the distribution of the soil and rock types present, the relief of the ground, the surface drainage and associated features, actual or likely landslip areas, ground cover and obstructions, earlier uses of the site such as tipping, or evidence of underground workings, etc. The inspection should not be restricted to the site but should examine adjacent areas to see how they affect or will be affected by construction on the site in question.

The importance of the preliminary investigation is that it should assess the suitability of the site for the proposed works and form the basis upon which the site exploration is planned. It also allows a check to be made on any conclusions reached in the desk study.

7.2 SITE EXPLORATION: DIRECT METHODS

The aim of a site exploration is to try to determine, and thereby understand, the nature of the ground conditions on and surrounding the site. The extent to which this is carried out depends upon the size and importance of the construction operation. A report embodying the findings can be used for design purposes and should contain geological plans of the site with accompanying sections, thereby conveying a three-dimensional picture of the subsurface strata.

The scale of the mapping will depend on the engineering requirement, the complexity of the geology, and the staff and time available. For example, on a large project geological mapping is frequently required on a large and detailed scale, for instance, thin, but suspect horizons such as clay bands should be recorded.

Rock and soil types should be mapped according to their lithology and if possible, presumed physical behaviour, that is, in terms of their engineering classification rather than age. Geomorphological conditions, hydrogeological conditions, landslips, subsidences, borehole and field test information can all be recorded on geotechnical maps. Particular attention should be given to the nature of the superficial deposits and, where present, made-over ground.

There are no given rules regarding the location of boreholes or drillholes or the depth to which they should be sunk. This depends upon two principal factors: (i) the geological conditions; and (ii) the type of structure concerned. The information provided by the preliminary reconnaissance and from any trial trenches should provide a basis for the initial planning and layout of the borehole programme. Holes should be located so as to detect the geological sequence and structure. Obviously, the more complex this is the greater the number of holes needed. In some instances it may be as well to start with a widely spaced network of holes. As information is obtained, further holes can be put down, if and where necessary.

Exploration should be carried out to a depth which includes all strata likely to be significantly affected by the structural load. Experience has shown that damaging settlement does not usually take place when the added stress in the soil due to the weight of a structure is less than 10% of the effective overburden stress. It would therefore seem logical to sink boreholes on compact sites to depths where the additional stresses do not exceed 10% of the stress due to the weight of the overlying strata. It must be borne in mind that if a number of loaded areas are in close proximity the effect of each is additive. Under certain special conditions boreholes may have to be sunk more deeply as, for example, when old voids due to mining operations are suspected, or when it is suspected that there are highly compressible layers, such as interbedded peats, at depth. If possible boreholes should be taken through superficial deposits to rockhead. In such instances adequate penetration of the rock should be specified to ensure that isolated boulders are not mistaken for the solid formation.

The results from a borehole or drillhole should be documented on a log (Fig. 7.1). Apart from the basic information such as number, location, elevation, date, client, contractor, and engineer responsible, the fundamental requirement of a borehole log is to show how the sequence of strata changes with depth. Individual soil or rock types are pre-sented in symbolic form on a borehole log. The material recovered must be adequately described, and in the case of rocks frequently includes an assessment of the degree of weathering, fracture index, and relative strength. The type of boring or drilling equipment should be recorded, the rate of progress made being a significant factor. Particular mention of the water level in the hole and any water loss, when it is used as a flush during rotary drilling, should be noted, as these reflect the mass permeability of the ground. If any *in situ* testing is done during boring or drilling operations, then the type(s) of test and the depth at which it/they were carried out must be recorded. When percussion methods are used the depths from which samples are taken must be noted. A detailed account of the logging of cores for engineering purposes is given in Anon, 1970.

Direct observation of strata, discontinuities, and cavities can be undertaken by cameras or closed-circuit television equipment and drillholes can be viewed either radially or axially. Remote focusing for all heads and rotation of the radial head through 360° are controlled from the surface. The television heads have their own light source. Colour changes in rocks can be detected as a result of the varying amount of light reflected from the drillhole walls. Discontinuities appear as dark areas because of the non-reflection of light. However, if the drillhole is deflected from the vertical, variations in the distribution of light may result in some lack of picture definition.

7.2.1 Subsurface exploration in soils

The simplest method whereby data relating to subsurface conditions in soils can be obtained is by hand augering. The two most frequently used types of augers are: (i) the post-hole auger; and (ii) the screw auger. These are principally used in cohesive soils.

Soil samples which are obtained by augering are badly disturbed and invariably some amount of mixing of soil types occurs. Critical changes in the ground conditions are therefore unlikely to be located accurately. Even in very soft soils it may be very difficult to penetrate more than 7 m with hand augers. The Mackintosh probe and Lang prospector are more specialized forms of hand tools.

Power augers are available as solid-stem or hollow-stem, both having an external continuous helical flight. The latter are used in those soils in which the borehole does not remain open. The hollow stem can be sealed at the lower end with a combined plug and cutting bit which is removed when a sample is required. Hollow-stem augers are useful for investigations where the requirement is to locate bedrock beneath overburden. Solid-stem augers are used in stiff clays which do not need casing. However, if an undisturbed sample is required they have to be removed. Disturbed samples taken from auger holes are often unreliable. In favourable ground conditions, such as firm and stiff homogeneous clays, auger rigs are capable of high output rates.

Pits and trenches allow the ground conditions in soils and highly weathered rocks to be examined directly although they are limited as far as their depth is concerned. Trenches, to a depth of some 5 m, can provide a flexible, rapid, and economic method of obtaining information. Groundwater conditions and stability of the sides obviously influence whether or not they can be excavated, and safety must at all times be observed. This at times necessitates shoring the sides. Pits and adits are expensive and should be considered only if the initial subsurface survey has revealed any areas of special difficulty. The soil conditions in pits and trenches also can be photographed. Undisturbed, as well as disturbed, samples can be collected. Such excavations are used to locate slip planes in landslides.

The development of large earth augers and patent piling systems has made it is possible to sink 1 m diameter boreholes in soils more economically than previously. The ground conditions can be inspected directly from such holes. Depending on the ground conditions, the boreholes may be either unlined, lined with steel mesh, or cased with steel pipe. In the latter case windows are provided at certain levels for inspection and sampling.

The light-cable and tool-boring rig is used for investigating soils (Fig. 7.2). The hole is sunk by repeatedly dropping one of the tools into the ground. A power winch is used to lift the tool, suspended on a wire, and by releasing the clutch of the winch the tool drops and cuts into the soil. Once a hole is established it is lined with casing, the drop tool operating within the casing. This type of rig is usually capable of penetrating about 60 m of soil. In doing so the size of the casing in the lower end of the borehole is reduced. The basic tools are the shell and the claycutter, which are essentially open-ended steel tubes to which cutting shoes are attached. The shell, which is used in granular soils, carries a flap valve at its lower end which prevents the material from falling out on withdrawal from the borehole.

The material is retained in the cutter by the adhesion of the clay.

For boring in stiff clays the weight of the claycutter may be increased by adding a sinker bar. In very stiff clays a little water is often added to assist boring progress. This must be done with caution to avoid possible changes in the properties of soil about to be sampled. In such clays the borehole can often be advanced without lining, except for a short length at the top to keep the hole stable. If cobbles or small boulders are encountered in clays, particularly tills, then these can be broken by using heavy chisels.

When boring in soft clays, although the hole may not collapse it tends to squeeze inwards and to prevent the cutter operating. The hole must therefore be lined. The casing is driven in and winched out; however, in difficult conditions it may have to be jacked out. Casing tubes have internal diameters of 150, 200, 250, and 300 mm, the most commonly used sizes being 150 and 200 mm (the large sizes are used in coarse gravels).

The usual practice is to bore ahead of the casing for about 1.5 m (the standard length of a casing section) before adding a new section of casing and surging it down. The reason for surging the casing is to keep it 'free' in the borehole so that it can be extracted more easily on completion. When the casing can no longer be advanced by surging, smaller-diameter casing is introduced. However, if the hole is near its allotted depth the casing may be driven into the ground for quickness. Where clays occur below granular deposits, the casing used as a support in the granular soils is driven a short distance in the clay to create a seal and the shell is used to remove any water which might enter the borehole.

Boreholes in sands or gravels almost invariably require lining. The casing should be advanced with the hole or overshelling is likely to occur, i.e. the sides collapse and prevent further progress. Because of the mode of operation of the shell, the borehole should be kept full of water so that the shell may operate efficiently. Where cohesionless soils are water bearing all that is necessary is for the water in the borehole to be kept topped up. If flow of water occurs, then it should be from the borehole to the surrounding soil. Conversely, if water is allowed to flow into the borehole, piping probably will occur. Piping can usually be avoided by keeping the head of water in the borehole above the natural head. To overcome artesian conditions the casing should be extended above ground and kept filled with water. The shell generally cannot be used in highly per-

| Name of company: A N Other Ltd. | Borehole No. 1 Sheet 1 of 1 |

Name of company:
A N Other Ltd.

Borehole No. 1
Sheet 1 of 1

Equipment & methods:
Light cable tool percussion rig. 200 mm dia. hole to 7.00 m. Casing 200 mm dia. to 6.00 m.

Location No: 6155

Carried out for:
Smith, Jones & Brown

Ground level:
9.90 m (Ordnance datum)

Coordinates:
E 350 N 901

Date:
17–18 June 1974

Description	Reduced level	Legend	Depth & thickness	Samples/tests				Field records
				Depth	Sample Type	No.	Test	
Made Ground (sand, gravel, ash, brick and pottery)	9.40		(0.50)	0.20	D	1		
Made Ground (red and brown clay with gravel)	9.10		0.50 (0.30) 0.80	0.70–1.15	U D	2 3		24 blows
Firm mottled brown silty CLAY (Brickearth)	7.90		(1.20) 2.0	1.15				
Stiff brown sandy CLAY with some gravel (Flood Plain Gravel)	6.25		(1.65) 3.65	2.10–2.55 2.55	U D	4 5		50 blows
Medium dense brown sandy fine to coarse GRAVEL (Flood Plain Gravel)	4.60		(1.65) 5.30	3.60–4.05 3.65 4.00–4.30 4.00–5.00 5.00–5.30 5.30	D U B D	6 7 8	S N27 S N15	No recovery
Firm becoming stiff to very stiff fissured grey silty CLAY with partings of silt (London Clay)	2.45		(2.15) 7.45	6.00–6.45 7.00–7.45	U U	9 10		Standpipe inserted 5.30 m below ground level 35 blows 44 blows

End of borehole

Water level observations during boring

Date	Time	Depth of hole, m	Depth of casing, m	Depth to water, m	Remarks
18 Jun	1615	7.00	0.00	3.65	casing withdrawn
24 Jun	1200	0.00	0.00	2.37	
27 Jun	0915	0.00	0.00	2.33	standpipe readings
27 Jun	1420	0.00	0.00	2.11	
28 Jun	1000	0.00	0.00	2.46	
1 Jul	1015	0.00	0.00	2.46	

SPT: Where full 0.3 m penetration has not been achieved, the number of blows for the quoted penetration is given (not N-value).

Depths: All depths and reduced levels in metres. Thicknesses given in brackets in depth column.

Water: Water level observations during boring are given on last sheet of log.

Sample/test key
D Disturbed sample
B Bulk sample
W Water sample
▮ Piston (P), tube (U) or core sample; length to scale
S Standard penetration test
V Vane test
C Core recovery (%)
r Rock Quality Designation (RQD %)

Remarks:
Water added to facilitate boring from 0.50 m to 7.00 m. Borehole back filled with natural spoil from 7.00 m to 5.30 m, gravel to 0.80 m, clay to 0.50 m, a concreted cock box to ground level.

Logged by:

Scale:

Fig. 7.1a Typical log of data from a light cable percussion borehole. (After *BS 5930* (Anon, 1981c).)

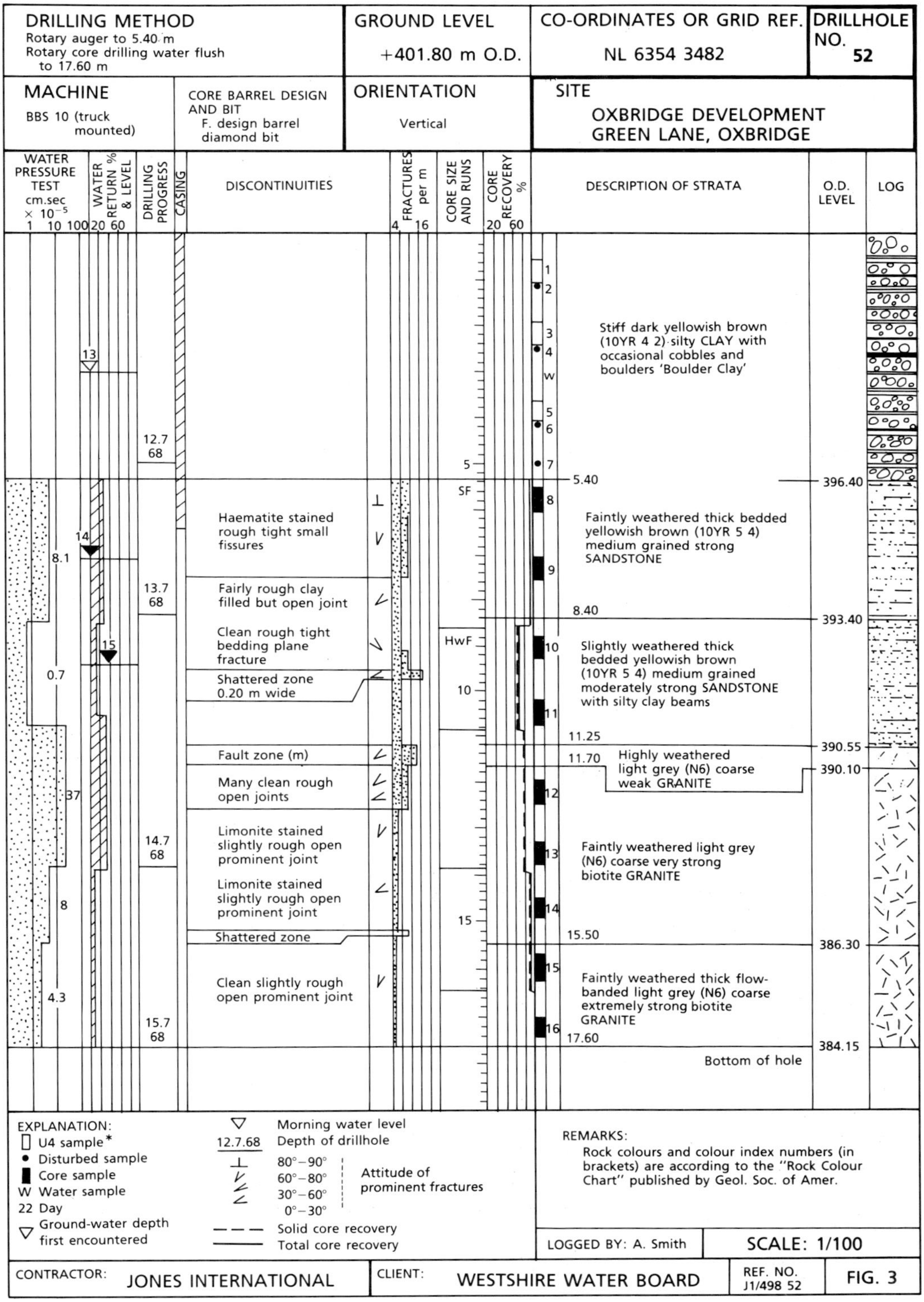

DRILLING METHOD Rotary auger to 5.40 m Rotary core drilling water flush to 17.60 m		GROUND LEVEL +401.80 m O.D.	CO-ORDINATES OR GRID REF. NL 6354 3482	DRILLHOLE NO. **52**
MACHINE BBS 10 (truck mounted)	CORE BARREL DESIGN AND BIT F. design barrel diamond bit	ORIENTATION Vertical	SITE OXBRIDGE DEVELOPMENT GREEN LANE, OXBRIDGE	

WATER PRESSURE TEST cm.sec $\times 10^{-5}$ 1 10 100	WATER RETURN % & LEVEL 20 60	DRILLING PROGRESS	CASING	DISCONTINUITIES	FRACTURES per m 4 16	CORE SIZE AND RUNS	CORE RECOVERY % 20 60	DESCRIPTION OF STRATA	O.D. LEVEL	LOG
13		12.7 68				1 2 3 4 W 5 6 7		Stiff dark yellowish brown (10YR 4 2) silty CLAY with occasional cobbles and boulders 'Boulder Clay'		
8.1	14	13.7 68		Haematite stained rough tight small fissures Fairly rough clay filled but open joint	SF	8 9		5.40 Faintly weathered thick bedded yellowish brown (10YR 5 4) medium grained strong SANDSTONE	396.40	
0.7	15			Clean rough tight bedding plane fracture Shattered zone 0.20 m wide	HwF	10 11		8.40 Slightly weathered thick bedded yellowish brown (10YR 5 4) medium grained moderately strong SANDSTONE with silty clay beams	393.40	
37		14.7 68		Fault zone (m) Many clean rough open joints Limonite stained slightly rough open prominent joint		12 13		11.25 11.70 Highly weathered light grey (N6) coarse weak GRANITE Faintly weathered light grey (N6) coarse very strong biotite GRANITE	390.55 390.10	
8				Limonite stained slightly rough open prominent joint Shattered zone		14			386.30	
4.3		15.7 68		Clean slightly rough open prominent joint		15 16		15.50 Faintly weathered thick flow- banded light grey (N6) coarse extremely strong biotite GRANITE 17.60	384.15	
								Bottom of hole		

EXPLANATION:
▢ U4 sample*
• Disturbed sample
■ Core sample
W Water sample
22 Day
▽ Ground-water depth
first encountered

▽ Morning water level
12.7.68 Depth of drillhole
⊥ 80°–90°
V 60°–80°
∠ 30°–60°
⟋ 0°–30°
Attitude of prominent fractures

- - - Solid core recovery
—— Total core recovery

REMARKS:
Rock colours and colour index numbers (in
brackets) are according to the "Rock Colour
Chart" published by Geol. Soc. of Amer.

LOGGED BY: A. Smith	SCALE: 1/100

CONTRACTOR: JONES INTERNATIONAL	CLIENT: WESTSHIRE WATER BOARD	REF. NO. J1/498 52	FIG. 3

* Now U100

Fig. 7.1b Drillhole log. (Courtesy of the Geological Society (Anon, 1970).)

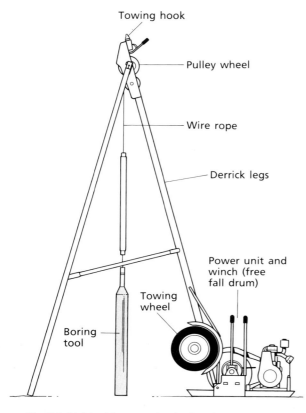

Towing hook

Pulley wheel

Wire rope

Derrick legs

Power unit and
winch (free
fall drum)

Towing
wheel

Boring
tool

Fig. 7.2 Light-cable percussion-boring rig.

meable coarse gravels since it is usually impossible to maintain a head of water in the borehole. Fortunately, these conditions often occur at or near ground level and the problem can sometimes be overcome by using an excavator to open a pit either to the water table or to a depth of 3 to 4 m. Casing can then be inserted, the pit backfilled, and boring can then proceed. Another method of penetrating gravels and cobbles above the water table is to employ a special grab with a heavy tripod and winch, and casing of 400 mm diameter or greater.

Rotary attachments are available which can be used with light-cable and tool rigs. However, they are much less powerful than normal rotary rigs and tend to be used only for short runs, for example, to prove rockhead at the base of a borehole.

In the wash-boring method the hole is advanced by a combination of chopping and jetting the soil or rock, the cuttings thereby produced being washed from the hole by the water used for jetting (Fig. 7.3). The method cannot be used for sampling and therefore its primary purpose is to sink the hole

between sampling positions. When a sample is required, the bit is replaced by a sampler. Nevertheless, some indication of the type of ground penetrated may be obtained from the cuttings carried to the surface by the wash water, from the rate of progress made by the bit, or from the colour of the wash water.

Several types of chopping bits are used. Straight and chisel bits are used in sands, silts, clays, and very soft rocks whilst cross-bits are used in gravels and soft rocks. Bits are available with the jetting points either facing upwards or downwards. The former type are better at cleaning the base of the hole than are the latter.

The wash-boring method may be used in both cased and uncased holes. Casing obviously has to be used in cohesionless soils to avoid the sides of the hole collapsing. Although this method of boring is commonly used in the USA, it has rarely been employed in the United Kingdom. This is mainly because wash-boring does not lend itself to many of the ground conditions encountered and also because of the difficulty of identifying strata with certainty.

Cable-and-tool or churn drilling is a percussion drilling method which can be used in more or less all types of ground conditions but the rate of progress tends to be slow. The rig can drill holes up to 0.6 m in diameter to depths of about 1000 m. The hole is advanced by raising and dropping heavy drilling tools which break the soil or rock. Different bits are used for drilling in different formations and an individual bit can weigh anything up to 1500 kg, so that the total weight of the drill string may amount to several thousand kilograms. A slurry is formed from the broken material and the water in the hole. The amount of water introduced into the hole is kept to the minimum required to form the slurry. The slurry is periodically removed from the hole by means of bailers or sand pumps. In unconsolidated materials, casing is kept near the bottom of the hole in order to avoid caving.

Changes in the type of strata penetrated can again be inferred from the cuttings brought to the surface, the rate of drilling progress, or the colour of the slurry. Sampling, however, has to be done separately. Unfortunately, the ground which has to be sampled may be disturbed by the heavy blows of the drill tools.

7.2.2 Sampling in soils

As far as soils are concerned samples may be divided into two types: (i) disturbed; (ii) undisturbed. Dis-

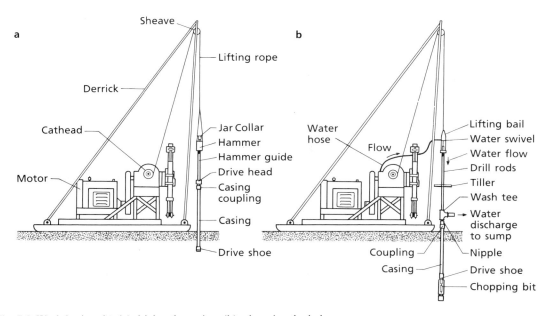

Fig. 7.3 Wash-boring rig: (a) driving the casing; (b) advancing the hole.

turbed samples can be obtained by hand, by auger, or from the claycutter or shell of a boring rig. Samples of cohesive soil should be approximately 0.5 kg in weight, this providing a sufficient size for index testing. They are sealed in jars. A larger sample is necessary if the particle size distribution of granular material is required and this may be retained in a tough plastic sack. Care must be exercised when obtaining such samples to avoid loss of fines.

An undisturbed sample can be regarded as one which is removed from its natural condition without disturbing its structure, density, porosity, moisture content, and stress condition. Although it must be admitted that no sample is ever totally undisturbed, every attempt must be made to preserve the original condition of such samples. Unfortunately, mechanical disturbances produced when a sampler is driven into the ground distort the soil structure. Furthermore, a change of stress condition occurs when a borehole is excavated.

Undisturbed samples may be obtained by hand from surface exposures, pits, and trenches. Careful hand trimming is used to produce a regular block, normally a cube of about 250 mm dimension. Block samples are waxed, together with reinforcing layers of thin cloth. Such samples are particularly useful when it is necessary to test specific horizons, such as shear zones.

The fundamental requirement of any undisturbed

sampling tool is that on being forced into the ground it should cause as little remoulding and displacement of the soil as possible. The amount of displacement is influenced by a number of factors: first, the cutting edge of the sampler — a thin cutting edge and sampling tube minimizes displacement but it is easily damaged, and it cannot be used in gravels and hard soils. The internal diameter of the cutting edge (D_i) should be slightly less than that of the sample tube, thus providing inside clearance which reduces drag effects due to friction. Similarly, the outside diameter of the cutting edge (D_o) should be from 1 to 3% larger than that of the sampler, again to allow for clearance. The relative displacement of a sampler can be expressed by the area ratio (A_r):

$$A_r = \frac{D_0^2 - D_i^2}{D_i^2} \times 100 \qquad (7.1)$$

This ratio should be kept as low as possible, e.g. according to Hvorslev (1949) displacement is minimized by keeping the area ratio below 15%. It should not exceed 25%. Friction can also be reduced if the tube has a smooth inner wall. A coating of light oil may also prove useful in this respect.

The standard sampling tube for obtaining samples from cohesive soils is referred to as the U100, this has a diameter of 100 mm, a length of approximately 450 mm and its walls are 1.2 mm thick (Fig. 7.4). The cutting shoe should meet the above require-

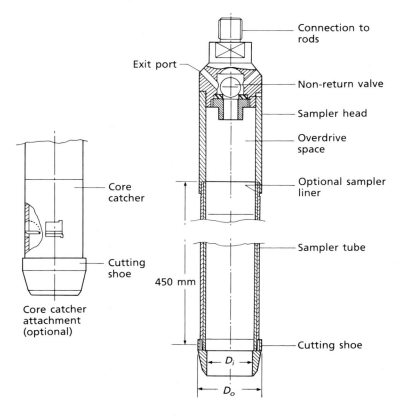

Fig. 7.4 The general purpose open-tube sampler (U100).

ments. The upper end of the tube is fastened to a check valve which allows air or water to escape during driving and helps to hold the sample in place when it is being withdrawn. On withdrawal from the borehole the sample is sealed in the tube with paraffin wax and the end caps screwed on. In soft materials two or three tubes may be screwed together to reduce disturbance in the sample.

The standard type of sampler is suitable for clays with a shear strength exceeding 50 kPa. A thin-walled piston sampler should be used for obtaining clays with a lower shear strength since soft clays tend to expand into the sample tube. Expansion is reduced by a piston in the sampler, a thin-walled tube being jacked down over a stationary internal piston, which, when sampling is complete, is locked in place and the whole assembly is then pulled (Fig. 7.5). Piston samplers range in diameter from 54 to 250 mm. A vacuum tends to be created between the piston and the soil sample, and thereby helps to hold it in place.

Where continuous samples are required, particularly from rapidly varying or sensitive soils, a Delft sampler may be used (Fig. 7.6). This can obtain a continuous sample from ground level to depths of about 20 m. The core is retained in a self-vulcanizing sleeve as the sampler is continuously advanced into the soil.

The most difficult undisturbed sample to obtain is that from saturated sand, particularly when it is loosely packed. In such instances the Bishop sand sampler, which makes use of compressed air and incorporates a thin-walled sampling tube, has been used (Fig. 7.7). The thin-walled sampling tube is housed in an outer tube. The inner tube is driven into the soil and compressed air introduced into the outer tube expels the water. Then the sampling tube is retracted into the outer tube, the air pressure creating capillary zones which retain the soil.

7.2.3 Subsurface exploration in rocks

Rotary drills (Fig. 7.8) are either skid-mounted, trailer-mounted, or, in the case of larger types, mounted on lorries. They are used for drilling through rock, although they can, of course, penetrate and take samples from soil.

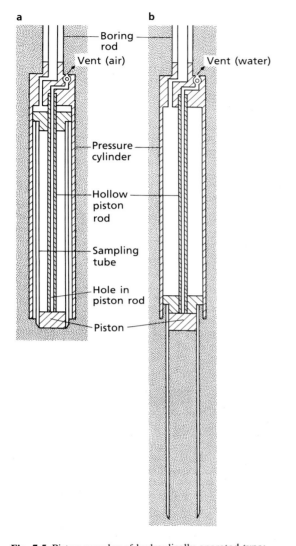

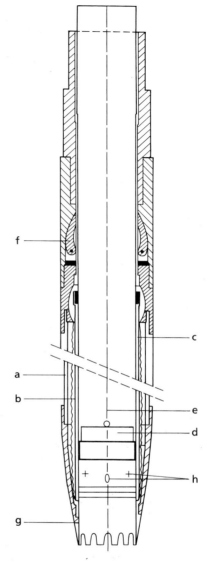

Fig. 7.5 Piston sampler of hydraulically operated type: (a) lowered to bottom of borehole, boring rod clamped in fixed position at ground surface; (b) sampling tube after being forced into soil by water supplied through bearing rod.

Fig. 7.6 Section through a 66 mm continuous sampling apparatus: (a) outer tube; (b) stocking tube over which precoated nylon stocking is slid; (c) plastic inner tube; (d) cap at top of sample; (e) steel wire to fixed point at ground surface (tension cable); (f) sample-retaining clamps; (g) cutting shoe; (h) holes for entry of lubricating fluid. (Courtesy of Delft Soil Mechanics Laboratory.)

Rotary–percussion drills are designed for rapid drilling in rock. The rock is subjected to rapid high-speed impacts whilst the bit (Fig. 7.9) rotates, which brings about compression and shear in the rock. The technique is most effective in brittle materials since it relies on chipping the rock. The rate at which drilling proceeds depends upon: (i) the type of rock, particularly on its strength, hardness, and fracture index; (ii) the type of drill and drill bit; (iii) the flushing medium and the pressures used; and (iv) the experience of the drilling crew. If the drilling operation is standardized, then differences in the rate of

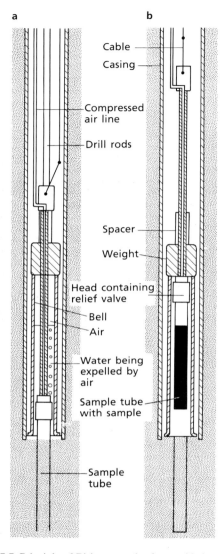

a   b

Cable

Casing

Compressed
air line

Drill rods

Spacer

Weight

Head containing
relief valve

Bell

Air

Water being
expelled by
air

Sample tube
with sample

Sample
tube

Fig. 7.7 Principle of Bishop sampler for sand below water table: (a) sampler forced into sand by drill rods and water in bell being displaced by compressed air; (b) sampler lifted by cable into air-filled bell.

Fig. 7.8 Medium size, skid-mounted rotary drill. (Courtesy of Atlas Copco, Ltd.)

penetration reflect differences in rock types. Drill flushings should be sampled at regular intervals, at changes in the physical appearance of the flushings, and at significant changes in penetration rates. Interpretation of rotary-percussion drillholes should be related to a cored drillhole nearby.

Because compressed air, water, or mud may be used as the flush, rock bit drilling may be used in any

type of material. However, a heavy rig is required to drill through rock. This method is sometimes used as a means of advancing a hole at low cost and high speed between intervals where core drilling is required.

For many engineering purposes a solid, and as near as possible continuous, rock core is required for examination. The core is cut with a bit (Fig. 7.10) and housed in a core barrel. The bit is set with diamonds or tungsten carbide inserts. In set bits diamonds are set on the face of the matrix. The coarser, surface-set bits tipped with diamond and tungsten carbide are used in softer formations. These bits are generally used with air rather than with water flush. Impregnated bits possess a matrix impregnated with diamond dust and their grinding action is suitable for hard and broken formations. Most core drilling is carried out using diamond bits, the type of bit used being governed by the rock type to be drilled, that is, the harder the rock the smaller the size and the higher the quality of the diamonds

Fig. 7.9 Rotary-percussion drill.

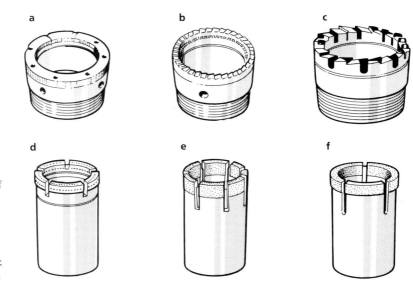

Fig. 7.10 Some common types of coring bits: (a) surface-set diamond bit (bottom discharge); (b) 'stepped' sawtooth bit; (c) tungsten carbide bit; (d) impregnated diamond bit; (e) 'Diadril' corebit impregnated; (f) 'Diadril' corebit impregnated.

that are required in the bit. Tungsten bits are not suitable for drilling in very hard rocks. Thick-walled bits are more robust but penetrate more slowly than thin-walled bits. The latter produce a larger core for a given hole size. This is important where several reductions in size have to be made. Core bits vary in size and accordingly core sticks range between 17.5 and 165 mm in diameter (Fig. 7.11) — generally, the larger the bit, the better the core recovery.

A variety of core barrels are available for rock sampling. The simplest type of core barrel is the single tube but because it is suitable only for hard

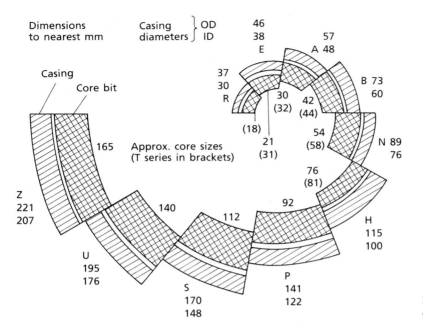

Dimensions to nearest mm

Casing diameters } OD / ID

46
38
E

57
A 48

Casing

Core bit

37
30
R

B 73
60

30
(32)

42
(44)

165

Approx. core sizes
(T series in brackets)

(18)

21
(31)

54
(58)

N 89
76

Z
221
207

140

112

92

76
(81)

U
195
176

S
170
148

P
141
122

H
115
100

Fig. 7.11 Core barrel and casing diameters.

massive rocks, it is rarely used. In the single-tube barrel, the barrel rotates the bit and the flush washes over the core. In double-tube barrels the flush passes between the inner and outer tubes. Double tubes may be of the rigid or swivel type. The disadvantage of the rigid barrel is that both the inner and outer tubes rotate together and in soft rock this can break the core as it enters the inner tube. It is therefore only suitable for hard rock formations.

In the double-tube swivel-type core barrel the outer tube rotates whilst the inner tube remains stationary (Fig. 7.12). It is suitable for use in medium and hard rocks and gives improved core recovery in soft, friable rocks. The face-ejection barrel is a variety of the double-tube swivel-type in which the flushing fluid does not affect the end of the core. This type of barrel is a minimum requirement for coring badly shattered, weathered, and soft rock formations. Triple-tube barrels have been developed for obtaining cores from very soft rocks and from highly jointed and cleaved rock.

Both the bit and core barrel are attached by rods

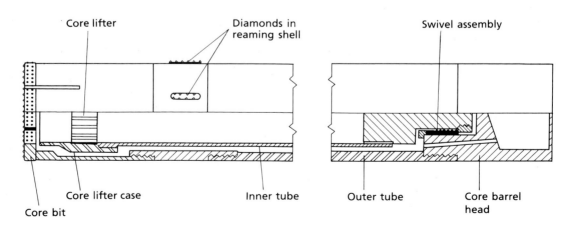

Core lifter

Diamonds in reaming shell

Swivel assembly

Core lifter case

Inner tube

Outer tube

Core barrel head

Core bit

Fig. 7.12 Double-tube swivel-type core barrel.

to the drill, by which they are rotated. Either water or air is used as a flush. This is pumped through the drill rods and discharged at the bit. The flushing agent serves to cool the bit and to remove the cuttings from the drillhole. Bentonite is sometimes added to water. It eases the running and pulling of casing by lubrication, it holds chippings in suspension, and promotes drillhole stability by the formation of a filter skin on the walls of the hole.

Disturbance of the core is likely to occur when it is removed from the core barrel. Most rock cores should be removed by hydraulic extruders whilst the tube is held horizontal. To reduce disturbance during extrusion, the inner tube of double-core barrels can be lined with a plastic sleeve before drilling commences. On completion of the core run the plastic sleeve containing the core is withdrawn from the barrel.

If casing is used for diamond drilling operations, then it is drilled into the ground using a tungsten carbide- or diamond-tipped casing shoe with air, water, or mud flush. The casing may be inserted down a hole drilled to a larger diameter to act as conductor casing when reducing and drilling ahead in a smaller diameter or it may be drilled or reamed in a larger diameter than the initial hole to allow continued drilling in the same diameter.

Many machines will core drill at any angle between vertical and horizontal. Unfortunately, inclined drillholes tend to go off line, the problem being magnified in highly jointed formations. In deeper drilling, the sag of the rods causes the hole to deviate. Drillhole deviation can be measured by an inclinometer.

The weakest strata are generally of the greatest interest but these are the very materials which are most difficult to obtain and, if recovered, most likely to deteriorate after extraction. Recently Hawkins (1986) introduced the concept of lithological quality designation (LQD), defining it as the percentage of solid core present greater than 100 mm in length within any lithological unit. The total core recovery and the maximum intact core length (MICL) should also be recorded. Shales and mudstones are particularly prone to deterioration and some may disintegrate completely if allowed to dry. If samples are not properly preserved they will dry out. Deterioration of suspect material may be reduced by wrapping the cores with aluminium foil or plastic sheeting. Core sticks may be photographed before they are removed from the site.

A simple but important factor is labelling. This must record the site, the drillhole number, and the position in the drillhole from which material was obtained. The labels themselves must be durable and properly secured. When rock samples are stored in a core box the depth of the top and bottom of the core contained and of the separate core runs should be noted both outside and inside the box. Zones of core loss should also be identified.

7.3 IN SITU TESTING

There are two categories of penetrometer tests: (i) the dynamic tests; and (ii) the static tests. Both methods measure the resistance to penetration of a conical point offered by the soil at any particular depth. Penetration of the cone creates a complex shear failure and thus provides an indirect measure of the *in situ* shear strength of the soil.

The most widely used dynamic method is the standard penetration test (SPT). This empirical test consists of driving a split-spoon sampler, with an outside diameter of 50 mm, into the soil at the base of a borehole. Drivage is accomplished by a trip hammer, weighing 63 kg, falling freely through a distance of 750 mm onto the drive head, which is fitted at the top of the rods (Fig. 7.13). First of all the split spoon is driven 150 mm into the soil at the bottom of the borehole. It is then driven a further 300 mm and the number of blows required to drive this distance is recorded. The blow count is referred to as the N value from which the relative density of the soil can be assessed (Table 7.2). Refusal is regarded as 100 blows. In deep boreholes the SPT suffers the disadvantage that the load is applied at the top of the rods so that some of the energy from the blow is dissipated in the rods. Hence, with increasing depth the test results become more suspect.

The results obtained from the SPT provide an evaluation of the degree of compaction of cohesionless sands and the N values may be related to the values of the angle of internal friction (ϕ) and the allowable bearing capacity.

Terzaghi and Peck (1968) suggested that for very fine or silty submerged sand with a standard penetration value, N', greater than 15, the relative density would be nearly equal to that of dry sand with a standard penetration value, N, where:

$$N = 15 + \tfrac{1}{2}(N' - 15) \qquad (7.2)$$

If this correction was not made, Terzaghi and Peck suggested that the relative density of even moder-

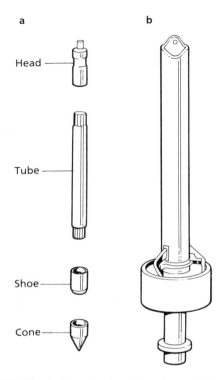

Fig. 7.13 Standard penetration test equipment: (a) split-spoon sampler; (b) trip hammer.

Table 7.2 Relative density and consistency of soil. (After Terzaghi and Peck, 1968; Sanglerat, 1972.)

(a) Relative density of sand and SPT values, and relationship to angle of internal friction

SPT (N)	Relative density (D_r)	Description of compactness	Angle of internal friction (ϕ)
4	0.2	Very loose	Under 30°
4–10	0.2–0.4	Loose	30°–35°
10–30	0.4–0.6	Medium dense	35°–40°
30–50	0.6–0.8	Dense	40°–45°
Over 50	0.8–1	Very dense	Over 45°

(b) N-values, consistency, and unconfined compressive strength of cohesive soils

N	Consistency	Unconfined compressive strength (kN/m^2)
Under 2	Very soft	Under 20
2–4	Soft	20–40
5–8	Firm	40–75
9–15	Stiff	75–150
16–30	Very stiff	150–300
Over 30	Hard	Over 300

$$D_r = \frac{e_{max} - e}{e_{max} - e_{min}}; \; e = \text{void ratio}$$

ately dense, very fine, or silty submerged sand might be overestimated by the results of SPTs. In gravel deposits care must be taken to determine whether a large gravel size may have influenced the results. Usually in the case of gravel, only the lowest values of N are taken into account. The lowest values of the angle of internal friction given in Table 7.2 are conservative estimates for uniform, clean sand and they should be reduced by at least 5° for clayey sand. The upper values apply to well graded sand and may be increased by 5° for gravelly sand. The SPT can also be employed in clays (Table 7.2), weak rocks, and in the weathered zones of harder rocks.

The most widely used static method employs the Dutch cone penetrometer (Fig. 7.14). It is particularly useful in soft clays and loose sands where boring operations tend to disturb *in situ* values. In this technique a tube and inner rod with a conical point at the base are hydraulically advanced into the ground. The cone has a cross-sectional area of 1000 mm^2 with an angle of 60°. At approximately every 300 mm depth the cone is advanced ahead of

the tube a distance of 50 mm and the maximum resistance is noted. The tube is then advanced to join the cone after each measurement and the process repeated. The resistances are plotted against their corresponding depths so as to give a profile of the variation in consistency (Fig. 7.15).

One type of Dutch cone penetrometer has a sleeve behind the cone which can measure side friction. The ratio of sleeve resistance to that of cone resistance is higher in cohesive than in cohesionless soils thus affording some estimate of the type of soil involved (Fig. 7.16).

In the piezocone a cone penetrometer is combined with a piezometer, the latter being located between the cone and the friction sleeve. The pore water pressure is measured at the same time as the cone resistance and sleeve friction. Because of the limited thickness of the piezometer (the filter is around 5 mm), much thinner layers can be determined with greater accuracy than with a cone penetrometer. If the piezocone is kept at a given depth so that the pore water pressure can dissipate with time, then this

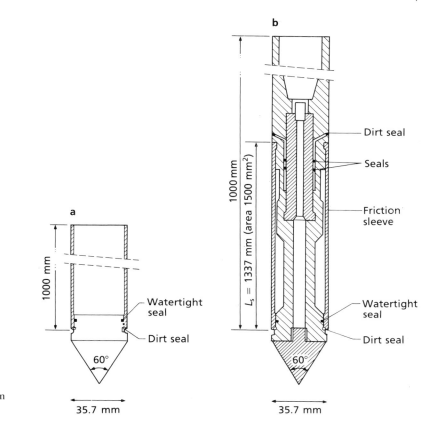

Fig. 7.14 An electric penetrometer tip: (a) without friction sleeve; (b) with friction sleeve.

allows assessment of the *in situ* permeability and consolidation characteristics of the soil to be made.

Because soft clays may suffer disturbance when sampled and therefore give unreliable results when tested for strength in the laboratory, a vane test is often used to measure the *in situ* undrained shear strength. Vane tests can be used in clays which have a consistency varying from very soft to firm. In its simplest form the shear vane apparatus consists of four blades arranged in cruciform fashion and attached to the end of a rod (Fig. 7.17). To eliminate the effects of friction of the soil on the vane rods during the test all rotating parts, other than the vane, are enclosed in guide tubes. The vane is normally housed in a protective shoe. The vane and rods are pushed into the soil from the surface or the base of a borehole to a point 0.5 m above the required depth. Then the vane is pushed out of the protective shoe and advanced to the test position. It is then rotated at a rate of 6° to 12° per minute. The torque is applied to the vane rods by means of a torque-measuring instrument mounted at ground level and clamped to the borehole casing or rigidly fixed to the ground. The maximum torque required for rotation is recorded. When the vane is rotated the soil falls along a cylindrical surface defined by the edges of the vane as well as along the horizontal surfaces at the top and bottom of the blades. The shearing resistance is obtained from the following expression:

$$\tau = \frac{M}{\pi \left(\dfrac{D^2 H}{2} + \dfrac{D^3}{6} \right)} \tag{7.3}$$

where τ is the shearing resistance, D and H are the diameter and height of the vane respectively, and M is the torque. Tests in clays with a high organic content or with pockets of sand or silt are likely to produce erratic results. The results should therefore be related to borehole evidence.

Loading tests can be carried out on loading plates (Fig. 7.18a). However, just because the ground immediatley beneath a plate is capable of carrying a heavy load without excessive settlement, this does

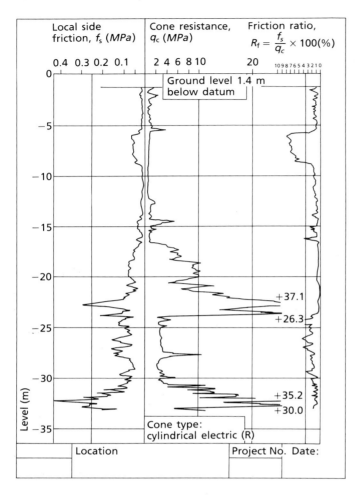

Fig. 7.15 Typical record of cone penetration test.

not necessarily mean that the ground will carry the proposed structural load. This is especially the case where a weaker horizon occurs at depth but is still within the influence of the bulb of pressure which will be generated by the structure (Fig. 7.18b). The plate-load test provides valuable information by which the bearing capacity and settlement characteristics of a foundation can be assessed. It is carried out in a trial pit, usually at excavation base level. Plates vary in size from 0.15 to 1.0 m in diameter, the size of plate used being determined by the spacing of discontinuities. The plate should be properly bedded and the test carried out on undisturbed material so that reliable results may be obtained. The load is applied by a jack, in increments, either of one-fifth of the proposed bearing pressure or in steps of 25 to 50 kPa (these are smaller in soft soils,

i.e. where the settlement under the first increment of 25 kPa is greater than 0.002D, D being the diameter of the plate). Successive increments should be made after settlement has ceased. The test is generally continued up to two or three times the proposed loading, or in clays until settlement equal to 10 to 20% of the plate dimension is reached or the rate of increase of settlement becomes excessive. Consequently, the ultimate bearing capacity at which settlement continues without increasing the load rarely is reached.

In clays when the final increment is applied the load should be maintained until the rate of settlement becomes less than 0.1 mm in 2 h. This can be regarded as the completion of the primary consolidation stage. With this information settlement curves can be drawn from which the ultimate loading can

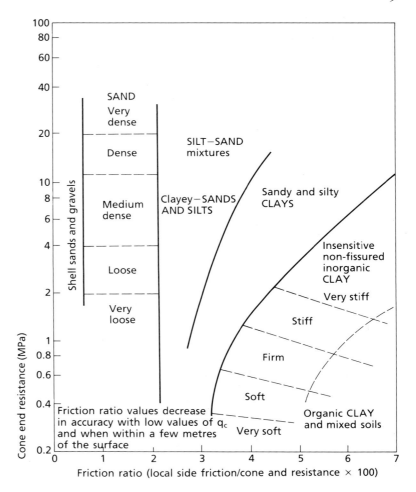

Fig. 7.16 Tentative evaluation of soil type from Dutch cone penetrometer. Some overlap between zones does occur. Local correlations are preferable. In sands and gravels, the relationship between compaction and cone resistance varies with grain size and grading.

be determined, and an evaluation of Young's modulus made. At the end of the consolidation stage the plate can be unloaded in the same incremental steps in order to obtain an unloading curve.

Large plate bearing tests are frequently used to determine the value of Young's modulus of the foundation rock at large civil engineering sites, such as dam sites. Loading of the order of several meganewtons is required to obtain measurable deformation of representative areas. The area of rock loaded is usually $1\,m^2$. Tests are usually carried out in specially excavated galleries in order to provide a sufficiently strong reaction point for the loading jacks to bear against.

The test programme usually includes cycles of loading and unloading. Such tests show that during loading a noticeable increase in rigidity occurs in the

rock mass and that during unloading a very small deformation occurs for the high stresses applied, with very large recuperation of deformations being observed for stresses near zero. This is due to joint closure. Once the joints are closed the adhesion between the faces prevents their opening until a certain unloading is reached. However, when brittle rocks like granite, basalt, and limestone have been tested they generally have given linear stress–strain curves and have not exhibited hysteresis.

Variations of this type of test include the freyssinet jack. This is placed in a narrow slit in the rock mass and then grouted into position so that each face is in uniform contact with the rock. Pressure is then applied to the jack. Unless careful excavation, particularly blasting, takes place in the testing area the results of a flatjack test may be worthless. All loose

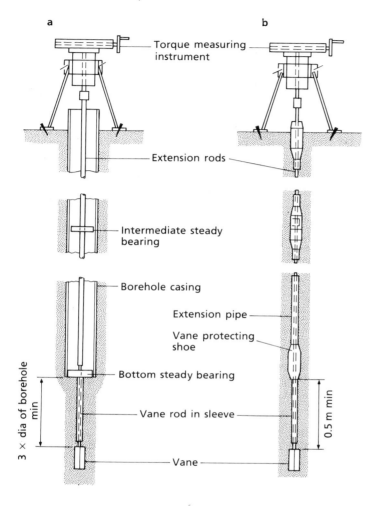

a

b

Torque measuring instrument

Extension rods

Intermediate steady bearing

Borehole casing

Extension pipe

Vane protecting shoe

Bottom steady bearing

Vane rod in sleeve

Vane

3 × dia of borehole min

0.5 m min

Fig. 7.17 Shear vane tests: (a) borehole vane test; (b) penetration vane test.

material must be removed before cutting the slot.

A dilatometer (Fig. 7.19) can be used in a drillhole to obtain data relating to the deformability of a rock mass. These instruments range up to about 300 mm in diameter and over 1 m in length and can exert pressures of up to 20 MPa on the drillhole walls. Diametral strains can be measured either directly along two perpendicular diameters or by measuring the amount of liquid pumped into the instrument. The last method is less accurate and is only used when the rock is very deformable.

The Menard pressuremeter (Fig. 7.20) is used to determine the *in situ* strength of the ground. It is particularly useful in those soils from which un-disturbed samples cannot readily be obtained. This pressuremeter consists essentially of a probe which

is placed in a borehole at the appropriate depth and then expanded. Where possible the test is carried out in an unlined hole but if necessary a special slotted casing is used to provide support. The probe consists of a cylindrical metal body over which are fitted three cylinders. A rubber membrane covers the cylinders and is clamped between them to give three independent cells. The cells are inflated with water and a common gas pressure is applied by a volumeter located at the surface, thus a radial stress is applied to the soil. The deformations produced by the central cell are indicated on the volumeter.

A simple pressuremeter test consists of ten or more equal pressure increments with corresponding volume-change readings, taken to the ultimate failure strength of the soil concerned. Four volume

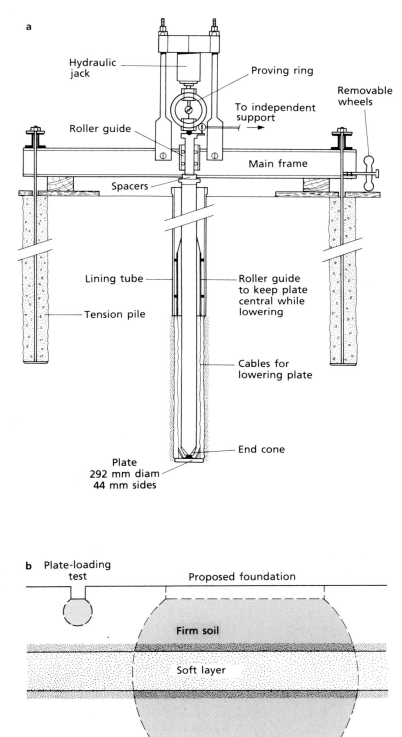

Fig. 7.18 (a) Plate-load test. (Courtesy of Building Research Establishment, Garston, Watford, WD2 7JR (from CP5/73).) (b) Bulb of pressure developed beneath a foundation compared with one developed beneath a plate-load test.

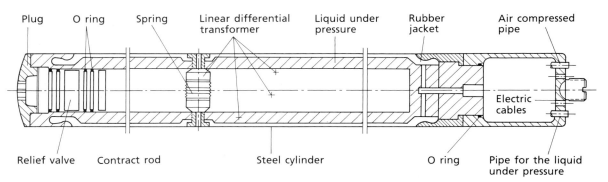

Plug   O ring   Spring   Linear differential transformer   Liquid under pressure   Rubber jacket   Air compressed pipe

Electric cables

Relief valve   Contract rod   Steel cylinder   O ring   Pipe for the liquid under pressure

Fig. 7.19 The dilatometer.

readings are made at each pressure step at time intervals of 15, 30, 60, and 120s after the pressure has stabilized. It is customary to unload the soil at the end of the elastic phase of expansion and to

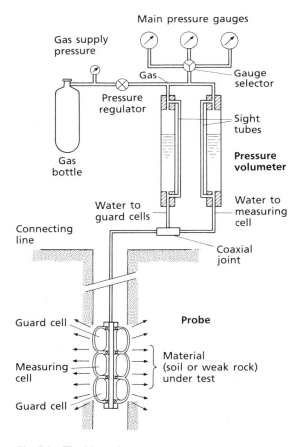

Main pressure gauges

Gas supply pressure

Gas

Gauge selector

Pressure regulator

Sight tubes

Gas bottle

Pressure volumeter

Water to guard cells

Water to measuring cell

Connecting line

Coaxial joint

Guard cell

Probe

Measuring cell

Material (soil or weak rock) under test

Guard cell

Fig. 7.20 The Menard pressuremeter.

repeat the test before proceeding to the ultimate failure pressure. This test thus provides the ultimate bearing capacity of soils as well as their deformation modulus. The test can be applied to any type of soil, and takes into account the influence of discontinuities. It can also be used in weathered zones of rock masses and in weak rocks such as shales and marls. It provides an almost continuous method of *in situ* testing.

The major advantage of a self-boring pressuremeter is that a borehole is unnecessary, hence the interaction of the probe and the soil is improved. Self-boring is brought about either by jetting or by using a chopping tool (Fig. 7.21). The camkometer (Fig. 7.21b) has a special cutting head so that it can be drilled into soft ground to form a cylindrical cavity of exact dimensions and thereby create a minimum of disturbance. The camkometer measures the lateral stress, undrained stress—strain properties, and the peak stress of soft clays and sands *in situ*. The tool may also be rotated into the soil by a motor in the probe itself.

In an *in situ* shear test a block of rock is sheared from the rock surface whilst a horizontal jack exerts a vertical load. It is advantageous to make the tests inside galleries, where reactions for the jacks are readily available (Fig. 7.22). The tests are performed at various normal loads and give an estimate of the angle of shearing resistance and cohesion of the rock. The value of this test in very jointed and heterogeneous rocks is severely limited both because of the difficulty in isolating undisturbed test blocks and because the results cannot be translated to the scale of conditions of the actual structure and its foundation. *In situ* shear tests are usually performed on blocks measuring 700×700 mm, cut in the rock. These tests can be made on the same rock where it shows different degrees of alteration and along dif-

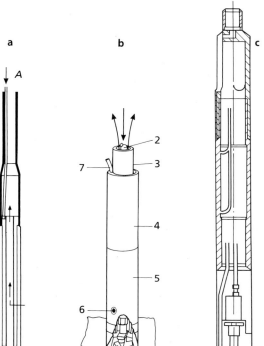

Fig. 7.21 (a) Schematic diagram of self-boring pressuremeter. This pressuremeter is available in three diameters (65, 100, and 132 mm). The injection-fluid flow which can be applied is limited by the section of the flexible pipe (*A*). This pressuremeter can consequently be used only in fine sandy or soft cohesive soils; (b) the Camkometer: the soil is broken up by a chopping tool rotated by a drill string driven from the surface. The chopping tool (1) is driven by a hollow middle rod (2) through which is injected water under pressure. This rod turns freely inside a tube (3) used for removing the sediment to the surface. There is also a tube (4) which carries the pressiometric cell (5) which may be equipped with a pore pressure tap (6). The pressiometric cell supply and measurement lines (7) run through the annulus between the two tubes; (c) self-boring pressuremeter probe with built-in motor.

ferent directions according to the discontinuity pattern.

7.4 FIELD INSTRUMENTATION

When some degree of risk is involved in construc-

tion, then some type of field instrumentation may be required in order to provide a continual check on the stability of the structure during its lifespan. Furthermore, field observations of both the magnitude and rate of subsurface ground movements are needed in connection with deep excavations, slope stability, and earth and rockfill dam construction. Such instrumentation needs to assess the pore water pressure, deformation, and stress and strain in the ground. However, an instrumentation programme does not usually constitute part of a site investigation.

Surface deformation either in the form of settlements or horizontal movements can be assessed by precise surveying methods, the use of EDM or laser equipment providing particularly accurate results. Settlement, for example, can be recorded by positioning reference marks on the structures concerned and readings can be taken by precise surveying methods. The observations are related to nearby bench marks. Vertical movements can also be determined by settlement tubes or by water-level or mercury filled gauges.

Drillhole extensometers are used to measure the vertical displacement of the ground at different depths. A single-rod extensometer is anchored in a drillhole, and movement between the anchor and the reference tube is monitored. Multiple-rod installations monitor displacements at various depths using rods of varying lengths. Each rod is isolated by a close-fitting plastic sleeve and the complete assembly is grouted into place, fixing the anchors to the ground while allowing free movement of each rod within its sleeve (Fig. 7.23a). A precise borehole extensometer consists of circular magnets embedded in the ground, which act as markers, and reed switch sensors move in a central access tube to locate the positions of the magnets (Fig. 7.23b).

An inclinometer (Fig. 7.24) is used to measure horizontal movements below ground and relies on the measurement of the angle a pendulum makes with the vertical at given positions in a specially cased borehole. The gravity operated pendulum transmits electrical signals to the recorder and a vertical profile is thereby obtained. Sets of readings over a period of time enable both the magnitude and rate of horizontal movement to be determined.

The measurement of stress, contact pressures, and stress change may be made in two ways. Strain may be measured and then converted to stress or stress may be measured directly by an earth-pressure cell such as the Glotzl cell. This is a hydraulic (flat-diaphragm) cell which has a high stiffness at constant

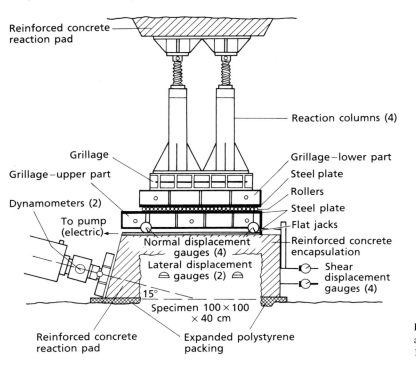

Reinforced concrete reaction pad

Reaction columns (4)

Grillage
Grillage–upper part
Dynamometers (2)
To pump (electric)

Grillage–lower part
Steel plate
Rollers
Steel plate
Flat jacks
Reinforced concrete encapsulation

Normal displacement gauges (4)
Lateral displacement gauges (2)

Shear displacement gauges (4)

15°
Specimen 100 × 100 × 40 cm

Reinforced concrete reaction pad

Expanded polystyrene packing

Fig. 7.22 The *in situ* shear test apparatus. (After Franklin *et al.*, 1974.)

temperature and is used for measuring contact pressures. An earth cell must be placed in position in such a way as to minimize disturbance of the stress and strain distribution. Ideally an earth pressure cell should have the same elastic properties as the surrounding soil. This, of course, cannot be attained and in order to minimize the magnitude of error (i.e. the cell-action factor) the ratio of the thickness to the diameter of the cell should not exceed 0.2 and the ratio of the diameter to the deflection of the diaphragm must be 2000 or greater. Most strain measurements in soils and soft rocks are deformation measurements which are interpreted in terms of strain. Strain is sometimes measured directly in hard rocks. This can be done by mounting strain gauges onto the ends of an insertion in a drillhole, so monitoring deformation of the drillhole.

A strain cell is required to move with the soil without causing it to be reinforced. To record strain it is necessary to monitor the relative movements of two fixed points at either end of a gauge length. Strain cells with a positive connection between their end plates have difficulty in measuring small strains. This can be overcome by substituting separated strain cells at either end of the gauge length.

7.5 GEOPHYSICAL METHODS: INDIRECT SITE EXPLORATION

A geophysical exploration may be included in a site investigation for an important engineering project in order to provide subsurface information over a large area at reasonable cost (Anon, 1988). The information obtained may help eliminate less favourable alternative sites, may aid the location of test holes in critical areas and may prevent unnecessary repetitive boring or drilling in fairly uniform ground. A geophysical survey helps to locate the position of holes and also detects variations in subsurface conditions between them. Test holes provide information about the strata where they are sunk but tell nothing about the ground in between. Nonetheless boreholes or drillholes to aid interpretation and correlation of the geophysical measurements are an essential part of any geophysical survey. Therefore an appropriate combination of direct and indirect methods often can yield a high standard of results.

Geophysical methods are used to determine the geological sequence and structure of subsurface rocks by the measurement of certain physical properties or forces. The properties which are made most use of in geophysical exploration are density, elas-

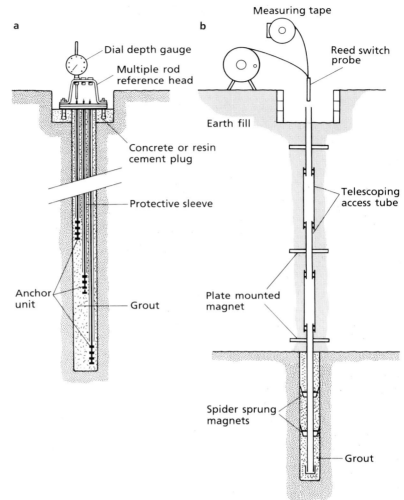

Fig. 7.23 (a) Multiple-rod extensometer; (b) magnetic-probe extensometer. (Courtesy of Soil Instruments, Ltd.)

ticity, electrical conductivity, magnetic susceptibility, and gravitational attraction, i.e. seismic and resistivity methods record the artificial fields of force applied to the area under investigation whilst magnetic and gravitational methods measure natural fields of force. The former techniques have the advantage over the latter in that the depth to which the forces are applied can be controlled. By contrast, the natural fields of force are fixed and can only be observed, not controlled. Seismic and resistivity methods are more applicable to the determination of horizontal or near-horizontal changes or contacts, whereas magnetic and gravimetric methods are generally used to delineate lateral changes or vertical structures.

In a geophysical survey, measurements of the variations in certain physical properties are usually taken in a traverse across the surface, although they may be made in order to log a hole. Anomalies in the physical properties measured generally reflect anomalies in the geological conditions. The ease of recognizing and interpreting these anomalies depends on the contrast in physical properties which, in turn, influences the choice of the method employed.

The actual choice of method to be used for a particular survey may not be difficult to make. The character and situation of the site also have to be taken into account, especially in built-up areas, which may be unsuitable for one or other of the

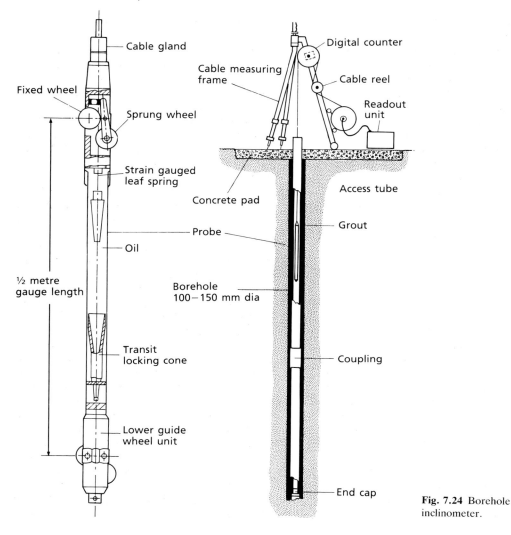

Fig. 7.24 Borehole inclinometer.

geophysical methods either because of the presence of old buildings or services on the site, interference from some source, or lack of space for carrying out the survey. When dealing with layered rocks, provided their geological structure is not too complex, seismic methods have a distinct advantage in that they give more detailed, precise, and unambiguous information than any other method. Electrical methods may be preferred for small-scale work where the structures are simple. On occasions more than one method may be used to resolve the same problem.

Generally speaking, observations should be close enough for correlation between them to be obvious, so enabling interpretation to be carried out without ambiguity, but an accurate and unambiguous interpretation of geophysical data is possible only where the subsurface structure is simple and even then there is no guarantee that this will be achieved.

7.5.1 Seismic methods

The sudden release of energy from the detonation of an explosive charge in the ground or the mechanical pounding of the surface generates shock waves which radiate out in a hemispherical wave front from the point of release. The waves generated are compressional (P), dilational shear (S), and surface waves. The velocities of the shock waves generally increase with depth below the surface since they increase with increasing elastic modulus and this increases with depth. The compressional waves travel faster and are more easily generated and recorded than shear waves. They are therefore used almost exclusively in seismic exploration. The shock-wave velocity depends on many variables, including rock fabric, mineralogy, and pore water. In general, velocities in crystalline rocks are high to very high (Table 7.3). Velocities in sedimentary rocks increase concomitantly with consolidation and decrease in pore fluids, and with increase in the degree of cementation and diagenesis. Unconsolidated sedimentary accumulations have maximum velocities varying as a function of the volume of voids (either air-filled or water-filled), mineralogy, and grain size.

When seismic waves pass from one layer to another some energy is reflected back towards the surface whilst the remainder is refracted. Thus, two methods of seismic surveying can be distinguished: (i) seismic reflection and (ii) seismic refraction. Measurement of the time taken from the generation of the shock waves until they are recorded by detector arrays forms the basis of the two methods.

The seismic reflection method is the most extensively used geophysical technique, its principal employment being in the oil industry. In this technique the depth of investigation is large compared with the distance from the shot to detector array. This is to exclude refraction waves. Indeed, the method is able to record information from a large number of horizons down to depths of several thousands of metres.

In the seismic refraction method (Fig. 7.25) one ray approaches the interface between two rock types at a critical angle which means that if the ray is passing from a low- (V_0) to a high-velocity (V_1) layer it will be refracted along the upper boundary of the latter layer. After refraction, the pulse travels along the interface with velocity V_1. The material at the boundary is subjected to oscillating stress from below. This generates new disturbances along the boundary which travel upwards through the low-velocity rock and eventually reach the surface.

At short distances from the point where the shock waves are generated the geophones record direct waves whilst at a critical distance both the direct and refracted waves arrive at the same time. Beyond this, because the rays refracted along the high-velocity layer travel faster than those through the low-velocity layer above, they reach the geophones first. In refraction work the object is to develop a time—distance graph which involves plotting arrival times against geophone spacing (Fig. 7.25). Thus the distance between geophones, together with the total length and arrangement of the array, has to be carefully chosen to suit each particular problem.

The most common arrangement in refraction work is profile shooting. Here the shot points and geophones are laid out in lines, the geophones receiving the refracted waves from the shots fired. For many surveys for civil engineering purposes where it is required to determine depth to bedrock it may be sufficient to record from two shot-point distances at each end of the receiving spread. By traversing in both directions the angle of dip can be determined.

In the simple case of refraction by a single high-velocity layer at depth, the travel times for the seismic wave which proceeds directly from the shot point to the detectors and the travel times for the critical refracted wave to arrive at the geophones, and plotted graphically against geophone spacing (Fig. 7.25). The depth, Z, to the high-velocity layer

Table 7.3 Velocities of compressional waves of some common rocks

	V_p (km/s)		V_p (km/s)
Igneous rocks		Sedimentary rocks	
Basalt	5.2−6.4	Gypsum	2.0−3.5
Dolerite	5.8−6.6	Limestone	2.8−7.0
Gabbro	6.5−6.7	Sandstone	1.4−4.4
Granite	5.5−6.1	Shale	2.1−4.4
Metamorphic rocks		Unconsolidated deposits	
Gneiss	3.7−7.0	Alluvium	0.3−0.6
Marble	3.7−6.9	Sands and gravels	0.3−1.8
Quartzite	5.6−6.1	Clay (wet)	1.5−2.0
Schist	3.5−5.7	Clay (sandy)	2.0−2.4

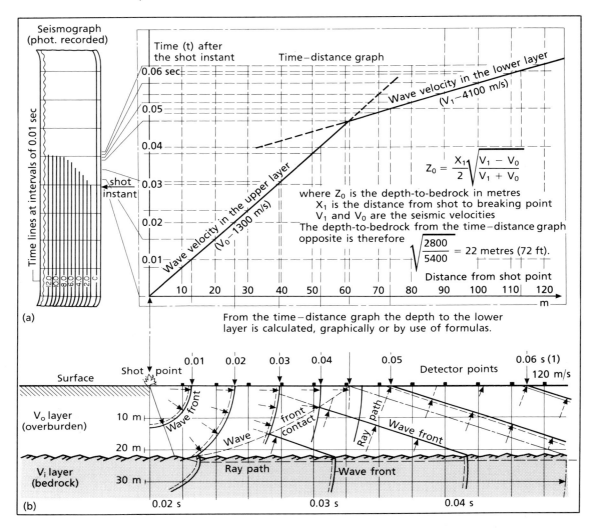

Fig. 7.25 Time–distance graphs for a theoretical single-layer problem, with parallel interface. With non-parallel interfaces, both forward and reverse profiles must be surveyed.

can then be obtained from the graph by using the expression:

$$Z = \frac{X}{2}\left(\frac{V_1 - V_0}{V_1 + V_0}\right) \tag{7.4}$$

where V_0 is the speed in the low-velocity layer, V_1 is the speed in the high-velocity layer, and X is the critical distance. The method also works for multi-layered rock sequences if each layer is sufficiently thick and transmits seismic waves at higher speeds than the one above it (Fig. 7.26a). However, in the refraction method a low-velocity layer underlying a high-velocity layer usually cannot be detected as in such an inversion the pulse is refracted into the low-velocity layer. Also a layer of intermediate velocity between an underlying refractor and overlying layers can be masked as a first arrival on the travel–time curve. The latter is known as a blind zone. The position of faults can also be estimated from the time–distance graphs.

The velocity of shock waves is closely related to the elastic modulus of rock or soil and can therefore provide data relating to the engineering performance

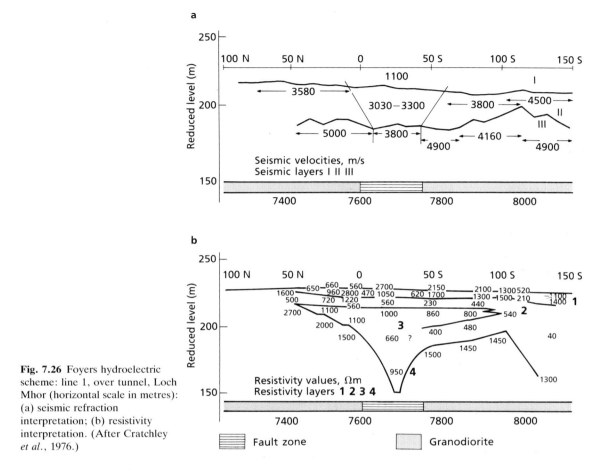

Fig. 7.26 Foyers hydroelectric scheme: line 1, over tunnel, Loch Mhor (horizontal scale in metres): (a) seismic refraction interpretation; (b) resistivity interpretation. (After Cratchley et al., 1976.)

of the ground. Young's modulus (E) and Poisson's ratio (v) can be derived if the density (ρ) and compressional (V_ρ) and shear (V_s) wave velocities are known by using the following expressions:

$$E = \rho V_\rho^2 \frac{(1 + v)\,(1 - 2_v)}{(1 - v)} \qquad (7.5)$$

or

$$E = 2V_s^2\,\rho(1 + v) \qquad (7.6)$$

or

$$E = \frac{V_s^2}{g}\,\rho\left[\frac{3(V_\rho/V_s)^2 - 4}{(V_\rho/V_s)^2 - 1}\right] \qquad (7.7)$$

$$v = \frac{0.5(V_\rho/V_s)^2 - 1}{(V_\rho/V_s)^2 - 1} \qquad (7.8)$$

where g is the acceleration due to gravity. These dynamic moduli correspond to the initial tangent moduli of the stress–strain curve for an instantaneously applied load and are usually higher than those obtained in static tests.

7.5.2 Resistivity methods

The resistivity of rocks and soils varies within a wide range. Since most of the principal rock forming minerals are practically insulators, the resistivity of rocks and soils is determined by the amount of conducting mineral constituents and the content of mineralized water in their pores. The latter condition is by far the dominant factor and in fact most rocks and soils conduct an electric current only because they contain water. The widely differing resistivity values of the various types of impregnating water

Table 7.4 Resistivity values of some different types of natural waters

Type of water	Resistivity (Ω-m)
Meteoric water derived from precipitation	30−1000
Surface waters in igneous rock districts	30−500
Surface waters in sedimentary rock districts	10−100
Groundwater in igneous rock districts	30−150
Groundwater in sedimentary rock districts	Larger than 1
Seawater	About 0.2

Table 7.5 Resistivity values of some common rock types

Rock type	Resistivity (Ω-m)
Topsoil	5−50
Peat and clay	8−50
Clay, sand, and gravel mixtures	40−250
Saturated sand and gravel	40−100
Moist to dry sand and gravel	100−3000
Mudstones, marls, and shales	8−100
Sandstones and limestones	100−1000
Crystalline rocks	200−10 000

can cause variations in the resistivity of rocks ranging from a few tenths of an ohm-metre to hundreds of ohm-metres (Ω-m) as can be seen from Table 7.4.

In the resistivity method an electric current is introduced into the ground by means of two current electrodes and the potential difference between two potential electrodes is measured. It is preferable to measure the potential drop or apparent resistance directly in ohms rather than observe both current and voltage. The ohm value is converted to apparent resistivity by use of a factor that depends on the particular electrode configuration in use (see p. 231).

The relation between the depth of penetration and the electrode spacing is given in Figure 7.27 from which it can be seen that 50% of the total current passes above a depth equal to about half the electrode separation and 70% flows within a depth equal to the electrode separation. Analysis of the

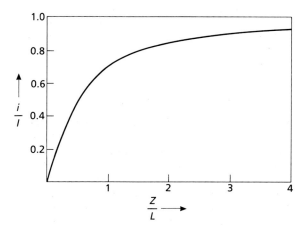

Fig. 7.27 Fraction of total current, I, which passes above a horizontal plane at depth, Z, as a function of the distance, L, between two current electrodes.

variation in the value of apparent resistivity with respect to electrode separation enables inferences to be drawn about the subsurface formations.

The resistivity method is based on the fact that any subsurface variation in conductivity alters the pattern of current flow in the ground and therefore changes the distribution of electric potential at the surface. Since the electrical resistivity of such features as superficial deposits and bedrock differ from each other (Table 7.5), the resistivity method may be used in their detection, and to give their approximate thicknesses, relative positions, and depths. The first step in any resistivity survey should be to conduct a resistivity depth sounding at the site of a borehole in order to establish a correlation between resistivity and lithological layers. If a correlation cannot be established then an alternative method is required.

The electrodes are normally arranged along a straight line, the potential electrodes being placed inside the current electrodes and all four are symmetrically disposed with respect to the centre of the configuration. The configurations of the symmetric type that are most frequently used are those introduced by Wenner and by Schlumberger. Other configurations include the dipole−dipole and the pole−dipole arrays. In the Wenner configuration the distances between all four electrodes are equal (Fig. 7.28). The spacings can be progressively increased, keeping the centre of the array fixed or the whole array, with fixed spacings, can be shifted along a given line. In the Schlumberger arrangement the potential electrodes maintain a constant separation about the centre of the station whilst if changes with depth are being investigated the current electrodes are moved outwards after each reading (Fig. 7.28). The expressions used to compute the apparent resistivity (ρ_a) for the Wenner and Schlumberger configurations are as follows:

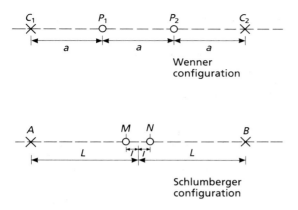

Fig. 7.28 Wenner and Schlumberger configurations.

Wenner

$$\rho_a = 2\pi a R \qquad (7.9)$$

Schlumberger:

$$\rho_a = \frac{\pi(L^2 - l^2)}{2l} \times R \qquad (7.10)$$

where a, L, and l are explained in Figure 7.28 and R is the resistance reading.

Horizontal profiling is used to determine variations in apparent resistivity in a horizontal direction at a preselected depth. For this purpose an electrode configuration, with fixed inter-electrode distances, is moved along a straight traverse, resistivity determinations being made at stations at regular intervals. The length of the electrode configuration must be carefully chosen because it is the dominating factor regarding depth penetration.

The data of a constant separation survey consisting of a series of traverses arranged in a grid pattern, may be used to construct a contour map of lines of equal resistivity. These maps are often extremely useful in locating areas of anomalous resistivity such as gravel pockets in clay soils and the trend of buried channels. Even so interpretation of resistivity maps as far as the delineation of lateral variations is concerned is mainly qualitative.

Electrical sounding furnishes information concerning the vertical succession of different conducting zones and their individual thicknesses and resistivities (Fig. 7.26b). For this reason the method is particularly valuable for investigations on horizontally stratified ground. In electrical sounding the midpoint of the electrode configuration is fixed at the observation station while the length of the configuration is increased in stages. As a result the current penetrates deeper and deeper, the apparent resistivity being measured each time the current electrodes are moved outwards. The readings therefore become increasingly affected by the resistivity conditions at advancing depths. The Schlumberger configuration is preferable to the Wenner configuration for depth sounding. The data obtained are usually plotted as a graph of apparent resistivity against electrode separation, in the case of the Wenner array, or half the current electrode separation for the Schlumberger array. The electrode separation (or half separation in the case of the latter array) at which inflection points occur in the graph provides an idea of the depth of interfaces. The apparent resistivities of the different parts of the curve provide some idea of the relative resistivities of the layers concerned.

If the ground approximates to an ideal condition, then a quantitative solution, involving a curve-fitting exercise, should be possible. The technique requires a comparison of the observed curve with a series of master curves prepared for various theoretical models.

Generally it is not possible to determine the depths to more than three or four layers. If a second layer is relatively thin and its resistivity much larger or smaller than that of the first layer, the interpretation of its lower contact will be inaccurate. For all depth determinations from resistivity soundings it is assumed that there is no change in resistivity laterally. This is not the case in practice. Indeed, sometimes the lateral change is greater than that occurring with increasing depth and so corrections have to be applied for the lateral effects when depth determinations are made.

7.5.3 Electromagnetic methods

In most electromagnetic methods, electromagnetic energy is introduced into the ground by inductive coupling and is produced by passing an alternating current through a coil. The receiver also detects its signal by induction, for example, the terrain conductivity meter measures the conductivity of the ground by such an inductive method. The conductivity meter is carried along traverse lines across a site and can provide a direct continuous readout. Hence, surveys can be carried out rapidly. Conductivity values are taken at positions set out on a grid pattern.

The ground probing radar method is based upon the transmission of pulsed electromagnetic waves in the frequency range 1 to 1000 MHz. In this method the travel times of the waves reflected from subsurface interfaces are recorded as they arrive at the surface and the depth (Z) to an interface is derived from:

$$Z = vt/2 \qquad (7.11)$$

where v is the velocity of the radar pulse and t is its travel time. The relative dielectric constant of the ground material (ε_r) is related to the radar velocity as follows:

$$\varepsilon_r = (c/v)^2 \qquad (7.12)$$

where c is the velocity of the electromagnetic waves in air (3×10^8 m/s). If the relative dielectric constant of this material is known, then the depth (Z) to the reflecting interface can be derived from

$$Z = t/2 \times c/v\varepsilon_r \qquad (7.13)$$

Darracott and Lake (1981) described a trolley-mounted system of ground probing radar which could be towed either by hand or behind a vehicle. Rates of towing usually vary between 1.5 and 6.5 km/h (the slower the progress, the better the resolution). They noted that the conductivity of the ground imposes the greatest limitation on the use of radar probing in site investigation, i.e. the depth to which radar energy can penetrate depends upon the effective conductivity of the strata being probed. This, in turn, is governed chiefly by the water content and its salinity (Table 7.6). The nature of the pore water exerts the most influence on the dielectric constant. Furthermore the value of effective conductivity is also a function of temperature and density as well as the frequency of the electromagnetic waves being propagated. The least penetration occurs in saturated clayey materials or where the moisture content is saline. For example, Leggo (1982) noted that useful data can be obtained from sites where clayey topsoil is more or less absent, as wet clay and silt, in particular, mean the greatest attenuation of electromagnetic energy so that depth of penetration frequently is less than 1 m. The technique appears to be reasonably successful in sandy soils and rocks in which the moisture content is non-saline. Rocks like limestone and granite can be penetrated for distances of tens of metres and in dry conditions the penetration may reach 100 m. Dry rock salt is radar-translucent, permitting penetration distances of hundreds of metres. The penetration of

radar energy can be increased by using a lower frequency but this unfortunately reduces its resolution which means that subsurface anomalies have to be correspondingly larger if they are to be detected.

Darracott and Lake (1981) recommended that the conductivity of ground material should be determined before a ground probing radar survey is carried out and data obtained from boreholes, trenches, and pits help to ensure more accurate interpretation. As few sites are completely radar opaque, trial runs yield useful data.

7.5.4 Magnetic and gravity methods

All rocks, mineral, and ore deposits are magnetized to a lesser or greater extent by the Earth's magnetic field. Thus, in magnetic prospecting accurate measurements are made of the anomalies produced in the local geomagnetic field by this magnetization. The intensity of magnetization, and hence the amount by which the Earth's magnetic field is changed locally, depends on the magnetic susceptibility of the material concerned. In addition to the magnetism induced by the Earth's field, rocks possess a permanent magnetism that depends upon their geological history.

Rocks have different magnetic susceptibilities related to their mineral content. Some minerals, for example, quartz and calcite, are magnetized reversely to the field direction and have therefore negative susceptibility and are described as dia-

Table 7.6 Approximate electromagnetic parameters of earth materials. (After Morey, 1974.)

Material	Approximate conductivity (mho/m)	Approximate dielectric constant
Air	0	1
Fresh water	10^{-4} to 3×10^{-2}	81
Sea water	4	81
Sand, dry	10^{-7} to 10^{-3}	4 to 6
Sand, saturated (fresh water)	10^{-4} to 10^{-2}	30
Silt, saturated (fresh water)	10^{-3} to 10^{-2}	10
Clay, saturated (fresh water)	10^{-1} to 1	8 to 12
Dry sandy coastal land	2×10^{-3}	10
Granite, dry	10^{-8}	5
Limestone, dry	10^{-9}	7

magnetic. Paramagnetic minerals, which are the majority, are magnetized along the direction of magnetic field so that their susceptibility is positive. The susceptibility of the ferromagnetic minerals, such as magnetite, ilmenite, pyrrhotite, and haematite, is a very complicated function of the field intensity. However, since the magnitudes of their susceptibility amount to 10 to 10^5 times the order of susceptibility of the paramagnetic and diamagnetic minerals, the ferromagnetic minerals can be found by magnetic field measurements.

If the magnetic field ceases to act on a rock, then the magnetization of paramagnetic and diamagnetic minerals disappears. However, in ferromagnetic minerals the induced magnetization is diminished only to a certain value. This residuum is called remanent magnetization and is of great importance in rocks. All igneous rocks have a very high remanent magnetization acquired as they cooled down in the Earth's magnetic field. In the geological past, during sedimentation in water, grains of magnetic material were orientated by ancient geomagnetic fields so that some sedimentary rocks show stable remanent magnetization.

The strength of a magnetic field is measured in oersteds, but as the average strength of the Earth's magnetic field is about 0.5 oersted the variations associated with magnetized rock formations are very much smaller than this. Consequently, the practical unit used in magnetic surveying is the gamma, which is 0.00001 oersted.

Aeromagnetic surveying has almost completely supplanted ground surveys for regional reconnaissance purposes. Accurate identification of the plan position of the aircraft for the whole duration of the magnetometer record is essential. The object is to produce an aeromagnetic map, the base map with transcribed magnetic values being contoured at 5 or 10 gamma intervals.

The aim of most ground surveys is to produce isomagnetic contour maps of anomalies (Fig. 7.29) to enable the form of the causative magnetized body to be estimated. Profiles are surveyed across the trend of linear anomalies with stations, if necessary, at intervals of as little as 1 m. A base station is set up beyond the anomaly where the geomagnetic field is uniform. The reading at the base station is taken as zero and all subsequent readings are expressed as plus or minus differences. Corrections need to be made for the temperature of the instrument as the magnets lose their effectiveness with increasing temperature. A planetary correction is also required

which eliminates the normal variation of the Earth's magnetic field with latitude. Large metallic objects like pylons are a serious handicap to magnetic exploration and must be kept at a sufficient distance, as it is difficult to correct for them.

A magnetometer may also be used for mapping geological structures, for example, in some thick sedimentary sequences it is sometimes possible to delineate the major structural features because the succession includes magnetic horizons. These may be ferruginous sandstones or shales, tuffs, or basic lava flows. In such circumstances anticlines produce positive and synclines negative anomalies.

Faults and dykes are indicated on isomagnetic maps by linear belts of somewhat sharp gradient or by sudden swings in the trend of the contours. However, in many areas the igneous and metamorphic basement rocks, which underlie the sedimentary sequence, are the predominant influence controlling the pattern of anomalies since they are usually far more magnetic than the sediments above. Where the basement rocks are brought near the surface in structural highs the magnetic anomalies are large and characterized by strong relief. Conversely, deep sedimentary basins usually produce contours with low values and gentle gradients on isomagnetic maps.

The Earth's gravity field varies according to the density of the subsurface rocks but at any particular locality its magnitude is also influenced by latitude, elevation, neighbouring topographical features, and the tidal deformation of the Earth's crust. The effects of these latter factors have to be eliminated in any gravity survey where the object is to measure the variations in acceleration due to gravity precisely. This information can then be used to construct isogals on a gravity map. The acceleration at the Earth's surface due to gravity is about 980 Gals, but the gravity variations which are measured are extremely small, of the order of 0.0001 Gal. Hence the unit of measurement is the milligal (mGal). Modern gravity meters used in exploration measure not the absolute value of the acceleration due to gravity but the small differences in this value between one place and the next.

Gravity methods are mainly used in regional reconnaissance surveys to reveal anomalies which may be subsequently investigated by other methods. Since the gravitational effects of geological bodies are proportional to the contrast in density between them and their surroundings, gravity methods are particularly suitable for the location of structures in

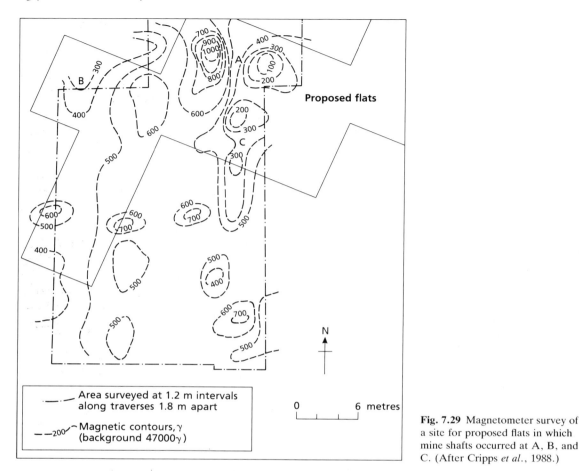

Fig. 7.29 Magnetometer survey of a site for proposed flats in which mine shafts occurred at A, B, and C. (After Cripps *et al.*, 1988.)

stratified formations. Gravity effects due to local structures in near surface strata may be partly obscured or distorted by regional gravity effects caused by large-scale basement structures. However, regional deep-seated gravity effects can be removed or minimized in order to produce a residual gravity map showing the effects of shallow structures which may be of interest.

A gravity survey is conducted from a local base station at which the value of the acceleration due to gravity is known with reference to a fundamental base where the acceleration due to gravity has been accurately measured. The way in which a gravity survey is carried out largely depends on the objective in view. Large-scale surveys covering hundreds of square kilometres, carried out in order to reveal major geological structures, are done by vehicle or helicopter with a density of only a few stations per

square kilometre. For more detailed work such as the delineation of ore bodies or basic minor intrusions or the location of faults, spacing between stations may be as small as 20 m. Because gravity differences large enough to be of geological significance are produced by changes in elevation of several millimetres and of only 30 m in north–south distance, the location and elevation of stations must be established with very high precision.

7.5.5 Drillhole logging techniques

Drillhole logging techniques can be used to identify some of the physical properties of rocks. The electrical resistivity method makes use of various electrode configurations down-the-hole. As the instrument is raised from the bottom to the top of the hole it provides a continuous record of the variations

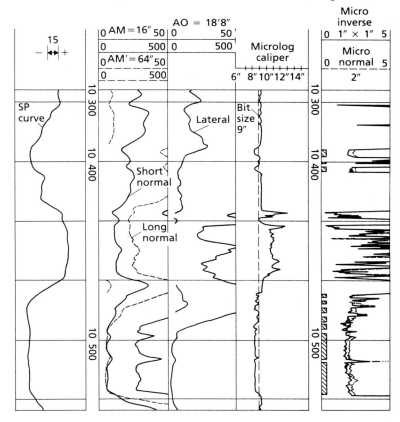

Fig. 7.30 Microlog curves. Microresistivity curves are shown on the right. Permeable portions of the section penetrated are indicated (cross-hatched bars) by extensions of the 50 mm micronormal curve beyond the microinverse. Note that the diameter of the bore, as recorded by the microlog caliper, is smaller than bit size where a mud filter cake is formed at the position of permeable beds. A standard electrical log of the same stratigraphic interval is shown on the left for comparison. (Courtesy Schlumberger Inland Services, Inc.)

in resistivity of the wall rock. In the normal or standard resistivity configuration there are two potential and one current electrodes in the sonde. The depth of penetration of the electric current from the drillhole is influenced by the electrode spacing. In a short normal resistivity survey spacing is about 400 mm, whereas in a long normal survey, spacing is generally between 1.5 and 1.75 m. Unfortunately in such a survey, because of the influence of thicker adjacent beds, thin resistive beds yield resistivity values which are much too low, whilst thin conductive beds produce values which are too high. The microlog technique may be used in such situations. In this technique the electrodes are very closely spaced (25 to 50 mm) and are in contact with the wall of the drillhole. This allows the detection of small lithological changes so that much finer detail is obtained than with the normal electric log (Fig. 7.30). A microlog is particularly useful in recording the position of permeable beds.

If, for some reason, the current tends to flow between the electrodes on the sonde instead of into the rocks, then the laterolog or guard electrode is used. The laterolog 7 has seven electrodes in an array which focuses the current into the strata of the drillhole wall. The microlaterolog, a focused microdevice, is used in such a situation instead of the microlog.

A dipmeter is a three, four, or six arm side-wall microresistivity device which measures small changes in resistivity, thereby permitting the relative shift of characteristic patterns of variation on the traces to be used to determine the attitudes of discontinuities in contact with the drillhole wall. In this way a fracture log can be produced.

Induction logging may be used when an electrical log cannot be obtained. In this technique the sonde sends electrical energy into the strata horizontally and therefore only measures the resistivity immediately opposite the sonde, unlike in normal electrical logging where the current flows between electrodes. As a consequence the resistivity is measured directly in an induction log whereas in a normal electrical log, since the current flows across the stratal boundaries, it is measured indirectly from the electrical log curves. A gamma-ray log (see below) is usually run with an induction log in order to reveal the boundaries of stratal units.

A spontaneous potential (SP) log is obtained by lowering a sonde down a drillhole which generates a small electric voltage at the boundaries of permeable rock units and especially between such strata and less permeable beds. For example, permeable sandstones show large SPs, whereas shales are typically represented by low values. If sandstone and shale are interbedded, then the SP curve has numerous troughs separated by sharp or rounded peaks, the widths of which vary in proportion to the thicknesses of the sandstones. Spontaneous potential logs are frequently recorded at the same time as resistivity logs. Interpretation of both sets of curves yields precise data on the depth, thickness, and position in the sequence of the beds penetrated by the drillhole. The curves also enable a semiquantitative assessment of lithological and hydrogeological characteristics to be made.

The sonic logging device consists of a transmitter–receiver system, transmitter(s) and receiver(s) being located at given positions on the sonde. The transmitters emit short high-frequency pulses several times a second, and differences in travel times between receivers are recorded in order to obtain the velocities of the refracted waves. The velocity of sonic waves propagated in sedimentary rocks is largely a function of the character of the matrix. Normally beds with high porosities have low velocities, and dense rocks are typified by high velocities. Hence the porosity of strata can be assessed. In the 3-D sonic log one transmitter and one receiver are used at a time. This allows both compressional and shear waves to be recorded, from which, if density values are available, the dynamic elastic moduli of the beds concerned can be determined. As velocity values vary independently of resistivity or radioactivity, the sonic log permits differentiation amongst strata which may be less evident on the other types of log.

The televiewer emits pulses of ultrasonic energy from a piezoelectric transducer which are then reflected by the fluid in the drillhole wall rock to be picked up by the transducer, which also acts as a receiver. The transducer is orientated relative to the

Earth's magnetic field by a down-the-hole magnetometer in the sonde and is rotated within the drillhole at 3 rev/s. In this way data can be obtained relating to the orientation of discontinuities within the drillhole wall (dip angles can be delineated down to 20° and dip directions down to values of 15°).

Radioactive logs include gamma-ray or natural gamma, gamma–gamma or formation density, and neutron logs. They have the advantage of being obtainable through the casing in a drillhole. On the other hand, the various electric and sonic logs, with the exception of interborehole acoustic scanning, can only be used in uncased holes. The natural gamma log provides a record of the natural radioactivity or gamma radiation from elements such as potassium 40, and uranium and thorium isotopes in rocks. This radioactivity varies widely among sedimentary rocks, being generally high for clays and shales and lower for sandstones and limestones. Evaporites give very low readings. The gamma–gamma log uses a source of gamma-rays which are sent into the wall of the drillhole. There they collide with electrons in the rocks and thereby lose energy. The returning gamma-ray intensity is recorded, a high value indicating low electron density and hence low formation density. The neutron curve is a recording of the effects caused by bombardment of the strata with neutrons. As the neutrons are absorbed by atoms of hydrogen, which then emit gamma-rays, the log provides an indication of the quantity of hydrogen in the strata around the sonde. The amount of hydrogen is related to the water (or hydrocarbon) content and therefore provides another method of estimating porosity. Since carbon is a good moderator of neutrons, carbonaceous rocks are liable to yield spurious indications as far as porosity is concerned.

The caliper log measures the diameter of a drillhole. Different sedimentary rocks show a greater or lesser ability to stand without collapsing from the walls of the drillhole. For example, limestones may present a relatively smooth face slightly larger than the drilling bit whereas soft shale may cave to produce a much larger diameter. A caliper log is obtained along with other logs to help interpret the characteristics of the rocks in the drillhole.

7.5.6 Cross-hole seismic methods

The cross-hole seismic method is based on the transmission of seismic energy between drillholes. Cross-hole seismic measurements are made between a seismic source in one drillhole (a small explosive charge, an air-gun, a drillhole hammer, or an electrical sparker) and a receiver at the same depth in an adjacent drillhole. The receiver can either be a three-component geophone array clamped to the drillhole wall or a hydrophone, as in interborehole acoustic scanning where a hydrophone is used in a liquid-filled drillhole to receive signals from an electric sparker in another drillhole similarly filled with liquid. The choice of source and receiver is a function of the distance between the drillholes, the required resolution and the properties of the rock mass. The best results are obtained with a high-frequency repetitive source (McCann et al., 1986).

Generally, the source and receiver in the two drillholes are moved up and down together. Drillholes must be spaced closely enough to achieve the required resolution of detail and be within the range of the equipment. This, Grainger and McCann (1977) noted, is up to 400 m in Oxford Clay, 160 m in the Chalk, and 80 m in sands and gravels. By contrast, because soft organic clay is highly attenuating, transmission is only possible over a few metres. These distances are for saturated material and the effective transmission is considerably reduced in dry superficial layers.

Cross-hole seismic measurements provide a means by which the engineering properties of the rock mass between drillholes can be assessed. For example, the dynamic elastic properties can be obtained from the values of the compressional and shear wave velocities, and the formation density. Other applications include assessment of the continuity of lithological units between drillholes, identification of fault zones and assessment of the degree of fracturing, and the detection of subsurface voids.

7.6 TERRAIN EVALUATION

Terrain evaluation is only concerned with the uppermost part of the land surface of the Earth, i.e. land at a depth of less than 6 m, excluding permanent masses of water. Mitchell (1973) described terrain evaluation as involving analysis (simplification of the complex phenomena which make up the natural environment), classification (organization of data in order to distinguish and characterize individual areas), and appraisal (manipulation, interpretation, and assessment of data for practical ends) of an area of the Earth's surface which is of interest to engineers.

In terrain evaluation the initial interpretation of landscape can be made from large-scale maps and aerial photographs. Observation of relief should give particular attention to direction (aspect) and angle of maximum gradient, maximum relief amplitude, and the proportion of the total area occupied by bare rock or slopes. Also, an attempt should be made to interpret the basic geology and the evolution of the landscape. An assessment of the risk of erosion (especially the location of slopes which appear potentially unstable) and the risk of excess deposition of water-borne or wind-blown debris should also be made. Terrain evaluation provides a method whereby the efficiency and accuracy of preliminary surveys can be improved. That is, it allows a subsequent site investigation to be directed towards the relevant problems. It also offers a rational means of correlating known and unknown areas, by applying information and experience gained on one project to a subsequent project. This is based on the fact that landscape systems of terrain evaluation have indicated that landscapes in different parts of the world are sufficiently alike to make predictions from the known to the unknown.

The following units of classification of land have been recognized for purposes of terrain evaluation, in order of decreasing size: (i) land zone; (ii) land division; (iii) land province; (iv) land region; (v) land system; (vi) land facet; and (vii) land element (Brink et al., 1966). The land system, land facet, and land element are the principal units used.

A land systems map shows the subdivision of a region into areas with common physical attributes which differ from those of adjacent areas. Land systems are usually recognized from aerial photographs, the boundaries between different land systems being drawn where there are distinctive differences between landform assemblages. Field work is necessary to confirm the landforms and to identify soils and bedrock.

In order to establish the pattern identified on the aerial photographs as a land system, it is necessary to define the geology and range of small topographic units referred to as land facets. A land system extends to the limits of a geological formation over which it is developed or until the prevailing landforming process gives way and another land system is developed. Land systems maps are usually prepared at scales of 1:500 000 or 1:1 000 000. More detailed maps may be required in complex terrain. They provide the engineer with background information which can be used in a preliminary assessment of the ground conditions in the area with which he is concerned and permit locations to be identified where detailed investigations may prove necessary.

A land system comprises a number of land facets. Each land facet possesses a simple form, generally being developed on a single rock type or superficial deposit. The soils, if not the same throughout the facet, at least vary in a consistent manner. An alluvial fan, a levee, a group of sand-dunes, or a cliff are examples of a land facet. Indeed, geomorphology frequently provides the basis for the identification of land facets. Land facets occur in a given pattern within a land system. They may be mapped from aerial photographs at scales between 1:10 000 and 1:60 000.

A land facet may, in turn, be composed of a small number of land elements, some of which deviate somewhat in a particular property, such as soils, from the general character. They represent the smallest unit of landscape that is normally significant, for example, a hill slope may consist of two land elements, an upper steep slope and a gentle lower slope. Other examples of land elements include small river terraces, gully slopes, and small outcrops of rock.

Although nearly all terrain evaluation mapping is carried out at the land system level, the land region may be used in a large feasibility study for some engineering operation. A land region consists of land systems which possess the basic geological composition and have an overall similarity of landforms. Land regions are usually mapped at a scale between 1:1 000 000 and 1:5 000 000.

7.7 MAPS FOR ENGINEERING PURPOSES

Engineering geological maps and plans are used mainly for planning and civil engineering purposes. They provide planners and engineers with information which assists them in the planning of land-use, and the location, construction, and maintenance of engineering structures of all types. Engineering geological maps usually are produced on the scale of 1:10 000 or smaller whereas engineering geological plans, being produced for a particular engineering purpose, have a larger scale (Anon, 1972).

Engineering geological maps may serve a special purpose or a multipurpose (Anon, 1976). Special-purpose maps provide information on one specific aspect of engineering geology such as grade of weathering, jointing patterns, mass permeability, or foundation conditions (Fig. 7.31). On the other hand, special purpose maps may serve one particular purpose, the engineering geological conditions at a dam site or along a routeway, or for zoning for land-use in urban development. Multipurpose maps cover various aspects of engineering geology and provide information for planning or engineering purposes.

In addition, engineering geological maps may be analytical or comprehensive. Analytical maps provide details, or evaluate individual components, of the geological environment. Examples of such maps include those showing the degree of weathering or seismic hazard. Comprehensive maps either depict all the principal components of the engineering geological environment or are maps of engineering geological zoning, delineating individual territorial units on a basis of uniformity of the most significant attributes of their engineering geological character.

Engineering geology maps frequently consist of basic geological maps on which some engineering geological data have been incorporated, the Belfast sheet produced by the British Geological Survey is such a map. In addition to the basic geology it includes isopachytes of the local estuarine clays and contours to rockhead. Detailed engineering geological information is given on such maps in tabular form (Fig. 7.32; Tables 7.7a and b).

Geotechnical maps and plans indicate the distribution of units, defined in terms of engineering properties, for example, they can be produced in terms of index properties, rock quality, or grade of weathering. A plan for a foundation could be made in terms of design parameters. The unit boundaries are then drawn for changes in the particular property. Frequently the boundaries of such units coincide with stratigraphical boundaries. In other instances, for example, where rocks are deeply weathered, they may bear no relation to geological boundaries. Unfortunately one of the fundamental difficulties in preparing geotechnical maps arises from the fact that changes in physical properties of rocks and soils are frequently gradational. As a consequence regular checking of visual observations by *in situ* testing or sampling is essential to produce a map based on engineering properties.

Frequently, it is impossible to represent all the engineering geological data obtained on one map. In such instances a series of overlays or an atlas of maps can be produced. For example, De Beer *et al.* (1980) described the production of a geotechnical 'atlas' of certain urban areas in Belgium. This 'atlas' comprised: a documentation map (this is a topographical map which shows the location of boreholes

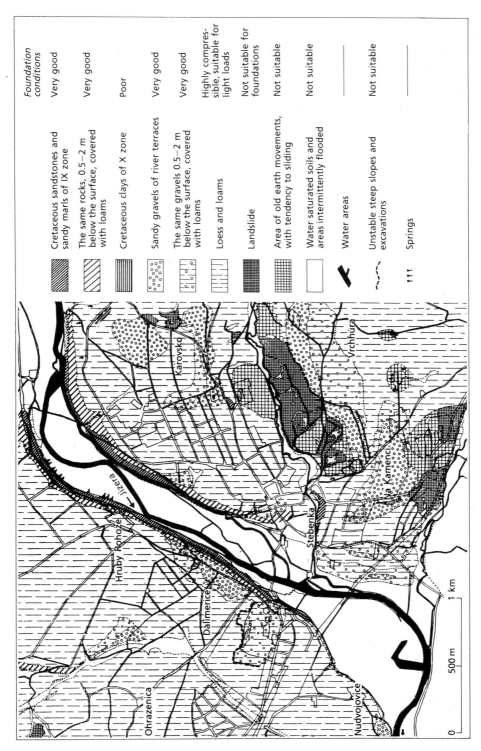

	Foundation conditions
Cretaceous sandstones and sandy marls of IX zone	Very good
The same rocks, 0.5–2 m below the surface, covered with loams	Very good
Cretaceous clays of X zone	Poor
Sandy gravels of river terraces	Very good
The same gravels 0.5–2 m below the surface, covered with loams	Very good
Loess and loams	Highly compressible, suitable for light loads
Landslide	Not suitable for foundations
Area of old earth movements, with tendency to sliding	Not suitable
Water saturated soils and areas intermittently flooded	Not suitable
Water areas	
Unstable steep slopes and excavations	Not suitable
Springs	

Fig. 7.31 Engineering geological map of Turnov, Czechoslovakia. (After Dearman, 1991.)

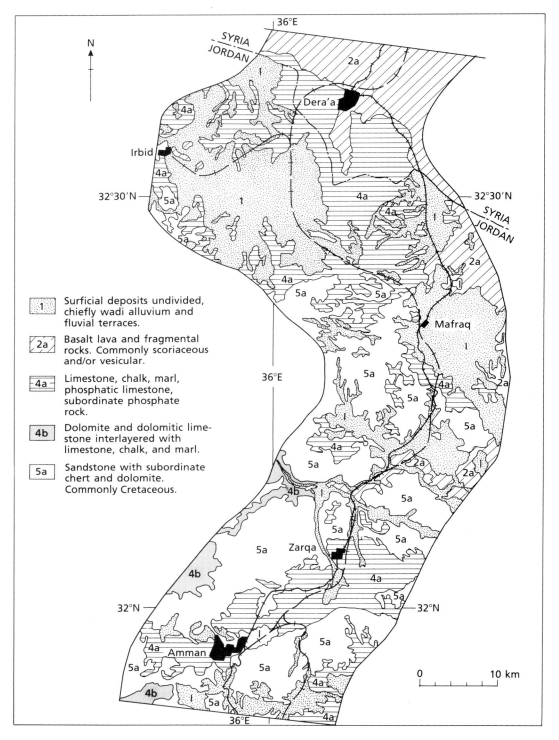

Fig. 7.32 Segment of the engineering geological map for the Hijaz railway in Jordan. Rock unit numbers are explained above. (After Briggs, 1987.)

Table 7.7(a) Excerpts from the engineering geology table illustrating the variety of materials in the study area for Hijaz railway. Symbols 1, 2a, 4a, 4b, and 5a occur in the area shown in Figure 7.32. **Table 7.7(b)** Key to the engineering characteristics column. (After Briggs, 1987.)

Map symbol	Geological description	Distribution	Map segments	Engineering characteristics	Suitability as source of material for:	Moderate water-supply favourability in shallow aquifers	Topographic expression
1	Surficial deposits undivided, chiefly wadi alluvium and fluvial and marine terraces	Most common in Saudi Arabia and southern Jordan	Present on most map segments	Excavation: easy Stability: poor Strength: fair Tunnel support: maximum	Ballast – 0 Coarse aggregate – + Sand – +++ Embankments – 0 Riprap – 0	Fair to good with seasonal fluctuations. Coastal areas poor	Generally flat, locally steeply dissected
2a	Basalt lava and fragmental rocks. Commonly scoriaceous and/or vesicular	Widespread in southern Syria. Locally elsewhere	01, 02, 13, 19, 20, and 28	Excavation: difficult Stability: good Strength: good Tunnel support: moderate	Ballast – + Coarse aggregate – + Sand – + Embankments – ++ Riprap – +	Generally poor. Locally fair to good, depending on interlayering	Flat to mountainous. Surfaces commonly bouldery
3c	Sandstone and conglomerate with limestone and marl. Loosely cemented. Locally hard	Along coastal plain between Al Wajh and Yanbu, Saudi Arabia	26, 27, 28, and 30	Excavation: intermediate Stability: fair Strength: fair Tunnel support: moderate to maximum	Ballast – 0 Coarse aggregate – 0 Sand – + Embankments – ++ Riprap – 0	Poor	Flat to rolling, locally hilly and dissected
4a	Limestone, chalk, marl, phosphatic limestone, subordinate phosphate rock	Widespread in Jordan	02–05 and 29	Excavation: difficult Stability: fair to good Strength: fair to good Tunnel support: moderate to minimum	Ballast – 0 Coarse aggregate – + Sand – 0 Embankments – + Riprap – +	Generally poor	Hilly, locally rolling or mountainous

Continued on p. 242

Table 7.7(a) *Continued*

Map symbol	Geological description	Distribution	Map segments	Engineering characteristics	Suitability as source of material for:	Moderate water-supply favourability in shallow aquifers	Topographic expression
4b	Dolomite and dolomitic limestone interlayered with limestone, chalk, and marl	Central Jordan	02 and 03	Excavation: moderately difficult Stability: fair to good Strength: fair to good Tunnel support: moderate to minimum	Ballast — + Coarse aggregate — + Sand — 0 Embankments — ++ Riprap — +	Generally poor	Hilly, locally mountainous
5a	Sandstone with subordinate chert and dolomite. Commonly calcareous	Widespread in Jordan	02–05 and 29	Excavation: moderately difficult Stability: fair to good Strength: good Tunnel support: moderate to minimum	Ballast — 0 Coarse aggregate — 0 Sand — + Embankments — ++ Riprap — 0	Poor to fair	Hilly to mountainous
6b	Chiefly andesite lava and fragmental rocks. Common medium-grade metamorphism, greenstone	Widespread in Hijaz Mountains	10–13, 18–25, 27, 30, and 31	Excavation: difficult Stability: good Strength: good Tunnel support: minimum	Ballast — +++ Coarse aggregate — +++ Sand — 0 Embankments — ++ Riprap — +++	Poor	Core of Hijaz Mountains. Relief locally greater than 2000 m
7b	Early and altered granites, granodiorite, quartz monzonite. Includes some gneiss	Common in the Hijaz Mountains and southern Jordan	10–13 and 19–31	Excavation: difficult Stability: good Strength: good Tunnel support: minimum to moderate	Ballast — + Coarse aggregate — + Sand — + Embankments — ++ Riprap — ++	Poor	Chiefly mountainous. Mostly more resistant than other intrusive rocks

Table 7.7(b) Key to the engineering characteristics column of the engineering geology table (Table 7.7a)

Excavation facility	Stability of cut slopes	Foundation strength	Tunnel support requirements
Easy — can be excavated by hand tools or light power equipment. Some large boulders may require drilling and blasting for their removal. Dewatering and bracing of deep excavation walls may be required	Good — these rocks have been observed to stand on essentially vertical cuts where jointing and fracturing are at a minimum. However, moderately close jointing or fracturing is common, so slopes not steeper than 4:1 (vertical: horizontal) are recommended. In deep cuts debris-catching benches are recommended	Good — bearing capacity is sufficient for the heaviest classes of construction, except where located on intensely fractured or jointed zones striking parallel to and near moderate to steep slopes	Minimum — support probably required for less than 10% of length of bore, except where extensively fractured
Moderately easy — probably rippable by heavy power equipment at least to weathered rock–fresh rock interface and locally to greater depth	Fair — cut slopes ranging from 2:1 to 1:1 are recommended; flatter where rocks are intensely jointed or fractured. Rockfall may be frequent if steeper cuts are made. Locally, lenses of harder rock may permit steeper cuts	Fair — choice of foundation styles is largely dependent on packing of fragments, clay content, and relation to the water table. If content of saturated clay is high, appreciable lateral movement of clay may be expected under heavy loads. If packing is poor, settling may occur	Moderate — support may be required for as much as 50% of length of bore, more where extensively fractured
Intermediate — probably rippable by heavy power equipment to depths chiefly limited by the manoeuvrability of the equipment. Hard rock layers or zones of hard rock may require drilling and blasting	Poor — flatter slopes are recommended. Some deposits commonly exhibit a deceptive temporary stability, sometimes standing on vertical or near-vertical cuts for periods ranging from hours to more than a year	Poor — foundations set in underlying bedrock are recommended for heavy construction, with precautions taken to guard against failure due to lateral stress	Maximum — support probably required for entire length of bore
Moderately difficult — probably require drilling and blasting for most deep excavations, but locally may be ripped to depths of several metres			
Difficult — probably require drilling and blasting in most excavations except where extensively fractured or altered			

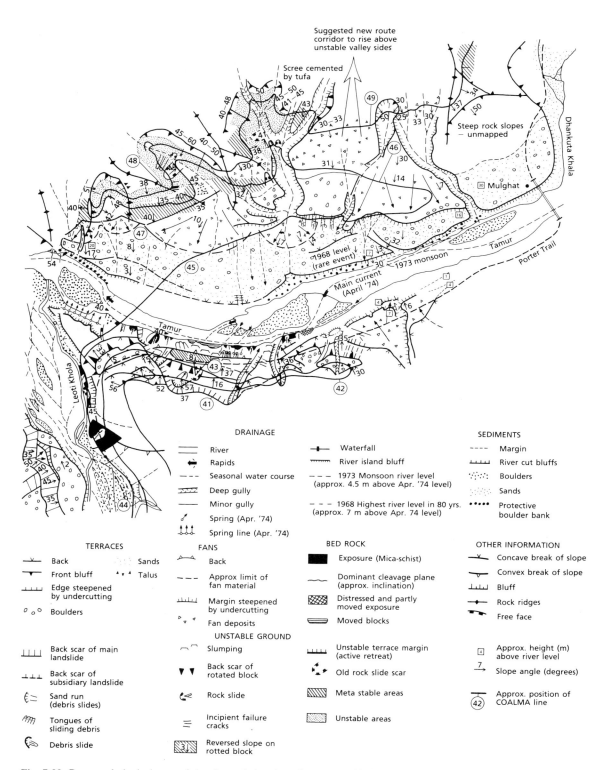

Fig. 7.33 Geomorphological map of the site and situation of a proposed bridge crossing on the Tamur River, eastern Nepal. Scale approximately 1:12 500. (After Brunsden *et al.*, 1975.)

and the data derived therefrom); an individual map showing the isopachytes of the upper surface of the formation in question; a hydrogeological map; a map of engineering geological zones; a number of engineering geological cross sections; and an explanatory key.

The purpose of engineering geomorphological maps is to portray the forms of the surface, the nature and properties of the materials of which the surface is composed, and to indicate the type and magnitude of the processes in operation. Surface form and aerial pattern of geomorphological processes often influence the choice of a site. Hence geomorphological maps give a rapid appreciation of the nature of the ground and thereby help the design of more detailed investigations, as well as focusing attention on problem areas. Such maps recognize landforms along with their delimitation in terms of size and shape (Fig. 7.33). The principal object during a reconnaissance survey is the classification of every component of the land surface in relation to its origin, present evolution, and likely material properties. Further precision can be afforded geomorphological interpretations by obtaining details from climatic, hydrological, or other records and by analysis of the stability of landforms. What is more an understanding of the past and present development of an area is likely to aid prediction of its behaviour during and after construction operations. Engineering geomorphological maps should therefore show how surface expression will influence an engineering project and should provide an indication of the general environmental relationship of the site concerned.

The aims of engineering geomorphological surveys are:

1 identification of the general characteristics of the terrain of an area, thereby providing a basis for evaluation of alternative locations and avoidance of the worst hazard areas;

2 identification of factors outside the site which may influence it, such as mass movement;

3 provision of a synopsis of geomorphological development of the area which includes:

(a) a description of the extent and degree of weathering;

(b) a classification of slopes based on their steepness, material composition, mode of development, and stability;

(c) a description of the location, pattern, and magnitude of the surface and subsurface drainage features (including karst development);

(d) definition of the shape and extent of geomorphological units such as fans, scree slopes, terraces, etc;

(e) recognition of specific hazards such as flooding and landslides;

4 location of suitable supplies of construction materials.

Obtaining such information should facilitate the planning of a subsequent site investigation, e.g. it should aid the location of boreholes, and these hopefully will confirm what has been discovered by the geomorphological survey.

If engineers are to obtain maximum advantage from a geomorphological survey, then derivative maps should be compiled from the geomorphological sheets. Such derivative maps generally are concerned with some aspect of ground conditions, such as landslip areas or areas prone to flooding or over which sand-dunes migrate (see Fig. 8.16).

8 Geology and planning

8.1 INTRODUCTION

The ultimate objective of planning is to determine a particular course of action and so involves attempting to resolve perceived problems. Although the policy which develops from planning embodies a particular course of action, planning proposals are often controversial in that they may offend one or more sections of the community. Hence, in the last analysis, planning policies are the prerogatives of government since legislation is necessary to put them into effect.

Land-use planning represents an attempt to resolve conflicts between man's need to utilize land and his need to protect the environment. Hence planners have to assess the advantages and disadvantages, costs and benefits of development. Therefore land-use planning involves the collection and evaluation of relevant data from which plans can be formulated. The policies which result depend on economic, sociological, and political influences in addition to the perception of the problem. In this context sufficient geological data should be provided to planners and engineers so that, ideally, they can develop the environment in harmony with nature. As indicated in Figure 8.1, geological information is required at all levels of planning and development from the initial identification of a social need to the construction stage. Even after construction, further involvement may be necessary in the form of advice on hazard monitoring, maintenance, or remedial works.

Over recent years public concern regarding the alteration and degradation of the environment has caused governmental and planning authorities to become more aware of the adverse effects of indiscriminate development. As a result laws have been passed to help protect the environment from spoliation. Most policies which deal with land use are concerned with either those processes which re-

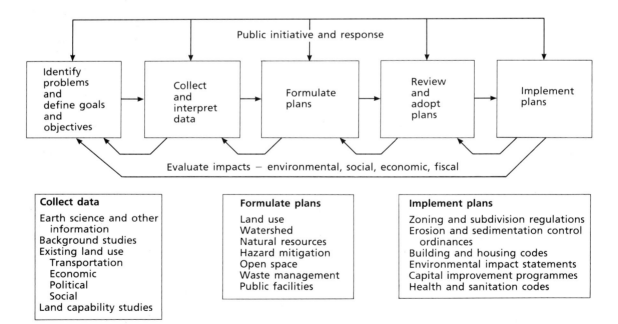

Fig. 8.1 Diagram of the land-use planning process.

present threats to life, health, or property including, for instance, hazardous events, pollution of air or water; or the exploitation, protection, or conservation of natural resources or the restoration of despoiled areas.

Since land use inevitably involves the different development of particular areas, some type of land classification(s) constitutes the basis on which land-use planning is carried out. However, land should also be graded according to its potential uses and capabilities; that is, indices are required to assess the environmental status of natural resources and their potential. Such indices should establish limits, trends, and thresholds, as well as providing insight that offers some measure of success of national and municipal programmes dealing with environmental problems.

Legget (1987) stressed the importance of geology in planning physical facilities and individual structures, and in the wise and best use of land. He pointedly noted the obvious, that land is the surface expression of underlying geology so that land-use planning can only be done with satisfaction if there is a proper understanding of the geology concerned. Legget further stated that the development of land must be planned with the full realization of the natural forces which have brought it to its present state, taking into account the dynamic character of nature so that development does not upset the delicate balance any more than is essential. Geology must therefore be the starting point of all planning.

It is therefore important that geological information should be readily understood by the planner. Unfortunately the conventional geological map often is inadequate for the needs of planners, developers, and civil engineers. Recently, however, various types of maps incorporating geological data have been produced for planning purposes. Such maps should be essentially simple and provide some indication of those areas where there are least constraints on development. Hazard maps (e.g. maps of rapid mass movement in relation to planning (Blikra, 1990)) and vulnerability maps (e.g. for recognition of areas which offer potential risk to groundwater supply if developed (Fobe and Goossens, 1990)) offer such examples.

Hazard zoning maps usually provide some indication of the degree of risk involved with a particular geological hazard (Seeley and West, 1990). An example of the production of hazard maps has been provided by Soule (1980). He outlined a method of mapping areas prone to geological hazards, by using map units based primarily on the nature of the potential hazards associated with them. The resultant maps, together with their explanation, are combined with a land-use/geological hazard-area matrix which provides some idea of the engineering problems which may arise in the area represented by the individual map. For instance, the matrix indicates the effects of any changes in slope or the mechanical properties of rocks or soils, and attempts to evaluate the severity of hazard for various land uses. As an illustration of this method, Soule used a landslide hazard map of the Crested Butte–Gunniston area, Colorado (Fig. 8.2). This map attempts to show which factors within individual map units have the most significance as far as potential hazard is concerned. The accompanying matrix outlines the problems likely to be encountered as a result of human activity (Fig. 8.2). In addition, environmental geology maps (EGMs) have been devised to meet the needs of planners (Robinson and Spieker. 1978; Culshaw et al., 1990). The production of comprehensive suites of EGMs in Britain began in the 1980s following a pilot study of the Glenrothes area of Fife, Scotland. This study produced 27 separate maps covering such aspects as stratigraphy and lithology of bedrock, superficial deposits, rockhead contours, engineering properties, mineral resources and workings, groundwater conditions, and landslip potential (Nickless, 1982). The maps primarily were for use of local and central planners but also were useful sources of information for civil engineers, developers, and mineral extraction companies. The main feature of the study was the presentation of each element of the geology on a separate map in a way that was easy for non-geologists to understand. Another feature was the interpretation of these elements so as to provide data other than mere outcrop distribution and lithology. Environmental potential maps are compiled from basic data maps and derived maps. They present, in general terms, the constraints on development such as areas with poor foundation conditions, land susceptible to land-slipping or subsidence, or land likely to be subjected to flooding (Fig. 8.3). They also can present those resources with respect to mineral, groundwater, or agricultural potential which might be used in development or which should not be sterilized by building over, or contamination from landfill sites. Again these maps can be readily understood by non-geologists.

Similar developments have taken place in the Netherlands, for instance, Mulder and Hillen (1990)

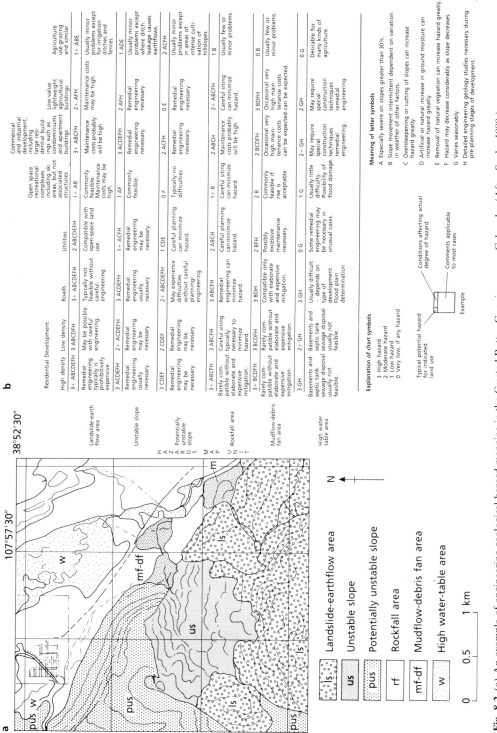

Fig. 8.2 (a) An example of engineering geological hazard mapping in the Crested Butte–Gunnison area, Colorado. (After Soule, 1980.) (b) The matrix is formated so as to indicate to the map user that several geological and geology-related factors should be considered when contemplating the land use in a given type of mapped area. This matrix can also serve to recommend additional types of engineering geological studies that may be needed for a site. Thus the map can be used to model or anticipate the kinds of problems that a land-use planner or land developer may have to overcome before a particular activity is permitted or undertaken.

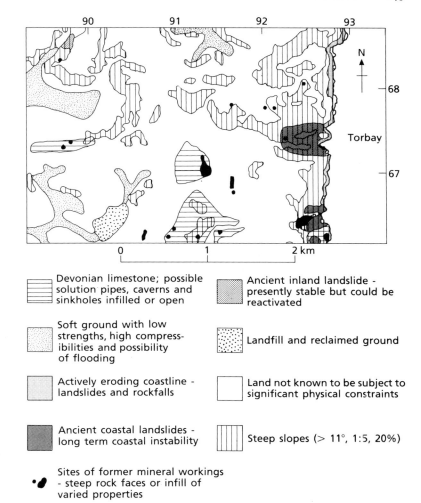

Fig. 8.3 Ground characteristics for planning and development for part of the Torbay area of south-west England. (After Culshaw *et al.*, 1990.)

Devonian limestone; possible solution pipes, caverns and sinkholes infilled or open

Soft ground with low strengths, high compress-ibilities and possibility of flooding

Actively eroding coastline - landslides and rockfalls

Ancient coastal landslides - long term coastal instability

Sites of former mineral workings - steep rock faces or infill of varied properties

Ancient inland landslide - presently stable but could be reactivated

Landfill and reclaimed ground

Land not known to be subject to significant physical constraints

Steep slopes (> 11°, 1:5, 20%)

described how sets of engineering geological maps of certain areas have been produced to help planning. These sets include a location map, a geotechnical cross-section of the area, multipurpose compre-hensive maps (e.g. sequence of soil layers as well as thickness and depth of pertinent individual layers (Fig. 8.4a and b)), and special maps (e.g. amount of settlement under load (Fig. 8.4c) foundation depth and hydrogeological environmental geology).

Wolden and Erichsen (1990) introduced the con-cept of geoplan maps which outline the relevant geological factors likely to influence planning (Fig. 8.5). These are similar to environmental potential maps. They maintained that such maps should help planners avoid making poor decisions at an early stage and would allow areas to be developed se-

quentially (e.g. mineral resources could be devel-oped before an area is built over). Such maps would facilitate the development of a multi-use plan of a particular area and avoid one factor adversely affecting another.

Many older industrial and urban areas are under-going redevelopment and therefore there is a need to plan the new environment so that land with difficult ground conditions is avoided by building development or the cost of building on it is fully appreciated, and that natural resources such as sand and gravel deposits are not sterilized or groundwater supplies contaminated. Hence planners require relevant information. In many urban areas data from site investigations are available and Forster and Culshaw (1990) have described how these can

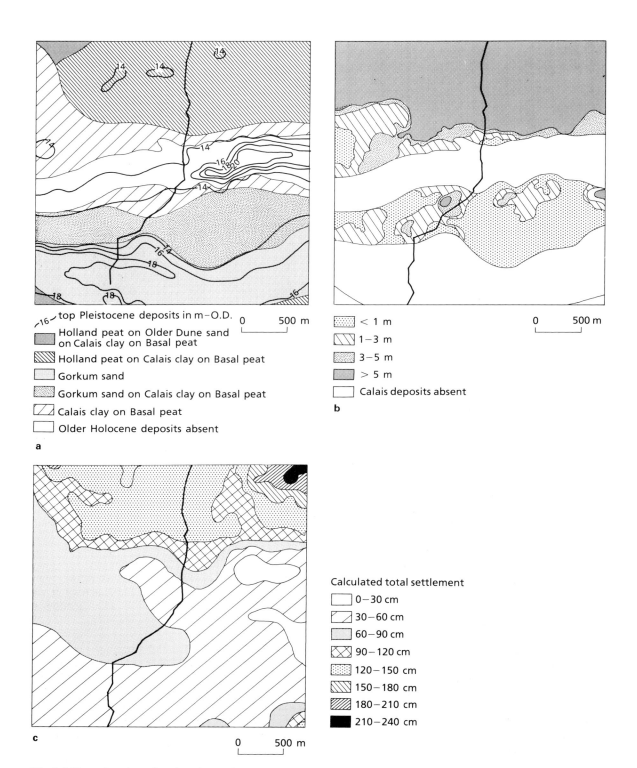

Fig. 8.4 Examples of set of engineering geology maps of Leiden, the Netherlands; (a) multipurpose comprehensive map of older Holocene deposits; (b) isopach map of Calais deposits (the most notable formation in the Leiden area); (c) induced settlement map for an imaginary surface load of 100 kPa. (After Mulder and Hillen, 1990.)

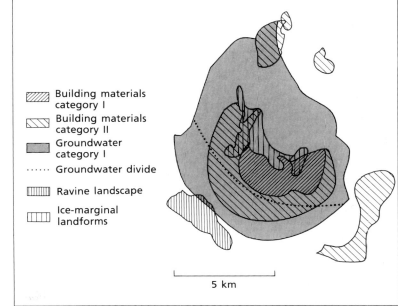

Fig. 8.5 Geoplan map. Building materials category I indicates that the sand and gravel deposits are of good quality and well documented. Building materials category II indicates that the deposits can be exploited but the area is not as well documented. Groundwater category I is an important exploitable resource with a capacity of 200 l/s or more. The ravine landscape and ice-marginal landforms represent areas recommended for conservation. (After Wolden and Erichsen, 1990.)

Legend:
- Building materials category I
- Building materials category II
- Groundwater category I
- Groundwater divide
- Ravine landscape
- Ice-marginal landforms

5 km

be made use of to prepare engineering geological maps for planners.

Laws are now in force in many countries which demand the investigation and evaluation of the consequences to the environment of all large engineering works. An environmental impact statement therefore usually involves a description of the proposed scheme, its impact on the environment, particularly noting any adverse effects, and alternatives to the proposed scheme. The aim of the environmental impact process is to improve the effectiveness of existing planning policy. It should bring together the requirements of the development and the constraints of the environment so that conflicts can be minimized. Leopold *et al.* (1971) devised a matrix system of analysis which acts as a super-checklist for those involved in the preparation of these statements and also tends to make assessment more objective. Their approach involved: (i) quantification of data; (ii) establishment of cause and effect relationships; and (iii) weighting of impacts. Geologists are invariably involved in environmental impact investigations.

As can be inferred from above, one of the aspects of planning which intimately involves geology is the control or reduction of the effects of geological processes or hazards which mitigate against the interests of man. The development of planning

policies for dealing with hazards in particular, requires an assessment of the severity, extent, and frequency of the hazard in order to evaluate the degree of risk. Once this has been accomplished methods whereby the risk can be reduced have to be investigated and evaluated in terms of public costs and benefits. The risks associated with geological hazards may be reduced, for instance, by control measures carried out against the hazard-producing agent; monitoring and warning systems which allow evacuation; restrictions on development of land; and the use of appropriate building codes together with structural reinforcement of property. In addition, the character of the ground conditions can affect both the viability and implementation of planning proposals.

The engineering design of structures is based on detailed measurement and interpretation of the behaviour of the soils and rocks liable to be affected by construction work. In this context probably the most significant features of soils and rocks are their capacity for bearing loads without excessive settlement, their volume change characteristics, their excavatability, and their intrinsic stability. The engineering behaviour of soils and rocks is a product of their composition, microfabric, and mass structure (Chapter 5). In turn, these features are functions of their mode of formation and subsequent history. It is important to appreciate that due to geological processes, especially weathering, the engineering properties of soils and rocks change with time. The nature, extent, and rate of occurrence of these changes are influenced by the character of the geological material concerned and the environmental conditions.

8.2 CONSERVATION, RESTORATION, AND RECLAMATION OF LAND

Conservation is concerned with safeguarding natural phenomena and the preservation and improvement of the quality of the environment. As such it is closely associated with the geology as geology and environment are intimately interrelated. However, the interests of conservation frequently are in conflict with social and economic pressures for development and so one of the roles of planning authorities is to balance the demands of urban, industrial, and infra-structural development with the need to preserve the countryside and areas of scientific and cultural interest. Hence, the geologist may find himself in conflict with the conservationist, especially in the case of the mineral extraction industry. On the other hand, geological knowledge should be made use of in conservation programmes since these should seek to make the wisest use of natural resources and in this regard there is no conflict with the geologist.

Conservation is not simply preservation. It seeks to improve existing conditions rather than simply maintaining the status quo. Hence, conservation does involve the reconciliation of differing views so that the best compromise can be reached. Obviously in this context land use is important. In some instances the same land can be used simultaneously to cater for several needs, whilst in others the uses to which it can be put are consecutive rather than concurrent and the final effect can be either to restore the land to its original use or to a new use which forms an acceptable part of the landscape. Thus, a sequence of events must be planned to ensure the greatest efficiency in the use of resources and that the most acceptable final state is achieved. Hence, in such situations, the geologist must be involved with planning from the onset otherwise natural resources may be sterilized as a consequence of premature, alternative development. The classic examples are the developments of urban areas over valuable surface mineral deposits, which could have been extracted and the areas then developed.

Both developed and developing societies make demands upon land which frequently mean that it is degraded. An appreciable contribution to the general economy can be made by bringing this derelict land back into worthwhile use. Whatever the ultimate use to which the land is put after restoration, it is imperative that it should fit the needs of the surrounding area and be compatible with other forms of land use that occur in the neighbourhood. Accordingly, the planning of the eventual land use must endeavour to include the integration of the restored area into the surrounding landscape.

Two of the major causes of the dereliction of land are the surface extraction of minerals and the disposal of waste products from mineral workings in the form of spoil heaps and tailings lagoons. Mineral extraction may also cause water pollution. In practice, the restoration and conservation of derelict sites involves the use of planning controls. Hence conditions are attached to planning permissions granting mineral developments and these apply not only to the form of development permitted but also to the ultimate state of the land after extraction ceases. For example, the after-use of some sites where minerals have been worked has caused problems, notably when pollution has resulted from their use for the disposal of toxic wastes. On the other hand, the use of such sites as recreational areas has provided additional amenities for the communities they serve. Again this emphasizes the need for thorough planning.

A preliminary reconnaissance of a derelict site is desirable to determine the sequence of work for the site survey and investigation. The exact boundaries of the site and the various physical features it contains are recorded during the survey and boreholes may be sunk to assess its geological character. The data gathered allows plans to be drawn and the restoration project to be designed.

The reclamation of land involves its upgrading to a use which is considered more beneficial to the community. The reclamation of swamplands and marshlands by drainage so that they can be used for agricultural or other purposes illustrates this. Some impressive examples of reclamation are provided by the various schemes undertaken in the Netherlands by which land has been reclaimed from the sea, the Zuider Zee scheme and the Delta scheme offering noteworthy examples of man's ingenuity.

8.3 GEOLOGICAL HAZARDS

Geological hazards can be responsible for devastating large areas of the land surface and so can pose serious constraints on development. However, geological processes such as volcanic eruptions, earthquakes, landslides, and floods cause disasters only when they impinge upon man or his activities. Nevertheless, both the number of recorded disaster events, together with the number of people killed, are increasing each year. As an example of the financial implications, Burton *et al.* (1978) estimated the cost of natural hazards to be approximately $25 billion a year for losses and $15 billion a year for costs involved in prevention and mitigation measures. They further estimated that some 250 000 lives were lost annually. Geological hazards pose formidable obstacles to economic growth, particularly in developing countries, and consequently preventive measures need to be included in national planning processes in order to avoid severe problems of economic and social dislocation. Unfortunately, however, informal development in many of these countries is placing increasing numbers of people and property at risk, often involving use of marginal and high-risk zones. In particular, overcultivation and deforestation aggravate the problems. Even so the associated problems of soil erosion and excessive runoff have not been confined to the developing world, for example, such problems occurred in the United States, notably in the 1930s.

The effects of a disaster may be lessened by reduction of vulnerability. Short-term forecasts a few days ahead of the event may be possible and complement relief and rehabilitation planning. In addition, it is possible to reduce the risk of disaster by a combination of preventive and mitigative measures. To do this successfully, the patterns of behaviour of the geological phenomena posing the hazards need to be understood and the areas at risk identified. Then the level of potential risk may be decreased and the consequence of disastrous events mitigated against by introducing regulatory measures or other inducements into the physical planning process. The impact of disasters may be reduced further by incorporating, into building codes and other regulations, appropriate measures so that structures will withstand or accommodate potentially devastating events.

Land-use planning for the prevention and mitigation of geological hazards should be based on criteria establishing the nature and degree of the risks present and their potential impact. Both the probable intensity and frequency of the hazard(s), and the susceptibility (or probability) of damage to

human activities in the face of such hazards are integral components of risk assessment. Vulnerability analyses comprising risk identification and evaluation should be carried out in order to make rational decisions on how best the effects of potentially disastrous events can be reduced or overcome through systems of permanent controls on land development. The geological data needed in planning and decision making, as summarized by Hays and Shearer (1981), are given in Table 8.1.

8.3.1 Volcanic activity

Volcanic eruptions and other manifestations of volcanic activity are variable in type, magnitude, duration, and significance as a hazard. Major zones of volcanic activity lie beneath the central parts of the oceans and other, more minor ones, transect certain of the continental plates (Fig. 1.5), for example, the oceanic rifts give rise to the volcanism which occurs in many ocean islands. Other oceanic islands are formed by volcanic activity associated with hot spots in the oceanic plates. Volcanic activity is also associated with continental rifts, notably the East African rift system. Volcanism also occurs as a consequence of mountain-building processes so that many currently active volcanoes are situated in recently formed mountain chains. Examples of volcanoes of this type include those around the Pacific Ocean and the Mediterranean Sea.

The lava produced by a volcano varies according to the geological situation so that although volcanic activity may take many forms, it is possible to distinguish two main styles of activity. Generally speaking, the activity associated with the oceanic rift systems entails the relatively gentle upwelling of lava from fissure vents and some central cone type volcanoes. Since the lava has low viscosity, the eruptions tend to be non-violent in nature although large quantities of ash and lava may be thrown into the air (Fig. 8.6) and the lava may flow large distances before it cools. In contrast, the activity associated with recently formed mountain ranges tends to be very explosive in nature. Owing to the high viscosity of the lava, volcanic vents can become blocked so that subterranean pressures build-up. The sudden release of these pressures during an eruption can give rise to *nuées ardentes*, localized seismicity, and the ejection of large quantities of material to great heights. Volcanoes of this type tend to attain a conelike form.

In any assessment of risk due to volcanic activity

Table 8.1 Data required to reduce losses from geological hazards. (After Hays and Shearer, 1981.)

Reduction decisions	Technical information needed about the hazards from earthquakes, floods, ground failures, and volcanic eruptions
Avoidance	Where has the hazard occurred in the past? Where is it occurring now? Where is it predicted to occur in the future?
Land-use zoning	What is the frequency of occurrence? Where has the hazard occurred in the past? Where is it occurring now? Where is it predicted to occur in the future? What is the frequency of occurrence? What is the physical cause? What are the physical effects of the hazard? How do the physical effects vary within an area? What zoning within the area will lead to reduced losses to certain types of construction?
Engineering design	Where has the hazard occurred in the past? Where is it occurring now? Where is it predicted to occur in the future? What is the frequency of occurrence? What is the physical cause? What are the physical effects of the hazard? How do the physical effects vary within an area? What engineering design methods and techniques will improve the capability of the site and the structure to withstand the physical effects of a hazard in accordance with the level of acceptable risk?
Distribution of losses	Where has the hazard occurred in the past? Where is it occurring now? Where is it predicted to occur in the future? What is the frequency of occurrence? What is the physical cause? What are the physical effects of the hazard? How do the physical effects vary within an area? What zoning has been implemented in the area? What engineering design methods and techniques have been adopted in the area to improve the capability of the structure to withstand the physical effects of a hazard in accordance with the level of acceptable risk? What annual loss is expected in the area? What is the maximum probable annual loss?

the number of lives at stake, the capital value of the property and the productive capacity of the area concerned have to be taken into account. Evacuation from danger areas is possible if enough time is available. However, the vulnerability of property is frequently close to 100% in the case of most violent volcanic eruptions. Hazard must also be taken into account in such an assessment. It is a complex function of the probability of eruptions of various intensities at a given volcano and of the location of the site in question with respect to the volcano. Hazard is the most difficult of factors to estimate, mainly because violent eruptions are rare events about which there are insufficient observational data for effective analysis, for instance, in the case of many volcanoes, large eruptions occur at intervals of hundreds or thousands of years.

Most dangerous volcanic phenomena happen very quickly, that is, the time interval between the beginning of an eruption and the appearance of the first *nuées ardentes* may be only a matter of hours. Fortunately, such events are usually preceded by visible signs of impending eruption (Schuster, 1983). No volcanic catastrophes have occurred at the very start of an eruption and this affords a certain length of time to take protective measures. Even so because less than one out of several hundred eruptions proves dangerous for a neighbouring population, evacuation, presuming that an accurate prediction could be made, would not take place before an eruption became alarming.

Baker (1979) estimated that the active lifespan of most volcanoes is probably between one and two million years. Since the activity frequently follows a

Fig. 8.6 Eruption of ash from Kirkefell volcano on the island of Heimaey, Iceland, in 1973. The ash entombed many houses. (Courtesy of the Icelandic Embassy.)

broadly cyclical pattern some benefit in terms of hazards assessment may accrue from the determination of the recurrence interval of particular types of eruption, the distribution of the resulting deposits, the magnitude of events, and recognition of any short-term patterns of activity. According to Booth (1979), four categories of hazard have been distinguished in Italy:

1 Very high frequency events with mean recurrence intervals (MRI) of less than 2 years. The area affected by such events is usually less than $1 km^2$.

2 High frequency events with MRI values of 2–200 years. In this category damage may extend up to $10 km^2$.

3 Low frequency events with MRI values 200–2000 years. Areal damage may cover over $1000 km^2$.

4 Very low frequency events which are associated with the most destructive eruptions and have MRI values in excess of 2000 years. The area affected may be greater than $10000 km^2$.

The return periods of particular types of activity of individual volcanoes or centres of volcanic activity can be obtained by thorough stratigraphic study and dating of the deposits so as to reconstruct past events. Used in conjunction with any available historical records, these data form the bases for assessing the degree of risk involved. However, it is unlikely that these studies could ever be refined sufficiently for the time of activity to be predicted to within a decade. Furthermore, because it is unlikely that man will be capable of significantly influencing the degree of hazard, then the reduction of risk can

only be achieved by reducing exposure of life and property to volcanic hazards.

Booth (1979) divided volcanic hazards into six categories: (i) premonitory earthquakes; (ii) pyroclast falls; (iii) pyroclast flows and surges; (iv) lava flows; (v) structural collapse; and (vi) associated hazards. Each type represents a specific phase of activity during a major eruptive cycle of a polygenetic volcano and may occur singly or in combination with other types. Damage resulting from volcanoseismic activity is rare although intensities on the Mercalli scale varying from VI to IX have been recorded over limited areas.

The likelihood of a certain location being inundated with lava at a given time can be estimated from information relating to the periodicity of eruptions in time and space, the distribution of rift zones on the flanks of a volcano, the topographic constraints on the directions of flow of lavas, and the rate of covering of the volcano by lava. The length of a lava flow is dependent upon the rate of eruption, the viscosity of the lava, and the topography of the area involved. Given the rate of eruption it may be possible to estimate the length of flow. Each new eruption of lava alters the topography of the slopes of a volcano to a certain extent and therefore flow paths may change. Also, prolonged eruptions of lava may eventually surmount obstacles which lie in their path and which act as temporary dams. This may then mean that the lava invades areas which were formerly considered safe.

Hazards associated with volcanic activity also

include destructive floods and mudflows (lahars) caused by sudden melting of snow and ice which cap high volcanoes, or heavy downfalls of rain (vast quantities of steam may be given off during any eruption), or the rapid collapse of a crater lake. Lahars can prove just as destructive as pyroclastic flows and can travel appreciable distances in a matter of minutes. Tsunamis generated by explosive eruptions can prove extremely dangerous. Poisonous or asphyxiating gases are a further threat to life. Air blasts, shock waves, and counter blasts are relatively minor hazards, although windows may be broken several tens of kilometres away from major eruptions.

Detailed geological mapping of the products of volcanic eruption is required in order to produce maps of volcanic hazard zones (Fig. 8.7). Thus, hazard zoning entails the identification of areas liable to be adversely affected by particular types of volcanic eruption during an episode of activity. Events with a mean recurrence interval of less than 5000 years should be taken into account in the production of maps of volcanic hazard zoning and data on any

events which have taken place in the last 50 000 years are probably significant. Two types of maps are useful for economic and social planning. One indicates areas liable to suffer total destruction by lava flows, nueés ardentes, and lahars. The other shows areas likely to be affected temporarily by damaging, but not destructive, phenomena including heavy falls of ash, toxic emissions, and the pollution of surface or underground waters. Examples of volcanic risk maps include expected ash-fall depths, lava and pyroclastic debris-flow paths, and the areal extent of lithic missile fallout. Such maps are needed by local and national governments so that appropriate land uses, building codes, and civil defence responses can be incorporated into planning procedures.

8.3.2 Earthquakes

Although some seismicity is caused by volcanic activity most is due to movements along faults within the Earth's crust. In fact earthquakes have been reported from all parts of the world but they primar-

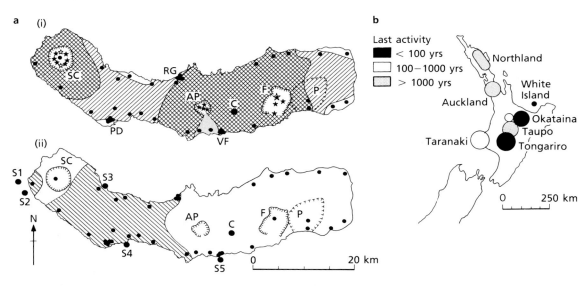

Fig. 8.7 (a) Volcanic hazard maps for Sao Miguel, Azores. (i) Hazard from trachytic eruptions. Light shaded areas liable to burial by over 0.25 m of air-fall pumice ($0.25 \, \mathrm{Mg/m^2}$ damp ash); cross-shaded areas liable to burial by over 1.0 m of air-fall pumice ($1 \, \mathrm{Mg/m^2}$ damp ash); dot-shaded area represents the probable route of hot lahars and/or pyroclast flows. Some principal towns: PD, Ponta Delgade; RG, Ribeira Grande; VF, Villa Franca. Other villages indicated by small black dots. Calderas; SC, Sete Cidades with its six internal craters and two crater lakes; AP, Agua de Pau with its four internal craters; C, Congro diatreme; F, Furnas with its four internal craters, one of which is occupied by Furnas village; P, Povoacao volcano, believed extinct. (ii) Hazard from basaltic eruptions. Shaded area is that in danger of burial by Strombolian ashes or lava extrusions. S1–S5 are the locations of ancient phreato-magmatic (Surtseyan) eruptions. (After Booth, 1979). (b) Volcanic hazard areas of North Island, New Zealand (From Dearman, 1991.)

ily are associated with the margins of the crustal plates which move with respect to each other. Earthquakes are a manifestation of this movement, displacements giving rise to elastic strain within the rock mass on either side of faults. Elastic-strain energy is released in the form of seismic waves.

Initially movement may occur over a small area of the fault plane, to be followed later by slippage over a much larger surface. Such initial movements account for the foreshocks which precede an earthquake. These are followed by the main movement, but complete stability is not restored immediately. The shift of rock masses involved in faulting relieves the main stress but develops new stresses in adjacent areas. Because stress is not relieved evenly everywhere, minor adjustments may arise along the fault plane and thereby generate aftershocks. The decrease in strength of the aftershocks is irregular and occasionally they may continue for a year or more.

The epicentre of an earthquake is located on the surface of the Earth immediately above the focus (i.e. the location of the origin) and shock waves radiate from the focus in all directions. Earthquake foci are confined within a limited zone of the upper Earth, the lower boundary occurring at 700 km depth from the surface. No earthquakes are known to have originated below this level. Moreover, earthquakes rarely originate at the Earth's surface. Because of its significance, the depth of foci has been used as the basis of a threefold classification of earthquakes:
1 those occurring within the upper 70 km are referred to as shallow;

2 those located between 70 and 300 km are intermediate; and
3 those between 300 and 700 km are deep.
Seventy per cent of all earthquakes are of shallow type.

An earthquake propagates three types of shock wave. The first pulses that are recorded on a seismograph are the P or compression waves. The next pulses recorded are the S or shear waves. S waves usually have a larger amplitude than the P waves but the latter travel almost three times as fast as the former. The third type of vibrations are known as the L waves. These waves travel from the focus of the earthquake to the epicentre above, and from there they radiate over the Earth's surface. P waves are not as destructive as S or L waves. This is because they have a smaller amplitude and the force their customary vertical motion creates rarely exceeds the force of gravity. On the other hand, S waves may develop violent tangential vibrations strong enough to cause great destruction. The intensity of an earthquake depends upon the amplitude and frequency of wave motion. S waves commonly have a higher frequency than L waves but the latter may be more powerful because of their larger amplitude.

The hazards due to seismicity include the possibility of a structure being severed by fault displacement but a much more likely event is damage due to shaking (Fig. 8.8). The destruction wrought by an earthquake depends on many factors. Of prime importance are the magnitude of the event, its dur-

Fig. 8.8 Collapse of cast steel workshop of rolling stock plant at Tangshan, China, Tangshan earthquake, 28 July 1976, magnitude 7.8.

ation, and the response of buildings and other elements of the infrastructure. Other hazards such as landslides, floods, subsidence, and secondary earthquakes may be triggered by a seismic event.

The strengths of earthquakes may be expressed in terms of intensity or magnitude. Earthquake intensity scales are a qualitative expression of the damage caused by an event. The most widely accepted intensity grading is the Mercalli scale given in Table 8.2.

The magnitude of an earthquake is an instrumentally measured expression of the energy liberated during the event. Richter (1935) devised a scale in which the maximum amplitude of the resulting seismic waves is expressed on a logarithmic scale. An earthquake of magnitude 2 is the smallest likely to be felt by humans and earthquakes of magnitude 5 or less are unlikely to cause damage to well-constructed buildings. The maximum magnitude of earthquakes is limited by the amount of strain energy which the rock mass can sustain before failure occurs, hence the largest tremors have had a magnitude of 8.9. Such events cause severe damage over wide areas.

The magnitude of an earthquake event depends on the length of the fault break and the amount of displacement which occurs. Generally speaking, movement only occurs on a limited length of a fault during one event. A magnitude 7 earthquake would be produced if a 150 km length of a fault underwent a displacement of about 1.0 m. The length of the fault break during a particular earthquake is generally only a fraction of the true length of the fault. Individual fault breaks during simple earthquakes have ranged in length from less than a kilometre to several hundred kilometres. What is more fault breaks do not only occur in association with large and infrequent earthquakes but also occur in association with small shocks and continuous slow slippage known as fault creep. Fault creep may amount to several millimetres per year and progressively deforms buildings located across such faults.

There is little information available on the frequency of breaking along active faults. All that can be said is that some master faults have suffered repeated movements — in some cases it has recurred in less than 100 years. By contrast much longer intervals, totalling many thousands of years, have occurred between successive breaks. Therefore, because movement has not been recorded in association with a particular fault in an active area, it cannot be concluded that the fault is inactive.

Many seismologists believe that the duration of an earthquake is the most important factor as far as damage or failure of structures, soils, and slopes are concerned. What is important in hazard assessment is the prediction of the duration of seismic shaking above a critical ground acceleration threshold. The magnitude of an earthquake affects the duration much more than it affects the maximum acceleration, since the larger the magnitude the greater the length of ruptured fault. Hence the more extended the area from which the seismic waves are emitted. With increasing distance from the fault the duration of shaking is longer but the intensity of shaking is less, the higher frequency waves being attenuated more than the lower ones.

The physical properties of the soils and rocks through which seismic waves travel, as well as the geological structure, also influence surface ground motion. For example, if a wave traverses vertically through granite overlain by a thick uniform deposit of alluvium, then theoretically the amplitude of the wave at the surface should be double that at the alluvium−granite contact. According to Ambraseys (1974) maximum acceleration within an earthquake source area may exceed 2 g for competent bedrock. However, normally consolidated clays with low plasticity are incapable of transmitting accelerations greater than 0.1 to 0.15 g to the surface. Clays with high plasticity allow accelerations of 0.25 to 0.35 g to pass through. Saturated sandy clays and medium dense sands may transmit 50 to 60% g, and in clean gravel and dry dense sand accelerations may reach much higher values.

The response of structures on different foundation materials has proved surprisingly varied. In general, structures not specifically designed for earthquake loadings have fared far worse on soft, saturated alluvium than on hard rock. This is because motions and accelerations are much greater on deep alluvium than on rock. By contrast, a rigid building may suffer less on alluvium than on rock. The explanation is attributable to the alluvium having a cushioning effect and the motion may be changed to a gentle rocking. This is easier on such a building than the direct effect of earthquake motions experienced on harder ground. Nonetheless alluvial ground beneath any kind of poorly constructed structure facilitates its destruction.

Ground vibrations often lead to the consolidation of cohesionless soils and associated settlement of the ground surface. Loosely packed saturated sands and silts tend to liquefy, thus losing their bearing capacity (Fig. 8.9). If liquefaction occurs in a sloping soil mass, the entire mass will begin to move as a flow slide. Such slides develop in loose, saturated co-

Table 8.2 Modified Mercalli scale (1956 version) with Cancani's equivalent acceleration

Degrees	Description	Acceleration* (mm s^{-2})
I	Not felt. Only detected by seismographs	Less than 2.5
II	Feeble. Felt by persons at rest, on upper floors, or favourably placed	2.5 to 5.0
III	Slightly felt indoors. Hanging objects swing. Vibration like passing of light trucks. Duration estimated. May not be recognized as earthquake	5.0 to 10
IV	Moderate. Hanging objects swing. Vibration like passing of heavy trucks, or sensation or a jolt like a heavy ball striking the walls. Standing motor cars rock. Windows, dishes, doors rattle. Glasses clink. Crockery clashes. In the upper range of IV wooden walls and frames creak	10 to 25
V	Rather strong. Felt outdoors, direction estimated. Sleepers wakened. Liquids disturbed, some spilled. Small unstable objects displaced or upset. Doors swing, close, open. Shutters and pictures move. Pendulum clocks stop, start, change rate	25 to 50
VI	Strong. Felt by all. Many frightened and run outdoors. Persons walk unsteadily. Windows, dishes, glassware broken. Ornaments, books, etc., fall off shelves. Pictures fall off walls. Furniture moved or overturned. Weak plaster and masonry cracked. Small bells ring (church, school). Trees, bushes shaken visibly or heard to rustle	50 to 100
VII	Very strong. Difficult to stand. Noticed by drivers of motor cars. Hanging objects quiver. Furniture broken. Damage to masonry D, including cracks. Weak chimneys broken at roof line. Fall of plaster, loose bricks, stones, tiles, cornices, also unbraced parapets and architectural ornaments. Some cracks in masonry C. Waves on ponds, water turbid with mud. Small slides and caving in along sand or gravel banks. Large bells ring. Concrete irrigation ditches damaged	100 to 250
VIII	Destructive. Steering of motor cars affected. Damage to masonry C, partial collapse. Some damage to masonry B, none to masonry A. Fall of stucco and some masonry walls. Twisting, fall of chimneys, factory stacks, monuments, towers, elevated tanks. Frame houses moved on foundations if not bolted down, loose panel walls thrown out. Decayed piling broken off. Branches broken from trees. Changes in flow or temperature of springs and wells. Cracks in wet ground and on steep slopes	250 to 500
IX	Ruinous. General panic. Masonry D destroyed, masonry C heavily damaged, sometimes with complete collapse, masonry B seriously damaged. General damage to foundations. Frame structures, if not bolted, shifted off foundations. Frames cracked, serious damage to reservoirs. Underground pipes broken. Conspicuous cracks in ground. In alluviated areas sand and mud ejected, earthquake fountains, sand craters	500 to 1000
X	Disastrous. Most masonry and frame structures destroyed with their foundations. Some well-built wooden structures and bridges destroyed. Serious damage to dams, dykes, embankments. Large landslides. Water thrown on banks of canals, rivers, lakes, etc. Sand and mud shifted horizontally on beaches and flat land. Rails bent slightly	1000 to 2500
XI	Very disastrous. Rails bent greatly. Underground pipelines completely out of service	2500 to 5000
XII	Catastrophic. Damage nearly total. Large rock masses displaced. Lines of sight and level distorted. Objects thrown into the air	Over 5000

* These are not peak accelerations as instrumentally recorded; A, B, C, D, USA masonry types.

Fig. 8.9 Intact blocks of flats which foundered into liquefied soils during the Niigata earthquake, Japan, 1964.

hesionless materials during earthquakes. Loose, saturated silts and sands often occur as thin layers underlying firmer materials. In such instances liquefaction of the silt or sand during an earthquake may cause the overlying material to slide over the liquefied layer. Structures on the main slide are frequently moved without suffering damage. However, a graben-like feature often forms at the head of the slide and buildings located in this area are subjected to large differential settlements and often are destroyed. Buildings near the toe of the slide are commonly heaved upwards or are even pushed over by the lateral thrust.

Clay soils do not undergo liquefaction when subjected to earthquake activity. However, it has been shown that under repeated cycles of loading large deformations can develop, although the peak strength remains about the same. Nonetheless these deformations can reach the point where, for all practical purposes, the soil has failed. Major slides can result from failure in deposits of clay.

Details about the occurrence, magnitude, and effects of earthquakes through time can be obtained by carrying out long-term seismic monitoring. In the absence of suitable instrumental data, the seismic history of an area may be established by interpretation of historical, archaeological, and geological evidence.

Maps can be drawn (Fig. 8.10) by using the seismic history of an area, which indicate the epicentral areas of earthquakes and these are zoned according to temporal activity and magnitude. Medvedev (1968) proposed that the seismic zoning of a region

should be based upon: (i) a study of the earthquakes which occur there; (ii) an investigation of the laws governing the occurrence of earthquakes of different intensity; (iii) an analysis of the geological conditions under which earthquakes occur; and (iv) the investigation of special features accompanying the occurrence of earthquakes. The production of maps of seismic zoning in the Soviet Union therefore takes account of the depth of earthquake foci, the relationship between energy at the focus and epicentral intensity, the correlation between magnitude and intensity, and information concerning attenuation of intensity with distance.

Specific conditions in each seismic region, peculiarities of the seismic regime, different types of seismogeological relationships, and the extent to which the area has been studied, mean that there can be no rigorous standard way of using the seismic data for zoning all regions. Furthermore, in the compilation of these maps, all engineering data concerning the surface manifestations of earthquakes should refer to identical ground conditions. Geological investigations may be used to establish the seismic history of an area, since differing seismic responses of geological materials allow the prediction of the effects of an earthquake and identify areas in which crustal strain is building up. However, geological data provide only a qualitative assessment of seismic risk. To obtain a quantitative expression the geological data must be examined in conjunction with the seismic response of particular engineering structures and seismic data.

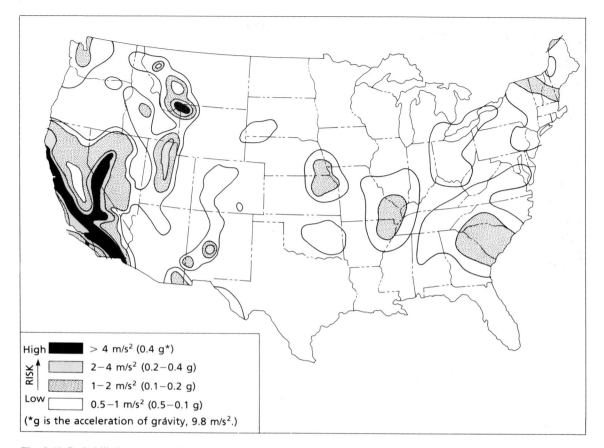

Fig. 8.10 Probabilistic representation of seismic hazard in the USA. Shaded areas give a probabilistic horizontal acceleration of rocks that, with a 90% probability, is not likely to be exceeded in 50 years. (After Algermissen and Perkins, 1976.)

8.3.3 Landslides and slope movements

Movements on slopes can range in magnitude from soil creep on the one hand to instantaneous and colossal landslides on the other (Chapter 3).

Because most landslides occur in areas previously affected by instability and because few occur without prior warning, Cotecchia (1978) emphasized the importance of carrying out careful surveys of areas which appear potentially unstable by making systematic records of relevant phenomena. He provided a review of the techniques involved in mapping mass movements, as well as itemizing which data should be included on such maps. He maintained that the ultimate aim should be the production of maps of landslide hazard zoning (Fig. 8.2).

A useful starting point in landslide investigations

is a checklist (Fig. 8.11) of the type suggested by Cooke and Doornkamp (1974). Each separate slope unit can be classified according to a stability rating and, furthermore, it provides for systematic examination of the main factors influencing mass movement. Aerial photographs, precise surveying, and geophysical methods have proved of value in landslide investigation. Computers can be used to produce maps of slope stability from the data obtained from aerial photographs and field investigations. A variety of types of landslide hazard maps have been produced (Bernknopf et al., 1988; Gupta and Joshi, 1990; Dearman, 1991).

As the same preventive or corrective work cannot always be applied to different types of slides it is important to identify the type of slide which is likely to take place, or which has taken place. Also, it is

Fig. 8.11 A checklist for sites liable to large scale instability. (After Cooke and Doornkamp, 1974.)

RELIEF

Valley depth	Small ☐	Moderate ☐	Large ☐	Very large ☐
Slope steepness	Low ☐	Moderate ☐	Steep ☐	Very steep ☐
Cliffs	Absent ☐			Present ☐
Height difference between different valleys	Small ☐	Moderate ☐	Large ☐	Very large ☐
Valley-side shape	Spur ☐	Straight ☐	Shallow cove ☐	Deep cove ☐

DRAINAGE

Drainage density	Low ☐	Moderate ☐	High ☐	Very high ☐
River gradient	Gentle ☐	Moderate ☐	Steep ☐	Very steep ☐
Slope undercutting	None ☐	Moderate ☐	Severe ☐	Very severe ☐
Concentrated seepage flow	Absent ☐	Present at local base level ☐	Present slowly draining ☐	Present rapidly draining ☐
Standing water	Absent ☐	Small ☐	Moderate ☐	Large ☐
Recent incision	Absent ☐	Small ☐	Moderate ☐	Large ☐
Pore-water pressure	Low ☐	Moderate ☐	High ☐	Very high ☐

BEDROCK

Jointing density	Low ☐	Moderate ☐	High ☐	Very high ☐
Direction of major joints (faults or bedding planes) with respect to steepest slopes	Away ☐	Normal ☐	Towards ☐	
Amount of dip for steepness of joint and/or fault planes	Horizontal ☐	Small ☐	Moderate ☐	Large ☐
Strong beds over weak beds	Absent ☐			Present ☐
Degree of weathering	None ☐	Small ☐	Moderate ☐	Large ☐
Compressive strength	High ☐	Moderate ☐	Low ☐	Very low ☐
Coherence (particularly of lower beds)	High ☐	Moderate ☐	Low ☐	Very low ☐

SOILS (incl. drift materials)

	Valley floor	Gentle slopes	Moderate slopes	Steep slope
Site	☐	☐	☐	☐
Coherent over incoherent beds	Absent ☐			Present ☐
Angle of rest	Low ☐	Moderate ☐	Steep ☐	Very steep ☐
Depth	Small ☐	Moderate ☐	Large ☐	Vert large ☐
Shear strength	High ☐	Moderate ☐	Low ☐	Very low ☐
Liquidity index	Low ☐	Moderate ☐	High ☐	Very high ☐

EARTHQUAKE ZONE

Tremors felt	Never ☐	Seldom ☐	Some ☐	Many ☐

LEGACIES FROM THE PAST

Fossil solifluction lobes and sheets	Absent ☐	Rare ☐	Some ☐	Many ☐
Previous landslides	Absent ☐	Rare ☐	Some ☐	Many ☐
Deep weathering	None ☐	Slight ☐	Moderate ☐	Much ☐

MAN-MADE FEATURES

Excavations—depth	None ☐	Small ☐	Moderate ☐	Large ☐
Excavations—position	Hillcrest ☐	High valley ☐	Low valley ☐	Bottom valley ☐
Reservoir	Absent ☐	Small ☐	Moderately deep ☐	Very deep ☐
Drainage diversion across hillside	Absent ☐			Present ☐
Lowering of reservoir level	None ☐	Small ☐	Moderate ☐	Large ☐
Loading of upper valley side	None ☐	Some ☐	Moderate ☐	Large ☐

important to bear in mind that landslides may change in character and that they are usually complex, frequently changing their physical characteristics as time proceeds. When it comes to the correction of a landslide, as opposed to its prevention, since the limits and extent of the slide are generally well defined, the seriousness of the problem can be assessed. Nevertheless, consideration must be given to the stability of the area immediately adjoining the slide. Obviously, any corrective treatment must not adversely affect the stability of the area about the slide.

If landslides are to be prevented, then areas of potential landsliding must first be identified, as must their type and possible amount of movement. Then, if the hazard is sufficiently real, the engineer can devise a method of preventive treatment (Kockelman, 1988). Economic considerations, however, cannot be disregarded. In this respect it is seldom economical to design cut slopes sufficiently flat to preclude the possibility of landslides.

Landslide prevention may be brought about by reducing the activating forces, by increasing the forces resisting movement or by avoiding or eliminating the slide. Reduction of the activating forces can be accomplished by removing material from that part of the slide which provides the force which will give rise to movement. Complete excavation of potentially unstable material from a slope may be feasible and such treatment is applicable to all types of mass movement. However, Baker and Marshall (1958) placed an upper limit of $50\,000\,m^3$ on the amount of material which can be removed economically. Hence the use of this form of treatment is limited.

Although partial removal is suitable for dealing with most types of mass movement, for some types it is inappropriate. For example, removal of head has little influence on flows or slab slides. On the other hand, this treatment is eminently suitable for rotational slips. Slope flattening, however, is rarely applicable to rotational or slab slides. Slope reduction may be necessary in order to stabilize the toe of a slope and so prevent successive undermining with consequent spread of failure upslope. Benching can be used on steeper slopes. It brings about stability by dividing a slope into segments.

Drainage is the most generally applicable preventive and corrective treatment for slides, regardless of type. Indeed drainage is the only economic way of dealing with landslides involving the movement of several million cubic metres of material. The surface of a landslide is generally uneven, hummocky, and traversed by deep fissures. This is particularly the case when the slipped area consists of a number of slices. Water collects in depressions and fissures, and pools and boggy areas are thus formed. In such cases the first remedial measure to be carried out is surface drainage (Chapter 9).

Restraining structures control landslides by increasing the resistance to movement. They include retaining walls, cribs, gabions, buttresses, piling, and rock bolts (Chapter 9). The following minimum information is required to determine the type and size of a restraining structure:

1 the boundaries and depth of the unstable area, its moisture content, and its relative stability;
2 the type of slide which is likely to develop or has occurred;
3 the foundation conditions since restraining structures require a satisfactory anchorage.

8.3.4 River action and flooding

All rivers form part of a drainage system, the form of which is influenced by rock type and structure, the nature of the vegetation cover, and the climate. An understanding of the processes which underlie river development forms the basis of proper river management.

Rivers form part of the hydrological cycle in that they carry precipitation runoff. This runoff is the surface water which remains after evapotranspiration and infiltration into the ground have taken place. Some precipitation may be frozen, only to contribute to runoff at some later time, while any precipitation that has soaked into the ground may reappear as springs where the water table meets the ground surface. Although, due to heavy rainfall or in areas with few channels, the runoff may occur as a sheet, usually it becomes concentrated into channels which become eroded by the flow of water and form valleys (Chapter 3).

Floods represent the commonest type of geological hazard (Fig. 8.12). They probably affect more individuals and their property than all the other hazards put together. However, the likelihood of flooding is more predictable than some other types of hazards such as earthquakes, volcanic eruptions, and landslides. Most disastrous floods are the result of excessive precipitation or snowmelt, that is, they are due to excessive surface runoff. It usually takes some time to accumulate enough runoff to cause a major disaster. This lag time is an important par-

Fig. 8.12 Collapse of the John Ross bridge due to flooding of the Tugela River, Natal, South Africa, September 1987.

ameter in flood forecasting. Flash floods prove the exception. In most regions floods occur more frequently in certain seasons than others. As the volume of water in a river is greatly increased during times of flood, so its erosive power increases accordingly. Thus, the river carries a much higher sediment load. Deposition of the latter where it is not wanted also represents a serious problem.

The influence of human activity can bring about changes in drainage basin characteristics, for example, removal of forest from parts of a river basin can lead to higher peak discharges which generate increased flood hazard. A most notable increase in flood hazard can arise as a result of urbanization; the impervious surfaces created mean that infiltration is reduced, and together with storm-water drains, they give rise to an increase in runoff. Not only does this produce higher discharges but lag times are also reduced.

Rivers may be considered in flood when their level has risen to an extent that damage occurs. The physical characteristics of a river basin together with

those of the stream channel affect the rate at which discharge downstream occurs and so calculation of the lag time is a complicated matter. Nonetheless once enough data on rainfall and runoff versus time have been obtained and analysed, an estimate of where and when flooding will occur along a river system can be made. Analyses of discharge are related to the recurrence interval to produce a flood frequency curve. The recurrence interval is the period of years within which a flood of a given magnitude or greater occurs and is determined from field measurements at gauging centres. It applies to those stations only. The flood frequency curve can be used to determine the probability of the size of discharge that could be expected during any given time interval.

The lower stage of a river, in particular, can be divided into a series of hazard zones based on flood stages and risk. Obviously a flood plain management plan involves the determination of such zones. These are based on the historical evidence related to flooding which includes: the magnitude of each flood and the elevation it reached as well as the recurrence intervals; the amount of damage involved; the effects of urbanization and any further development; and an engineering assessment of flood potential. Maps then are produced from such investigations which, for example, show the zones of most frequent flooding and the elevation of the flood waters. Flood hazard maps provide a basis for flood management schemes, Kenny (1990) recognized four geomorphological flood hazard map units in central Arizona, which formed the basis of a flood management plan (Table 8.3).

Flood plain zones can be designated for specific types of land use, that is, in the channel zone water should be allowed to flow freely without obstruction, for example, bridges should allow sufficient waterway capacity. Another zone could consist of that area with recurrence intervals of 1 to 20 years which could be used for agricultural and recreational purposes. Buildings would be allowed in the zone encompassing the 20 to 100-year recurrence interval, but they would have to have some form of protection against flooding.

River control refers to projects designed to hasten the runoff of flood waters or confine them within restricted limits, to improve drainage of adjacent lands, to check stream-bank erosion or to provide deeper water for navigation (Rahn, 1984). What has to be borne in mind, however, is that a river in an alluvial channel is continually changing its position due to the hydraulic forces acting on its banks and bed. As a consequence any major modifications to the river regime which are imposed without consideration of the channel will give rise to a prolonged and costly struggle to maintain the change.

Artificial strengthening and heightening of levees or constructing artificial levees are frequent measures employed as protection against flooding. Because these measures confine a river to its channel its efficiency is increased. The efficiency of a river is also increased by cutting through the constricted loops of meanders. River water can be run off into the cutoff channels when high floods occur. Canalization or straightening of a river can help to regulate flood flow, and improves the river for navigation. However, canalization in some parts of the world has only proved a temporary expedient. This is because canalization steepens the channel, as sinuosity is reduced. This, in turn, increases the velocity of flow and therefore the potential for erosion. Moreover, the base level for the upper reaches of the river has, in effect, been lowered, which means

Table 8.3 Generalized flood hazard zones and management strategies. (After Kenny, 1990.)

Flood hazard zone I (Active floodplain area)

Prohibit development (business and residential) within floodplain
    Maintain area in a natural state as an open-area or for recreational uses only

Flood hazard zone II (Alluvial fans and plains with channels less than a metre deep, bifurcating, and intricately interconnected systems subject to inundation from overbank flooding)

Flood proofing to reduce or prevent loss to structures is highly recommended
    Residential development densities should be relatively low; development in obvious drainage channels should be prohibited
Dry stream channels should be maintained in a natural state and/or the density of native vegetation should be increased to facilitate superior water drainage retention and infiltration capabilities
Installation of upstream storm water retention basins to reduce peak water discharges
Construction should be at the highest local elevation site where possible

Flood hazard zone III (Dissected upland and lowland slopes; drainage channels where both erosional and depositional processes are operative along gradients generally less than 5%)

Similar to flood hazard zone II
    Roadways which traverse channels should be reinforced to withstand the erosive power of a channelled stream flow

Flood hazard zone IV (Steep gradient drainages consisting of incised channels adjacent to outcrops and mountain fronts characterized by relatively coarse bedload material)

Bridges, roads, and culverts should be designed to allow unrestricted flow of boulders and debris up to a metre or more in diameter
    Abandon roadways which presently occupy the wash flood plains
    Restrict residential dwelling to relatively level building sites
    Provisions for subsurface and surface drainage on residential sites should be required
    Stormwater retention basins in relatively confined upstream channels to mitigate against high peak discharges

that channel incision will begin there. Although the flood and drainage problems are temporarily solved, the increase in erosion upstream increases sediment load. This is then deposited in the canalized sections which can lead to a return of the original flood and drainage problems.

Diversion is another method used to control flooding. This involves opening a new exit for part of the river water. But any diversion must be designed in such a way that it does not cause excessive deposition in the main channel otherwise it defeats its purpose. If, in some localities, little damage will be done by flooding, then these areas can be inundated during times of high flood and so act as safety valves.

Channel regulation can be brought about by training dams or walls which are used to deflect channels into a more desirable alignment or to confine them to lesser widths. Walls and dams can be used to close secondary channels and thus divert or concentrate the river into a preferred course. In some cases ground sills or weirs need to be constructed to prevent undesirable deepening of the bed by erosion. Bank revetment by pavement, riprap gabions, or protective mattresses (e.g. fascines or geomats) to retard erosion is usually carried out along with channel regulation or it may be undertaken independently for the protection of the lands bordering the river.

Control of the higher stretches of a river in those regions prone to soil removal by gullying and sheet erosion is very important. The removal of the soil mantle means that runoff becomes increasingly more rapid and consequently, the problem of flooding is aggravated. These problems were tackled in the Tennessee valley by establishing a well-planned system of agriculture so that soil fertility was maintained, and the valley slopes were reafforested. The control of surface water thereby was made more effective. Gullies were filled and small dams were constructed across the headstreams of valleys to regulate runoff. Larger dams were erected across tributary streams to form catchment basins for flood waters. Finally, large dams were built across the main river to smooth flood flow. Nonetheless the construction of a dam across a river can lead to other problems. It decreases the peak discharge and reduces the quantity of bedload through the river channel. The width of a river may be reduced downstream of a dam as a response to the decrease in flood peaks. Moreover, scour normally occurs immediately downstream of a dam. Removal of the finer fractions of the bed material by scouring action may cause armouring of the channel.

River channels may be improved for navigation purposes by dredging. When a river is dredged its floor should not be lowered so much that the water level is appreciably lowered. In addition, the nature of the materials occupying the floor and their stability should be investigated. This provides data from which the stability of the slopes of the projected new channel can be estimated and indicates whether blasting is necessary. The rate at which sedimentation takes place provides some indication of the regularity at which dredging should be carried out.

8.3.5 Marine action

Waves, acting on beach material, are a varying force. They vary with time and place due to: (i) changes in wind force and direction over a wide area of sea; and (ii) changes in coastal aspects and offshore relief. This variability means that the beach is rarely in equilibrium with the waves, in spite of the fact that it may only take a few hours for equilibrium to be attained under new conditions. Such a more or less constant state of disequilibrium occurs most frequently where the tidal range is considerable, as waves are continually acting at a different level on the beach.

The rate at which coastal erosion proceeds is influenced by the nature of the coast itself and is most rapid where the sea attacks soft, unconsolidated sediments (Fig. 8.13). When soft deposits are being actively eroded the cliff displays signs of landsliding together with evidence of scouring at its base. For erosion to continue the debris produced must be removed by the sea. This is usually accomplished by longshore drift. If, on the other hand, material is deposited to form extensive beaches and the detritus is reduced to a minimum size, then the submarine slope becomes very wide. Wave energy is dissipated as the water moves over such beaches and cliff erosion ceases.

Dunes are formed by onshore winds carrying sand sized material landward from the beach along lowlying stretches of coast where there is an abundance of sand on the foreshore. Leatherman (1979) maintained that dunes act as barriers, energy dissipators, and sand reservoirs during storm conditions, for example, the broad sandy beaches and high dunes along the coast of the Netherlands present a natural defence against inundation during storm surges. Because dunes provide a natural defence against erosion, once they are breached the ensuing coastal changes may be long-lasting. On the other hand, along parts of the coast of North Carolina where dune protection is limited, washovers associated

Fig. 8.13 Marine erosion of till along the Holderness coast near Skipsea, England, causing the destruction of holiday cottages. (Courtesy of Alan Forster.)

with storm tides have been responsible for high rates of erosion (Cleary and Hosier, 1987).

In spite of the fact that dunes inhibit erosion, Leatherman (1979) stressed that without beach nourishment they cannot be relied upon to provide protection, in the long term, along rapidly eroding shorelines. Beach nourishment widens the beach and maintains the proper functioning of the beach − dune system during normal and storm conditions.

The amount of longshore drift along a coast is influenced by coastal outline and wave length. Short waves can approach the shore at a considerable angle, and generate consistent downdrift currents. This is particularly the case on straight or gently curving shores and can result in serious erosion where the supply of beach material reaching the coast from updrift is inadequate. Conversely, long waves suffer appreciable refraction before they reach the coast.

Before any project or beach planning and management scheme can be started a complete study of the beach must be made. The preliminary investigation of the area concerned should first consider the landforms and rock formations along the beach and adjacent rivers, giving particular attention to their durability and stability. In addition, consideration must be given to the width, slope, composition, and state of accretion or erosion of the beach, the presence of bluffs, dunes, marshy areas, or vegetation in the backshore area, and the presence of beach structures such as groynes. Estimates of the rates of erosion, and the proportion and size of eroded material deposited on the beach must be made, as well as whether these are influenced by seasonal effects (Martinez *et al.*, 1990). The behaviour of many unconsolidated materials, which form cliffs, when weathered together with their slope stability and likelihood of sliding has to be taken into account. Samples of the beach and underwater material have to be collected and analysed for such factors as their particle size distribution and mineral content. Mechanical analyses may prove useful in helping to determine the amount of material which is likely to remain on the beach, for beach sand is seldom finer than 0.1 mm in diameter. The amount of material moving along the shore must be investigated in as much as the effectiveness of the structures erected may depend upon the quantity of drift available.

Topographic and hydrographic surveys of an area allow the compilation of maps and charts from which a study of the changes along a coast may be made. Observations should be taken of winds, waves, and currents, and information gathered on streams which enter the sea in or near the area concerned. Inlets across a beach need particular evaluation. During normal times there may be relatively little longshore drift but if upbeach breakthroughs occur in a bar off the inlet mouth, then sand is moved downbeach and is subjected to longshore drift. The contributions made by large streams may vary, for example, material brought down by large floods may cause a temporary, but nevertheless appreciable, increase in the beach width around the mouths of rivers. The effects of any likely changes in sea level must also be taken into consideration (Clayton, 1990). The data collected can be used to plan a system of coastal defences which should also incorporate a scheme of coastal management.

The groyne is the most important structure used to stabilize or increase the width of the beach by arresting longshore drift (Fig. 8.14). Consequently, they are constructed at right angles to the shore. Groynes should be approximately 50% longer than the beach on which they are erected. Standard types usually slope at about the same angle as the beach. Permeable groynes have openings which increase in size seawards and thereby allow some drift material to pass through them. The common spacing rule for groynes is to arrange them at intervals of one to three groyne lengths. The selection of the type of groyne and its spacing depends upon the direction and strength of the prevailing or storm waves, the amount and direction of longshore drift, and the relative exposure of the shore. With abundant long-shore drift and relatively mild storm conditions almost any type of groyne appears satisfactory, whilst when the longshore drift is lean, the choice is much more difficult. Groynes, however, reduce the amount of material passing downdrift and therefore can prove detrimental to those areas of the coastline. Their effect on the whole coastal system should therefore be considered.

Artificial replenishment of the shore by building beach fills is used either if it is economically prefer-able or if artificial barriers fail to defend the shore adequately from erosion. In fact, beach nourishment represents the only form of coastal protection which does not adversely affect other sectors of the coast. Unfortunately, it is often difficult to predict how frequently a beach should be renourished. Ideally, the beach fill used for renourishment should have a similar particle size distribution to the natural beach material.

Seawalls (Fig. 8.15) and bulkheads are protective waterfront structures. The former range from a simple riprap deposit (frequently underlain by geo-textile which acts as a separator and filter) to a regular, masonry retaining wall. Bulkheads are vertical walls either of timber boards or of steel-sheet piling. Foundation ground conditions for these retaining structures must be given careful attention and due consideration must be given to the likelihood of scour occurring at the foot of the wall and to changes in beach conditions. Because walls are impermeable they can increase the backwash and therefore its erosive capability.

Offshore breakwaters and jetties are designed to protect inlets and harbours. Breakwaters disperse the waves of heavy seas whilst jetties impound long-shore drift material upbeach of the inlet and thereby prevent sanding of the channel they protect. Off-shore breakwaters commonly run parallel to the shore or at slight angles to it, chosen with respect to the direction from which storm waves approach the coast. Although long, offshore breakwaters shelter their leeside, they also cause wave refraction and so may generate currents in opposite directions along the shore towards the centre of the sheltered area with resultant impounding of sand. Jetties are usually built at right angles to the shore although their outer segments may be set at an angle. Two parallel jetties may extend from each side of a river for some distance out to sea and because of its confinement, the velocity of river flow is increased, which in turn lessens the amount of deposition which takes place. Like groynes, such structures inhibit the downdrift movement of material; the deprivation of downdrift beaches of sediment may result in serious erosion.

Marine inundation of low-lying coastal areas may

Fig. 8.14 Groynes along the foreshore near Cromer, Norfolk, England.

Fig. 8.15 Seawall at Newbiggin-by-the-Sea, Northumberland, England.

occur as a result of high spring tides if the coast is not protected adequately. The Delta Scheme in the Netherlands and the Thames barrier in England, are designed to prevent such disasters occurring. An even more terrifying event to occur along coasts is inundation by large masses of water called tsunamis. Most tsunamis originate as a result of earthquakes on the sea floor, though they can also be developed by submarine landslides or volcanic activity. Seismic tsunamis usually develop where submarine faults have a significant vertical movement, and faults of this type most commonly occur along the coasts of South America, the Aleutian Islands, and Japan. Horizontal fault movements, such as occur along the California coast, do not result in tsunamis.

In the open ocean tsunamis normally have a very long wavelength and their amplitude is hardly noticeable. Successive waves may be from 5 min to an hour apart. They travel at speeds of around 650 km/h. However, their speed is proportional to the depth of water which means that in coastal areas the waves slow down and increase in height, rushing onshore as highly destructive breakers. Waves have been recorded up to nearly 20 m in height above normal sea level. Large waves are most likely when tsunamis move into narrowing inlets. Such waves cause terrible devastation, for example, if the wave is breaking as it crosses the shore, it can destroy houses merely by the weight of water. The subsequent backwash may carry many buildings out to sea and may remove several metres depth of sand from dune coasts. Structures which offer protection against potential damage caused by tsunamis include breakwaters, revetments, etc. Design against tsunami damage involves raising buildings off the ground and orienting shear walls normal to the direction of the wavefront. Reinforced concrete structures having adequate foundations do not usually suffer noticeable damage.

The Pacific Tsunami Warning System (PTWS) is a communications network covering all the countries bordering the Pacific Ocean and is designed to give advance warning of dangerous tsunamis. Clearly, the PTWS cannot provide a warning of an impending tsunami to those areas which are very close to the earthquake epicentre which is responsible for the generation of the tsunami. On the other hand, waves generated off the coast of Japan will take 10 h to reach Hawaii. Evacuation of an area depends on estimating just how destructive any tsunami will be when it arrives on a particular coast.

8.3.6 Wind action

In arid regions, in particular, because there is little vegetation, wind action is much more significant than elsewhere. By itself, wind can only remove uncemented rock debris but once armed with rock particles, the wind becomes a noteworthy agent of abrasion. The movement of windblown material, not always in dune form, often gives rise to problems in many arid regions as far as settlements and agricultural land are concerned. Urbanization of drylands has accentuated the problem of windblown sand and silt because of the adverse effect of human activity on sensitive desert soils. A review of the problems involved, the methods of data collection, and suggested management strategies has been provided by Jones *et al.* (1986). An analysis of meteorological data provides an indication of the potential of wind to move sand and silt, but the actual patterns of movement are influenced by topography, mean ground wind speed and turbulence, and the availability of the material involved. Location of sediment sources and drift zones allows supply areas to be stabilized and development to be planned. Kerr and Nigra (1952) presented a review of the objectives and methods of control of windblown sand (Table 8.4). Hazard maps (Fig. 8.16), showing unstable sand dunes and their migration, have been produced for development purposes in arid regions (Cooke *et al.*, 1978).

8.4 GEOLOGICAL RELATED HAZARDS INDUCED BY MAN

8.4.1 Soil erosion, desertification, and salinization

Soil erosion refers to the removal of loose surface material by water or wind. It is a normal aspect of landscape development but in some regions of the world it is the dominating process of denudation. Frequently, this is because man has upset the balance of nature. Nonetheless, once slopes are reduced to gradients that are relatively stable as far as mass movements are concerned, erosion becomes dominant.

Table 8.4 Objectives and methods of aeolian sand control. (After Kerr and Nigra, 1952.)

Objectives
1 The destruction or stabilization of sand accumulations in order to prevent their further migration and encroachment
2 The diversion of wind-blown sand around features requiring protection
3 The direct and permanent stoppage or impounding of sand before the location or object to be protected
4 The rendition of deliberate aid to sand movement in order to avoid deposition over a specific location, especially by augmenting the saltation coefficient through surface smoothing and obstacle removal

Methods
The above objectives are achieved by the use of one or more types of surface modification.
1 *Transposing*. Removal of material (using anything from shovels to bucket cranes) — rarely economical or successful, and does not normally feature in long-term plans
2 *Trenching*. Cutting of transverse or longitudinal trenches across dunes destroys their symmetry and may lead to dune destruction. Excavation of pits in the lee of sand mounds or on the windward side of features to be protected will provide temporary loci for accumulation
3 *Planting* of appropriate vegetation is designed to stop or reduce sand movement, bind surface sand, and provide surface protection. Early stages of control may require planting of sand-stilling plants (e.g. *Ammophila arenaria*, beach grass), protection of surface (e.g. mulching), seeding, and systematic creation of surface organic matter. Planting is permanent and attractive, but expensive to install and maintain
4 *Paving* is designed to increase the saltation coefficient of wind transported material by smoothing or hard-surfacing a relatively level area, thus promoting sand migration and preventing its accumulation at undesirable sites. Often used to leeward of fencing, where wind is unladen of sediment, and paving prevents its recharge. Paving may be with concrete, asphalt or wind-stable aggregates (e.g. crushed rock)
5 *Panelling* in which solid barriers are erected to the windward of areas to be protected, is designed either to stop or to deflect sand movement (depending largely on the angle of the barrier to wind direction). In general, this method is inadequate, unsatisfactory, and expensive, although it may be suitable for short-term emergency action
6 *Fencing*. The use of relatively porous barriers to stop or divert sand movement, or destroy or stabilize dunes. Cheap, portable and expendable structures are desirable (using, for example, palm fronds or chicken wire)
7 *Oiling* involves the covering of aeolian material with a suitable oil product (e.g. high-gravity oil) which stabilizes the treated surface and may destroy dune forms. It is, in many deserts, a quick, cheap and effective method

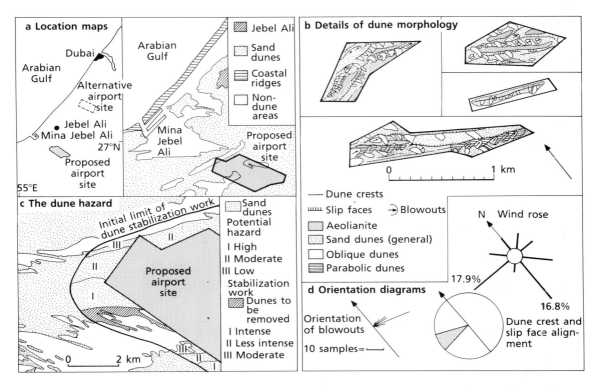

Fig. 8.16 Geomorphological analysis of a proposed airport site in Dubai with respect to the threat from mobile sand dunes. (After Cooke *et al.*, 1978.)

Soil erosion is most active where it is difficult for most rainfall to infiltrate into the ground, so that most flows over the surface and in so doing removes soil. Some sheet flow can have catastrophic effects. These conditions are most frequently met in semi-arid areas. It is difficult to separate natural from man-induced changes in erosion rates in these regions. Nevertheless, soil erosion occurs in many different climatic regions where vegetation has been removed. Severe soil erosion, associated with the formation of gullies (Fig. 5.11), can give rise to mass movements on the steepened slopes at the sides of these gullies. Soil erosion removes the topsoil, which contains a high proportion of the organic matter and the finer mineral fractions in soil which provide water and nutrient supplies for plant growth.

Sheet flow may cover up to 50% of the surface of a slope during a heavy rainfall but erosion does not take place uniformly across a slope. Linear concentrations of flow may occur within sheet wash. The depth of sheet wash, up to 3 mm, and the velocity of flow are such that both laminar and turbulent flow take place. Erosion due to turbulent flow only occurs

where flow is confined. Hence the flow elsewhere is laminar and non-erosive. The velocity of sheet flow ranges between 15 and 300 mm/s. Velocities of 160 mm/s are required to erode soil particles of 0.3 mm diameter but velocities as low as 20 mm/s will keep these particles in suspension.

Rills and gullies begin to form when the velocity of flow increases to speeds in excess of 300 mm/s and flow is turbulent. Whether they form depends on soil factors as well as velocity and depth of water flow. Rills and gullies remove much larger volumes of soil per unit area than sheet flow.

Generally, an increase of erosion occurs with increasing rainfall and erosion decreases with increasing vegetation cover. However, the growth of natural vegetation depends on rainfall, so producing a rather complex variation of erosion with rainfall. Agriculture can make rainfall and vegetation partially independent of one another. Accordingly, increased erosion resulting from farming practices depends on the change in vegetation cover, the total rainfall at periods of low cover, and the intensity of rainfall. Semi-arid regions show greater proportionate changes in erosion rates with rainfall than other environments and are very sensitive to small changes in climate.

The exhaustion of the organic matter in soil and soil nutrients used by plants is closely allied to soil erosion. Organic matter in the soil fulfils a similar role to clays in holding water and both inorganic and organic nutrients. Organic matter is also very important in maintaining soil aggregates and in providing a moist soil which remains highly permeable. High permeability and aggregate strength minimize the risk of overland flow. Loss of organic matter depends largely on the vegetation cover and its management. Partial removal of vegetation or wholesale clearance prevents the addition of plant debris as a source of new organic material for the soil. Over a period of years this results in a loss of plant nutrients and in a dry climate there is a significant reduction in soil moisture. The process can turn a semi-arid area into a desert in less than a decade. This organic depletion leads to lower infiltration capacity and increased overland flow with consequent erosion on slopes of more than a few degrees.

Erosion is greatest when intense rains fall before the vegetation cover is properly established. Two regions appear to be at great risk: (i) the arid and semi-arid zones in which exhaustion of organic matter through removal of vegetation either as a crop or through grazing may play an important part in the erosion of usable land; and (ii) those humid areas which have been seasonally stripped of vegetation for crop cultivation.

Wind erosion is most common in desert regions. Like water erosion, it depends upon the force that the wind can exert on soil particles. In addition, the roughness of the surface over which wind blows has an important influence on erosion, for example, where the surface is rough because of large stones the wind speed near the surface is low and little erosion occurs. By contrast, uncultivated fields are susceptible to wind erosion, especially when the soil contains appreciable silt size material. Losses of up to 10 mm/year were experienced in the dustbowl of Kansas.

The redistribution and resorting of particles by wind erosion may have profound effects on the soils affected, their related microtopography and any agricultural activity associated with them. The process operates in a variety of natural environments that lack a protective cover of vegetation. Its human consequences are most serious in agricultural areas which experience low, variable, and unpredictable rainfall, high temperatures and rates of evaporation, and high wind velocity, as in semi-arid areas, as well as some of the more humid regions that experience periodic droughts. In such areas the natural process of wind erosion may be accelerated by imprudent agricultural practices leading to soil damage and related problems.

As far as arid and semi-arid lands are concerned crop yields are generally low near desert margins so that it is rarely worth spending large sums of money on conservation. Good management is necessary in these regions and on low slopes it involves controlling yields at low, sustainable levels, and varying production with wet and dry years. On steep slopes or where there are migrating dunes, vegetation must be established and grazing discouraged. However, such policies are difficult to put into effect.

Soil erosion in humid areas can be tackled by contour ploughing, terracing, strip cropping, and mulching. However, it is possible that the control of soil erosion in these regions is not effective enough and there is a danger that because some areas of spectacular erosion have been dealt with, complacency may allow less suspect land gradually to be removed.

The world loses some 20 million hectares per year to desertification. This is a process of environmental degradation due to excessive human activity and demand in regions with fragile ecosystems. Deserts are encroaching into semi-arid regions largely as a consequence of poor farming practices which include overstocking and deforestation. The improper use of water resources leading to inefficient use and even to streams drying up aggravates the problem still further. Excessive abstraction of water from wells lowers the water table, which adversely affects plant growth. Desertification can occur within a short time, that is, in 5 to 10 years.

Desertification brings with it associated problems such as: removal of soil as well as reduction in its fertility; deposition of windblown sand and silt, which can bury young plants and block irrigation canals and rivers; and moving sand dunes. Also, when rain does fall a greater proportion of it contributes towards runoff so that erosion becomes more aggressive. This, in turn, means that the amount of sediment carried by streams and rivers increases. It also means that there is less water infiltrating into the ground for plant growth.

To counter desertification, land use and water resources need to be managed effectively. Such management must take account of the cyclic nature of semi-arid climates in that wetter and drier periods tend to alternate with each other.

Irrigation is used in semi-arid and arid regions to increase agricultural production. There are several methods of irrigation and some are more efficient than others. For example, conveying water via canals and ditches into furrows is thought, at most, to be only 60% efficient, sprinkler systems are about 75% efficient, and drip irrigation (a low water pressure system for delivering small amounts of water directly to plant roots) is 90% efficient. However, ditch irrigation, in particular, has given rise to problems of salinization and waterlogging.

Many soils in semi-arid and arid regions contain significant salt contents so that the water, when used for irrigation, becomes highly mineralized. Capillary action also brings salts to the surface. Inefficient irrigation, especially when accompanied by poor drainage, leads to salinization of soil, that is, the accumulation of salts near the ground surface. Generally soil with more than 0.1% soluble salts within 0.2 m of the surface is regarded as salinized. Heavily salinized land has been abandoned to agriculture in many parts of the world (e.g. in parts of China, Pakistan, and the United States). To avoid salinization an adequate system of drainage must be installed prior to the commencement of irrigation. This lowers the water table as well as conveying water away more quickly. Subsurface drainage can be used together with controlled surface drainage. Excess water can be applied to fields during the non-growing season to flush salts from the soil. Water use must be managed, and planting trees for shelter belts and rotating crops, where possible, helps. The other problem which can result from inadequate drainage is waterlogging, which generally is brought about by a rising water table caused by irrigation.

8.4.2 Waste disposal

Waste disposal is one of the most expensive environmental problems (Langer, 1989). Many types of waste material are produced by society of which domestic waste, hazardous waste, and radioactive waste are probably the most notable.

At the present time, increasing concern is being expressed over the disposal of domestic waste products. Although domestic waste is disposed of in a number of ways, quantitatively the most important method is placement in a sanitary landfill. In the United Kingdom, of approximately 18.6 million tonnes of domestic solid waste produced each year, about 76% is disposed of in landfills.

Domestic refuse is a heterogeneous collection of almost anything, much of which is capable of reacting with water to give a liquid rich in organic matter, mineral salts, and bacteria. The organic carbon content is especially important since this influences the growth potential of pathogenic organisms. Leachate is formed when rainfall infiltrates a landfill and dissolves the soluble fraction of the waste, and from the soluble products formed as a result of the chemical and biochemical processes occurring within the decaying wastes. Generally, the conditions within a landfill are anaerobic, so leachates often contain high concentrations of dissolved organic substances. Barber (1982) estimated that a small landfill site with an area of 1 ha located in southern England could produce up to 8 m^3 of leachate per day. A site with an area ten times as large would produce a volume of effluent with approximately the same biochemical oxygen demand (BOD) per year as a small, rural sewage treatment works. Hence, the location and management of these sites must be carefully controlled. Clearly, the production of leachate may represent a health hazard, for example, by contaminating a groundwater supply which would then require costly treatment prior to use and, since it can also threaten the surface water resource potential of a region, an economic problem of considerable magnitude may be created.

Barber (1982) identified three classes of landfill site based upon hydrogeological criteria (Table 8.5). When assessing the suitability of a site, two of the principal considerations are the ease with which the pollutant can be transmitted through the substrata and the distance it is likely to spread from the site. Consequently, the primary and secondary permeability of the formations underlying the landfill area are of major importance. It is unlikely that the first type of site mentioned in Table 8.5 would be considered suitable. There also would be grounds for an objection to a landfill site falling within the second category of Table 8.5 if the site was located within the area of diversion to a water supply well. Generally, the third category, in which the leachate is contained within the landfill area, is to be preferred. Since all natural materials possess some degree of permeability, total containment can only be achieved if an artificial impermeable lining is provided over the bottom of the site. However, there is no guarantee that clay, soil cement, asphalt, or plastic linings will remain impermeable perma-

Table 8.5 Classification of landfill sites based upon their hydrogeology. (After Barber, 1982.)

Designation	Description	Hydrogeology
Fissured site, or site with rapid subsurface liquid flow	Material with well developed secondary permeability features	Rapid movement of leachate via fissures, joints, or through coarse sediments. Possibility of little dispersion in the groundwater, or attenuation of pollutants
Natural dilution, dispersion, and attenuation of leachate	Permeable materials with little or no significant secondary permeability	Slow movement of leachate into the ground through an unsaturated zone. Dispersion of leachate in the groundwater, attenuation of pollutants (sorption, biodegradation, etc.) probable
Containment of leachate	Impermeable deposits such as clays or shales, or sites lined with impermeable materials or membranes	Little vertical movement of leachate. Saturated conditions exist within the base of the landfill

nently. Thus, the migration of materials from the landfill site into the substrata will occur eventually, although the length of time before this happens may be subject to uncertainty. In some instances the delay will be sufficiently long for the pollution potential of the leachate to be greatly diminished. One of the methods of tackling the problem of pollution associated with landfills is by dilution and dispersal of the leachate. Otherwise leachate can be collected by internal drains within the landfill and conveyed away for treatment.

Selection of a landfill site for a particular waste or a mixture of wastes involves a consideration of economic and social factors as well as the hydrogeological conditions. As far as the latter are concerned, most argillaceous sedimentary, massive igneous, and metamorphic rock formations have low intrinsic permeability and therefore afford the most protection to water supply. By contrast, the least protection is provided by rocks intersected by open discontinuities or in which solution features are developed. Granular materials may act as filters leading to dilution and decontamination. Hence, sites for disposal of domestic refuse can be chosen where decontamination has the maximum chance of reaching completion and where groundwater sources are located far enough away to enable dilution to be effective. The position of the water table is important as it determines whether wet or dry tipping is involved, as is the thickness of unsaturated material underlying a potential site. Generally, unless waste is inert, wet tipping should be avoided. The hydraulic gradient determines the direction and velocity of the flow of

leachates when they reach the water table and also is related to both the dilution which the leachates undergo, and to the points at which flow is discharged. Gray *et al.* (1974) recommended that at dry sites, tipping should take place on granular material which has a thickness of 15 m or more, while any water wells should be located at least 0.8 km away.

Carbon dioxide and methane are generated by the decomposition of organic materials within sanitary landfills. However, the amount of gas produced by domestic waste varies appreciably and a site investigation is required to determine the amount if such information is required, but it has been suggested that between 2.2 and 250 l/kg dry weight may be produced (Oweis and Khera, 1990). Both the gases mentioned are toxic and methane forms a highly explosive mixture with air. Unfortunately, there are numerous cases on record of explosions occurring in buildings due to the ignition of accumulated methane derived from underlying or nearby landfills (Williams and Aitkenhead, 1991). Accordingly, planners of residential developments should avoid such sites. Proper closure of a landfill site can require gas management to control methane gas by passive venting, power-operated venting, or the use of an impermeable barrier (Raybould and Anderson, 1987).

When sites have to cope with liquid or industrial wastes, or wastes of indeterminate composition, the danger is increased (Bell and Wilson, 1988). Chemical and biochemical hazardous wastes include those which are toxic, infectious, corrosive, and/ or ignitable. Obviously their uncontrolled dumping can pollute or contaminate soil and groundwater resources.

Protection of groundwater from the disposal of toxic waste can be brought about by containment. A number of containment systems have been developed which isolate wastes and include compacted clay barriers, slurry trench cutoff walls, geomembrane walls, diaphragm walls, sheet piling, grout curtains, and hydraulic barriers (Mitchell, 1986). A compacted clay barrier consists of a trench which has been backfilled with clay compacted to give a low hydraulic conductivity. Slurry trench cutoff walls are narrow trenches filled with soil—bentonite mixtures which extend downwards into an impermeable layer. Again they have a low hydraulic conductivity. In a geomembrane wall a 'U'-shaped geomembrane is placed in a trench and filled with sand so that its shape conforms to that of the trench. Monitoring wells are placed in the trench so that they can detect and abstract any leakage from the fill. Diaphragm walls are constructed in a similar manner to slurry trenches but are an expensive form of containment. Their use is therefore restricted to situations where high structural stability is required. If leakage occurs via interlocks in steel sheet pile cutoffs, then their water-tightness is impaired. In such instances wells may be needed to ensure effectiveness. Grout curtains may be used in certain situations and are likely to consist of three rows of holes, the outer rows being grouted first, followed by the inner to seal any voids. Extraction wells are used to form hydraulic barriers and are located so that the contaminant plume flows towards them. These systems are used to contain existing waste disposal sites.

When a new site has been chosen Mitchell (1986) maintained that a properly designed and constructed liner and cover offer long-term protection for ground and surface water. Clay liners are suitable for the containment of many wastes because of the low hydraulic conductivity of clay and its ability to adsorb some wastes. They are constructed by compaction in lifts of about 150 mm thickness. Care must be taken during construction to ensure that the clay is placed at the specified moisture content and density, and to avoid cracking due to drying out during and after construction (Quigley et al., 1988). Much care is also needed in the construction of geomembrane liners so that the seams are properly welded and the geomembrane is not torn. Composite liners incorporate both clay blankets and geomembranes. Although resistant to many chemicals, some geomembranes are susceptible to degradation by organic solvents.

Hence the US Environmental Protection Agency recommends the use of double-liner systems (Fig. 8.17). An alternative requirement recommended by Gray et al. (1974) was that a site handling toxic wastes should be underlain by at least 15 m of impermeable strata. Any well abstracting groundwater for domestic use and confined by such impermeable strata should be more than 2 km away.

Disposal of liquid hazardous waste has also been undertaken by injection into deep wells located in rock below freshwater aquifers, thereby ensuring that contamination or pollution of underground water supplies does not occur. In such instances the waste generally is injected into a permeable bed of rock several hundreds or even thousands of metres below the surface which is confined by relatively impervious formations. However, even where geological conditions are favourable for deep well disposal, the space for waste disposal frequently is restricted and the potential injection zones usually are occupied by connate water. Accordingly, any potential formation into which waste can be injected must possess sufficient porosity, permeability, volume, and confinement to guarantee safe injection. The piezometric pressure in the injection zone influences the rate at which the reservoir can accept the liquid waste. A further point to consider is that induced seismic activity has been associated with the disposal of fluids in deep wells (section 8.4.6). Two important geological factors relating to the cost of construction of a well are its depth and the ease with which it can be drilled.

Monitoring is especially important in deep-well disposal that involves toxic or hazardous materials. Effective monitoring requires that the geological and hydrogeological conditions are accurately evaluated and mapped before the disposal programme is started. A system of observation wells sunk into the subsurface reservoir concerned in the vicinity of the disposal well allows the movement of waste to be monitored. In addition, shallow wells sunk into freshwater aquifers permit monitoring of water quality so that any upward migration of the waste can be readily noted.

Radioactive waste may be of low or high level. Low-level waste contains small amounts of radioactivity and so does not present a significant environment hazard if properly dealt with. The dilute and disperse method frequently has been used to dispose of this material. Although many would not agree, it would appear that low-level radioactive waste can be disposed of safely by burying in carefully

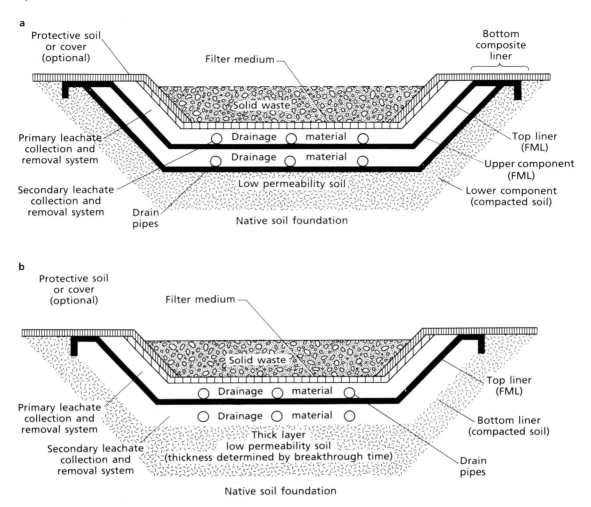

Fig. 8.17 Double-liner systems proposed by the 1985 United States Environmental Protection Agency guidelines. The leachate collection layer is considered also to function as the geomembrane protection layer: (a) FML/composite double liner; (b) FML/compacted-soil double liner. FML, flexible membrane liner. (After Mitchell, 1986.)

controlled and monitored sites where the hydro-logical and geological conditions severely limit the migration of radioactive material.

High-level radioactive materials need to be separated from biological systems for hundreds or even thousands of years before they no longer represent hazards to health. Their disposal therefore presents one of the most acute problems of the present day. The disposal of high-level liquid waste can be achieved by solidifying and mixing it with inert, non-leachable material. It is then placed in steel and concrete containers which may be stored in under-ground caverns (Morfeldt, 1989) or dumped into the oceans. Permanent storage in caverns (either modi-fied old mines or purpose-built) in thick impermeable rock formations such as salt, shale, granite, or basalt more than 500 m below the surface is usually re-garded as the most feasible means of disposal. In particular, thick deposits of salt have certain advan-tages: (i) salt has a high thermal conductivity and so will rapidly dissipate any heat associated with high-level nuclear waste; (ii) it is 'plastic' at proposed repository depths so that any fractures which may develop due to construction operations will 'self-heal'; (iii) it possesses gamma-ray protection similar to concrete; (iv) it undergoes only minor changes when subjected to radioactivity; and (v) it tends not to provide paths of escape for fluids (Langer and Wallner, 1988). The location of caverns should be in geologically stable areas with a minimum risk of seismic disturbance. Deep structural basins are considered as possible locations for disposal (Baillieul, 1987; Ericksson, 1989). Placement of encapsulated nuclear waste in deep holes in stable rock formations cannot dispose of high-volume wastes. Injection of liquid waste into porous or fractured strata, at depths from 1000 to 5000 m, which are suitably isolated, relies on the dispersal and diffusion of the waste through the host rock. The limits of diffusion need to be well defined. Alternatively, thick beds of shale, at depths between 300 and 500 m, can be fractured by high-pressure injection and then waste mixed with cement or clay grout can be injected into the fractured shale.

Waste from mines has been deposited on the surface as large heaps which disfigure the landscape. Spoil heaps associated with coal mines consist of run-of-mine material which reflects the various rock types which are extracted during mining operations. This material contains varying amounts of coal which has not been separated by the preparation process. Some spoil heaps, especially those with high contents of coal, may be partially burnt or burning. The spontaneous combustion of carbonaceous material is frequently aggravated by the oxidation of pyrite. Spontaneous combustion can give rise to hot spots in a spoil heap where the temperatures may average around 600°C but temperatures as high as 900°C may occur. Obviously, this can make for difficulties when restoring spoil heaps. In addition, noxious gases are emitted from burning spoil including carbon monoxide, carbon dioxide, sulphur dioxide, and, less frequently, hydrogen sulphide.

Slips in spoil heaps represent a hazard, one of the most notable being that which occurred at Aberfan, South Wales, in 1966 resulting in 144 deaths. The landslide occurred after heavy rain and turned into a mudflow which moved at about 32 km/h and engulfed several houses and a school (Bishop, 1973).

The fine discard from the preparation of coal consists of tailings or slurry and is pumped from the washery into lagoons. Tailings lagoons may be surrounded by spoil heaps, enclosed by embankments or impounded by dams. They are also associated with metalliferous mining. Obviously, the construction of safe tailings dams and embankments is im-

portant as far as public safety is concerned. A catastrophic failure of a tailings dam used for the disposal of slurry from coal mining occurred at Buffalo Creek, West Virginia, in 1972. After heavy rain, water overtopped the dam and it failed, releasing $7 \times 10^5 \, m^3$ of tailings and water within a few minutes. More than 1500 houses were destroyed and 118 people killed.

The tailings themselves may also represent a hazard, for example, the tailings from uranium mines still contain radioactive elements such as thorium-230 and its daughter product radium-226. Radon, which is a carcinogenic gas, is given off as radium decays. In the United States when a uranium tailings lagoon is filled it has to be covered with 3 m of earth so that the emission of radon does not exceed $2 \, pCi/m^2/s$.

8.4.3 Groundwater pollution

Pollution can be defined as an impairment of water quality by chemicals, heat, or bacteria to a degree that does not necessarily create an actual public health hazard, but does adversely affect waters for normal, domestic, farm, municipal, or industrial use. Contamination denotes impairment of water quality by chemical or bacterial pollution to a degree that creates an actual hazard to public health.

The greatest danger of groundwater pollution is from surface sources including animal manure, sewage sludge, leaking sewers, polluted streams, and refuse disposal sites. Areas with a thin cover of superficial deposits or where an aquifer is exposed, such as a recharge area, are the most critical from the point of view of pollution potential. Any possible source of contamination in these areas should be carefully evaluated, both before and after any groundwater supply well is constructed and the viability of groundwater protection measures considered. One approach to groundwater quality management is to indicate areas with high pollution potential on a map and to pay particular attention to activities within these vulnerable areas.

The attenuation of a pollutant as it enters and moves through the ground occurs as a result of biological, chemical, and physical processes. Hence the self-cleansing capacity of a soil-aquifer system depends upon the physical and chemical form of the pollutant, the nature of the soil/rock comprising the aquifer, and the way in which the pollutant enters the ground. In general, the concentration of a pollutant decreases as the distance it has travelled through the ground increases. Thus the greatest pollution potential exists for wells tapping shallow aquifers that intersect or lie near ground level.

The form of the pollutant is clearly an important factor with regard to its susceptibility to the various purifying processes, for example, pollutants which are soluble, such as fertilizers and some industrial wastes, cannot be removed by filtration. Metal solutions may not be susceptible to biological action. Solids, on the other hand, are amenable to filtration provided that the transmission media are not coarse-grained, fractured, or cavernous. Karst or cavernous limestone areas pose particular problems in this respect. Insoluble liquids such as hydrocarbons are generally transmitted through porous media, although some fraction may be retained in the material. Usually, however, the most dangerous forms of groundwater pollution are those which are miscible with the water in the aquifer.

Concentrated sources of pollution are most undesirable because the self-cleansing ability of the soil in that area is likely to be exceeded. As a result the 'raw' pollutant may be able to enter an aquifer and travel some considerable distance from the source before being reduced to a negligible concentration. A much greater hazard exists when the pollutant is introduced into an aquifer beneath the soil horizon since the purifying processes that take place within the soil are bypassed and attenuation of the pollutant is reduced. This is most critical when the pollutant is added directly to the zone of saturation, because in most soils and rocks the horizontal component of permeability is usually much greater than the vertical one. Consequently, a pollutant then can travel a much greater distance before significant attenuation occurs. This type of hazard often arises from poorly maintained domestic septic tanks and soakaways, from the discharge of quarry wastes, farm effluents, and sewage into surface water courses, and from the disposal of refuse and commercial wastes.

It is generally assumed that bacteria move at a maximum rate of about two-thirds the water velocity. Since most groundwaters only move at a rate of a few metres per year, the distances travelled by bacteria are usually quite small and, in general, it is unusual for bacteria to spread more than 33 m from the source of the pollution. However, Brown et al. (1972) suggested that viruses are capable of spreading over distances which exceed 250 m, although 20 to 30 m may be a more typical figure. Of course, in porous gravel, cavernous limestone, or fissured rock, bacteria and viruses may spread over distances of many kilometres.

Induced infiltration occurs where a stream is hydraulically connected to an aquifer and lies within the area of influence of a well. When the well is overpumped, a cone of depression develops and spreads. Eventually the aquifer may be recharged by the influent seepage of surface water, so that some proportion of the pumpage from the well is then obtained from the surface source. Induced infiltration is significant from the point of view of groundwater pollution in two respects: (i) hydraulic gradients may result in pollutants travelling in the opposite direction from that normally expected; (ii) surface water resources are often less pure than the underlying groundwater so the danger of contamination is introduced. However, induced infiltration does not automatically cause pollution, and it is a common method of augmenting groundwater supplies.

A list of potential groundwater pollutants would be almost endless, although one of the most common sources is sewage sludge. This material arises from the separation and concentration of most of the waste materials found in sewage. Since sludge contains nitrogen and phosphorus it has a value as a fertilizer. While this does not necessarily lead to groundwater pollution, the presence in sludge of contaminants such as metals, nitrates, persistent organic compounds, and pathogens, does mean that the practice must be carefully controlled. The widespread use of chemical and organic pesticides or herbicides is another possible source of groundwater contamination. Runoff from roads can contain chemicals from many sources, and therefore represents yet another potential source of pollution, as do cemeteries.

Nitrate pollution is basically the result of intensive cultivation due to the large quantity of synthetic nitrogenous fertilizer used, although overmanuring with natural organic fertilizer can have the same result. Rapid transformation into nitrate results in an ion which since it is neither adsorbed nor precipitated in the soil becomes easily leached by heavy rainfall and infiltrating water (Foster and Crease, 1974). However, the nitrate does not have an immediate affect on groundwater quality, possibly because most of the leachate that percolates through the unsaturated zone as intergranular seepage has a typical velocity of about a metre or so per year. Thus, there may be a considerable delay between the application of the fertilizer and the subsequent increase in the concentration of nitrate in the groundwater. This time-lag, which is frequently of the order of 10 years or more, makes it very difficult to correlate fertilizer application with increased concentration of nitrate in groundwater. Hence, although nitrate levels are unacceptably high now, they may worsen in the future because of the increasing use of nitrogenous fertilizer.

There are at least two ways in which nitrate pollution of water is known or suspected to be a threat to health. First, the build-up of stable nitrate compounds in the bloodstream reduces its oxygen-carrying capacity. Infants under a year old are most at risk. A limit of 500 mg/l of nitrate (NO_3) has been recommended by the World Health Organization (WHO) for European countries. Second, the possible combination of nitrates and amines through the action of bacteria in the digestive tract results in the formation of potentially carcinogenic nitrosamines.

Measures that can be taken to alleviate nitrate pollution include better land-use management, mixing of water from various sources, or the treatment of high-nitrate water before it is put into supply. In general, the ion-exchange process has been recommended as the preferred means of treating groundwaters, although this may not be considered cost-effective for all sources.

Excessive lowering of the water table along a coast as a consequence of over abstraction can lead to saline intrusion, the salt water entering the aquifer via submarine outcrops thereby displacing fresh water. However, the fresh water still overlies the saline water and continues to flow from the aquifer to the sea. The encroachment of salt water may extend several kilometres inland. Typically, chloride levels may increase from a normal value of around 25 mg/l to something approaching 19 000 mg/l, compared with a recommended limit for drinking water in Europe of 200 mg/l. Once intrusion develops it is not easy to control. The slow rates of groundwater flow, the density differences between fresh and salt waters, and the flushing required, usually mean that contamination, once established, may take many years to remove under natural conditions. Reduction of pumping to eliminate overdraft or artificial recharging have been used as methods of controlling saline intrusion.

The routine monitoring of groundwater level and water quality provides an early warning of pollution incidents. The first important step in designing an efficient groundwater monitoring system is the proper understanding of the mechanics and dynamics of contaminant propagation, the nature of the controlling flow mechanism and the aquifer characteristics. There should be a sufficient number of wells to allow the extent, configuration, and concentration of a contamination plume to be determined. Further-

more, construction of a water quality monitoring well must be related to the geology of the site, in particular the well structure should not react with the groundwater if water quality is being monitored. These wells are frequently constructed using inert plastic casings and screens. Monitoring also can be carried out by using geophysical methods, especially resistivity surveys (Oteri, 1983).

8.4.4 Effluents from coal mines

A variety of waste waters and process effluents arise during coal-mining operations. These may be produced by the actual extraction process, by the subsequent preparation of the coal, or from the disposal of colliery spoil. The strata from which the waste is derived, the mineralogical character of the coal and the colliery spoil, and the washing processes employed all affect the type of effluent produced.

The major pollutants generally associated with coal mining are suspended solids, dissolved salts (particularly chlorides), acidity, and iron compounds (Bell and Kerr, 1993). Colliery discharges have little oxygen demand with BOD values generally being very low. Elevated levels of suspended matter are associated with most coal mining effluents. In fact the increase in the concentration of suspended solids in a receiving water course is frequently the most common pollutant associated with coal mining. The blanketing effect of coal-slurry particles on the bed of a river is unacceptable both in terms of appearance and its influence on the flora and fauna of the stream. The extraction of coal, particularly from opencast sites and from drift mines, may lead to the discharge of high loads of silt and fine coal particles into rivers.

A high level of mineralization is characteristic of many coal-mining discharges and is reflected in high values of electrical conductivity (values of 335 000 µS/cm have been recorded). Not all mine waters, however, are highly mineralized. The principal groups of salts in mine drainage waters are chlorides and sulphates. The former occur in the waters lying in the confined aquifers between the coal seams and are released into the workings by mining operations. In general, the salinity increases with depth and with distance from the outcrop or incrop. The more saline waters contain significant concentrations of barium, strontium, ammonium, and manganese ions. Dissolved sulphates occur only in trace concentrations in the confined aquifer waters but mine drainage waters at the point of discharge almost invariably contain sulphates. These are either present in the waters lying in the more shallow un-confined aquifers, or are generated in the workings by the action of atmospheric oxygen on pyrite. The primary oxidation products of pyrite are ferrous and ferric sulphates and sulphuric acid. These react with clay and carbonate minerals to form secondary prod-ucts including manganese and aluminium sulphates. Further reactions with these minerals and incoming waters produce tertiary products such as calcium and magnesium sulphates.

The composition of waters derived from sulphide oxidation due to the breakdown of pyrite is extremely variable, depending principally upon the constituents of the strata through which the waters percolate, the availability of oxygen, and residence time. At one extreme drainage waters may be very acid (pH < 3) with high concentrations of metals in solution, es-pecially ferrous sulphate and, at the other, water may have become neutralized and contain hydrox-ides of iron, aluminium, and other metals.

Another major source of acid mine water results from the closure of mines and the cessation of pumping (Bell and Kerr, 1993). The water table rises and groundwater re-occupies the strata. It will drain to the surface where conditions are favour-able (e.g. within the catchment of a drainage adit, where a shaft intercepts rock containing water under artesian pressure or via springs). Hence those streams receiving drainage from abandoned mines frequently are chronically polluted.

The high level of dissolved solids often present in mine waters presents the most intractable water-pollution problem connected with coal mining as they are not readily susceptible to treatment or removal. The only treatment normally possible is dilution and dispersion in the receiving water course.

Although modern tipping techniques may render colliery spoil impervious, surface water runoff can leach out soluble mineral salts, particularly the chlor-ide ion. This may result in the loss of up to 1 tonne of chloride per hectare of exposed tip per annum under average rainfall conditions (tips may extend to many hundreds of hectares). The runoff from the tips may discharge directly into drainage ditches or to land around the periphery of the tip and soak into an aquifer below or run into streams.

8.4.5 Surface subsidence

Surface subsidence occurs as a consequence of the extraction of mineral deposits or the abstraction of water, oil, or natural gas from the ground. The subsidence effects of mineral extraction depend on the type of deposit, the geological conditions, in particular the nature and structure of the overlying rocks or soils, the mining methods, and any mitigat-ive action. In addition to subsidence due to the removal of support, mining often entails the lowering of groundwater levels which, since the vertical effec-tive stress is raised, causes consolidation of the overburden.

Many ore deposits are concentrated in veins or ovoid bodies, the mining of which may cause local-ized subsidence. Transmission of a ground loss to the surface entails spreading its effect beyond the edges of the mined area and indeed can fracture the surface. Hence the removal of a deep ore body is liable to cause subsidence over a wider area than one of equal thickness at shallow depth although, in the former case, the lowering of the ground surface will be less severe. On the other hand, the mining of stratiform deposits, especially the total extraction of seams of coal, can create subsidence problems over wide areas.

The mining method is a very important consider-ation, not least because of the control it exerts over the amount of mineral actually extracted during mining operations. The mining of coal or iron ore prior to the sixteenth century in Britain usually took the form of outcrop workings or bell-pitting (Fig. 8.18). The latter are shaft-like excavations up to about 12 m deep which extend down into the mineral seam. Once in the seam the workings extend out-wards for a short distance. The extent of lateral workings depends on the stability of the excavation. Bell-pitting leaves an area of highly disturbed ground which generally requires treatment or complete ex-cavation before it is suitable for use for any other purpose.

Later coal workings were usually undertaken by the pillar and stall method. This method entails leav-ing pillars of the mineral in place to support the roof of the workings. The method also is used at the present day to work stratiform deposits including limestone, gypsum, anhydrite, salt, and sedimentary iron ores (Fig. 8.19). The amount of mineral actually mined depends primarily on maintaining the stability of the excavation so that in weaker rock masses large pillars and small stalls are formed. However, in many cases in old coal workings pillars were robbed prior to abandonment of the mine, which increases the possibility of pillar collapse. Slow de-terioration and failure of pillars may take place years after mining operations have ceased, although

Fig. 8.18 Bell pits in the German Ironstone Band exposed at Sproats opencast site, Northumberland, England. These bell pits generally were 4.6 to 6.1 m in diameter at their base.

Fig. 8.19 Pillar and stall workings in salt, Meadowbank Mine, Winsford, Cheshire, England. (Courtesy of ICI, Ltd.)

observations in coal mines at shallow depth together with the resistance of coal to weathering, suggest that this is a relatively uncommon feature at depths less than about 30 m. On the other hand, small pillars may be crushed out once the overburden exceeds 50 to 60 m. Old pillars at shallow depth have occasionally failed near faults and they may fail if they are subjected to the effects of subsequent long-wall mining of coal. The yielding of a large number of pillars can bring about a shallow, broad trough-like subsidence over a large area.

Squeezes or crushes sometimes occur in a coal mine as a result of the pillars being punched into either the roof or floor bed, which might have become weakened or altered by the action of water or weathering. Once again surface subsidence adopts a trough-like or basin form and minor strain and tilt problems occur around the periphery of the basin thereby produced.

Even if pillars in old shallow workings are relatively stable the surface can be affected by void migration. Void migration develops if roof rock falls into the worked-out zones and represents the main problem in areas of shallow, abandoned mines. It

Fig. 8.20 This crown hole suddenly appeared in a garden in Gateshead, Tyne and Wear, England, it being due to void migration from abandoned shallow workings.

can occur within a few months, or a very long period of years after mining has ceased. The material involved in the fall bulks, so that migration is eventually arrested, although the bulked material never completely fills the voids. Nevertheless, the process can, at shallow depth, continue upwards to the surface leading to the sudden appearance of a crown hole (Fig. 8.20).

The mining of coal by longwall methods is a more recent innovation which has developed as a mechanized mining system during the twentieth century. It involves total extraction of panels in a seam. The working face is temporarily supported, the support being moved as the face advances, leaving the roof from which support has been withdrawn to collapse. The resulting subsidence is largely contemporaneous with mining, producing more or less direct effects on any surface development (Bell, 1988a).

The surface effects of longwall mining include not only lowering but also tilting and both compressive and tensile ground strains (Fig. 8.21). As longwall mining proceeds the ground is subject to tilting accompanied by tension and then compression. Once the subsidence front has passed by, the ground attains its previous slope and the ground strain returns to zero. However, permanent ground strains affect the ground above the edges of the extracted panel. For total mineral extraction and a low depth-to-width extraction ratio, the maximum amount of subsidence will be up to about 0.9 times the thickness of the seam. Normally, however, the amount of surface subsidence is significantly less than this amount.

Structural damage is not simply a function of ground strain but the shape, size, and form of construction are also important controls. In many instances subsidence effects also have been affected by the geological structure, notably the presence of faults and the character of the rocks and soils above the workings (Bell and Fox, 1991). A review of measures which can be taken to mitigate the effects of subsidence due to coal working is provided by Anon (1975b) and Anon (1977b). Methods of dealing with old shafts are outlined by Anon (1982).

Deposits that readily go into solution, in particular salt, can be extracted by solution mining, for example, salt has been obtained by brine pumping in a number of areas in the United Kingdom, Cheshire being by far the most important. Subsidence due to salt extraction by wild brine pumping still continues to be an inhibiting factor, on a very much reduced scale, as far as developments in Cheshire are concerned (Bell, 1992c). Wild brine pumping is carried out on the major natural brine runs and active subsidence is normally concentrated at the head and sides of a brine run where fresh water first enters the system. Hence, serious subsidence has occurred at considerable distances, up to 8 km from pumping centres, and is of unpredictable nature. In addition, tension cracks and small fault scars have formed in the surface tills on the convex flanks of subsidence hollows (Fig. 8.22). Because the exact area from which salt is extracted is not known, the magnitude of subsidence developed cannot be related to the volume of salt worked. Consequently, there is no accurate means of predicting the amount of ground

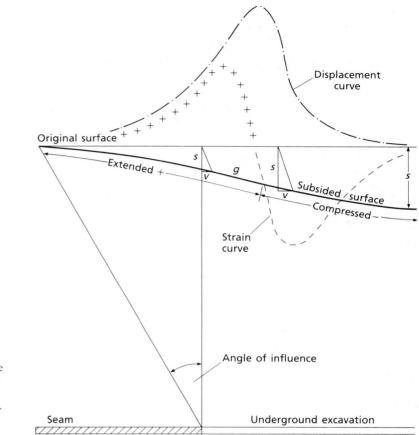

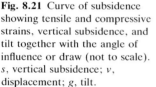

Fig. 8.21 Curve of subsidence showing tensile and compressive strains, vertical subsidence, and tilt together with the angle of influence or draw (not to scale). s, vertical subsidence; v, displacement; g, tilt.

movement or strain. Sulphur is mined in Texas and Louisiana by the Frasch process which involves pumping hot water into the beds of sulphur and then pumping the brine to the surface. Subsidence troughs with associated small faults at the peripheries of the basins have formed.

In addition to the subsidence caused by the shrinkage of peat following surface drainage (Fig. 8.23), surface subsidence also occurs in areas where there is intensive abstraction of fluids from the ground, especially oil or groundwater, as well as natural gas (Bell, 1988b). Subsidence is attributed to the consolidation of the fluid-bearing formations which results from the increase in vertical effective stress (Chapter 4). In most cases, and particularly in clayey deposits, the subsidence does not occur simultaneously with the abstraction of fluid. A reduction in the rate of abstraction can lead to a rise in groundwater levels. This, in turn, can lead to a rise in the ground

surface, as in the Venice area where there has been some 20 mm of rebound since 1970. Although controlled withdrawal of groundwater permits the re-establishment of the natural hydraulic balance, recovery uplift is never complete.

More than 40 known examples of differential subsidence, horizontal displacement, or surface faulting have been associated with 27 oil and gas fields in California and Texas. The most spectacular case of subsidence is that of the Wilmington oilfield near Long Beach, California (Fig. 8.24). By 1966, after 30 years of production and 8 of repressurizing by injection, an elliptical area of more than 75 km^2 had subsided more than 8.8 m.

8.4.6 Induced sinkholes

Rapid subsidence can take place due to the collapse of holes and cavities within limestone which has

Fig. 8.22 Subsidence trough caused by wild brine pumping of salt in Cheshire, England. Note the tension scars on the cambered flanks. The trough is now occupied in part by a flash and the road has been made up periodically.

Fig. 8.23 Damage caused to Benwick church, Huntingdonshire, England, due to subsidence of peat caused by drainage. The church was subsequently demolished.

been subjected to prolonged solution, this occurring when the roof rocks are no longer thick enough to support themselves (Beck, 1984). It must be emphasized, however, that the solution of limestone is a very slow process (Chapter 3). Nevertheless, solution may be accelerated by man-made changes in the groundwater conditions. In fact many sinkholes develop as a result of man's activities, especially as a result of declines in groundwater level due to excess-ive abstraction. Jammal (1986) recorded that 70 sinkholes had appeared in Orange and Seminole counties in central Florida during the previous 20 years. Most developed in those months of the year when rainfall was least (i.e. April and May) and withdrawal of groundwater was high. In fact, throughout the southeast United States there are thousands of sinkholes of many different sizes and shapes. They range from 1 to 2 m in diameter to

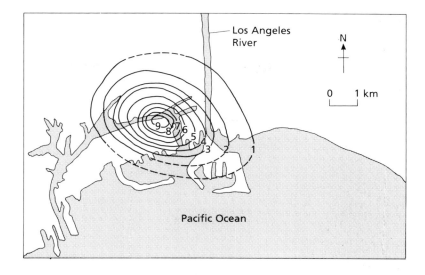

Fig. 8.24 Subsidence due to the abstraction of oil in the coastal region of Los Angeles. Contours in metres.

some more than 3 km. Depths of a few metres are common, whilst the largest sinkholes exceed 40 m in depth. Whereas it takes thousands of years to create natural sinkholes, those created by man largely have occurred since the early 1900s. More than 4000 sinkholes have been catalogued in Alabama as being caused by man's activity with the great majority of these developing since 1950.

Some induced sinkholes develop within hours of the effects of man's activity being imposed upon the geological and hydrogeological conditions. Several collapse mechanisms have been suggested, including: (i) the loss of buoyant support to roofs of cavities or caverns in bedrock previously filled with water and to overlying unconsolidated deposits; (ii) increases in the amplitude of water level fluctuations; (iii) increases in the velocity of movement of groundwater; and (iv) movement of water from the land surface to openings in underlying bedrock where most recharge previously had been rejected since the openings were occupied by water.

Areas underlain by highly cavernous limestones possess most dolines, hence doline density has proven a useful indicator of potential subsidence, as has sinkhole density. As there is preferential development of solution voids along zones of high secondary permeability because these concentrate groundwater flow, data on fracture orientation and density, fracture intersection density, and the total length of fractures have been used to model the presence of solution cavities in limestone. Therefore, the location of areas of high risk of cavity collapse has been estimated by using the intersection of

lineaments formed by fracture traces and lineated depressions (dolines). Aerial photographs have proven particularly useful in this context. Brook and Alison (1986) described the production of subsidence susceptibility maps for karst terrain which incorporate most of the data mentioned.

Most collapses forming sinkholes result from roof failures of cavities in unconsolidated deposits overlying limestones and represent the most dangerous subsidence phenomena associated with these rocks. The phenomenon is referred to as ravelling and occurs when solution enlarged openings extend upward from the rock into the unconsolidated deposits. The openings should be interconnected and lead into channels through which the deposits can be eroded by groundwater flow. Initially, the deposits arch over the openings but as they are enlarged a stage is reached when the deposits above the roof can no longer support themselves and therefore collapse. A number of conditions accelerate the development of cavities in the deposits and initiate collapse. Rapid changes in moisture content lead to aggravated slabbing in clays and flow in cohesionless sands. Lowering the water table: (i) increases the downwards seepage gradient and accelerates downward erosion; (ii) reduces capillary attraction in sand and increases instability of flow through narrow openings; and (iii) gives rise to shrinkage cracks in highly plastic clays which weaken the mass in dry weather and produce concentrated seepage during rains.

Spectacular sinkhole development due to groundwater abstraction has occurred in the Hershey Valley, Pennsylvania, where the water table was lowered some 50 m to allow the extension of quarrying operations. Similarly, dewatering associated with mining in the gold-bearing reefs of the Far West Rand, South Africa, which underlie dolostone and unconsolidated deposits, has led to the formation of sinkholes and produced differential subsidence over large areas (Bezuidenhout and Enslin, 1970). Hence, certain areas became unsafe for occupation and were evacuated. Subsidence was initially noticed in 1959 and the seriousness of the situation was highlighted in December 1962 when a sinkhole appeared at the West Driefontein Mine and engulfed a three-storey crusher plant with the loss of 29 lives. Then in August 1964 two houses and parts of two others disappeared into a sinkhole in Blyvooruitzicht Township with the loss of five lives (Fig. 8.25). Consequently, it became a matter of urgency that the areas which were subject to subsidence or to the occurrence of sinkholes be delineated.

Sinkholes formed concurrently with the lowering of the water table in areas which formerly, in general, had been free from sinkholes. Normally, sinkholes develop where the cover of unconsolidated material is less than 15 m, otherwise it chokes the cavity on collapse. Bezuidenhout and Enslin (1970) distinguished three areas. First, there were those areas where the original water table was less than 15 m below, and less frequently within 30 m of the surface. Second, sinkholes formed in the scarp zones which bordered deep buried valleys. Third, sinkholes occurred in narrow buried valleys where the limited width meant that initially unconsolidated material bridged sinks in the dolostone below. Bezuidenhout and Enslin found that sinkholes of larger dimensions than would be expected to occur, develop in such valleys. The thickness of the unconsolidated deposits varies laterally, thereby giving rise to differential subsidence which, in turn, causes large fissures to occur at the surface. In fact the most prominent fissures frequently demarcate the areas of subsidence.

8.4.7 Induced seismicity

Induced seismicity occurs where changes in the local stress conditions brought about by man give rise to changes in strain and deformation in rock masses causing movements along discontinuities. These movements may be on a microscopic or macroscopic scale. Most instances of this type have been associated with large reservoirs, underground liquid waste disposal, hydrocarbon extraction, underground mining, or large explosions. Such activities through the single or combined effect of unloading, loading, or increased pore pressures in the rock masses concerned allow the release of stored strain energy. The earthquakes produced tend to have magnitudes of less than 5, although more serious events have occurred. For instance, a 6.5 magnitude earthquake associated with the Koyna reservoir in India resulted in loss of life and damage to the dam itself. The area in which Koyna reservoir is located is characterized

Fig. 8.25 Sinkhole developed at Blyvooruitzicht in the Transvaal. It engulfed a house, claiming five lives. (Courtesy of Dr Brian Gregory.)

by wrench faulting. Evidence suggests that it was already in a critical state of stress before the dam was constructed and impounding of the reservoir provided the necessary triggering mechanism for movements to occur along the faults due to the increase in pore pressure. Evans (1966) documented induced seismicity in the Denver area which was attributed to the disposal of liquid waste down a deep well. Over 700 minor earthquakes of magnitudes up to 4.3 were recorded. Water injection to enhance the recovery of oil has also led to the occurrence of seismic events.

Seismic emissions may be brought about as a result of the fluid acting as a lubricant in microcracks and fissures on the one hand, and by increasing the pore pressures thereby reducing the shear strength of the rock mass on the other. Both may trigger the release of strain energy from the rocks and facilitate movements along faults. It has been suggested that there is a critical pore-fluid pressure required for earthquake activity to rise above background levels.

Violent rock failures (i.e. rock bursts) are associated with deep mining. The latter causes stress changes which lead to the sudden release of stored strain energy. In the gold mines of the Witwatersrand system in South Africa such release of strain energy is responsible for the generation of seismic events which range up to magnitudes around 4. The sources of most of the seismic events are located in the rock mass in the immediate vicinity of mining activity. Minor seismicity has also been caused by coal mining (Kusznir et al., 1980). In the latter part of the 1970s minor earth tremors were recorded in the neighbourhood of Stoke-on-Trent, England, and were linked with mining at Hem Heath Colliery. The presence of old workings in the same area appeared to be an additional causative factor.

9   Geology and construction

9.1 OPEN EXCAVATION

Open excavation refers to the removal of material, within certain specified limits, for construction purposes. In order to accomplish this economically and without hazard the character of the rocks and soils involved and their geological setting must be investigated. Indeed, the method of excavation and the rate of progress are very much influenced by the geology on site (Kummerle and Benvie, 1988). Furthermore, the position of the water table in relation to the base level of the excavation is of prime importance, as are any possible effects of operations on the surrounding ground and/or buildings.

9.1.1 A note on slope stability

The stability of slopes is a critical factor in open excavation. This is particularly the case in cuttings (e.g. for roads, canals, and railways) where slopes should be designed to resist disturbing forces over long periods. In other words, a stability analysis should determine under what conditions a proposed slope will remain stable.

Instability in a soil mass occurs when slip surfaces develop and movements are initiated within it. Undesirable properties in a soil such as low shearing strength, development of fissures, and high water content tend to encourage instability and are likely to lead to deterioration of slopes. In the case of open excavation, removal of material can give rise to the dissipation of residual stress which can aid instability.

There are several methods available for analysis of the stability of slopes in soils (Chowdhury, 1978). Most of these may be classed as limit equilibrium methods in which the basic assumption is that the failure criterion is satisfied along the assumed path of failure. Starting from known or assumed values of the forces acting upon the soil mass, calculation is made of the shear resistance required for equilibrium. This shearing resistance is then compared with the estimated or available shear strength of the soil to give an indication of the factor of safety. The analysis gives a conservative result.

The design of a slope excavated in a rock mass requires as much information as possible on the character of the discontinuities within the rock mass, since its stability is frequently dependent upon the nature of the discontinuities (Hoek and Bray, 1981). Information relating to the spatial relationships between discontinuities affords some indication of the modes of failure which may occur and information relating to the shear strength of the rock mass, or more particularly the shear strength along discontinuities, is required for use in the stability analysis. The joint inclination is always the most important parameter for slopes of medium and large height, whereas density is more important for small slopes than friction. Cohesion becomes less significant with increasing slope height whilst the converse is true as far as the effects of water pressure are concerned.

9.1.2 Excavations in rocks and soils

Excavation in fresh, massive, plutonic igneous rocks such as granite and gabbro can be left more or less vertical after removal of loose fragments. On the other hand, volcanic rocks such as basalts and andesites are generally bedded and jointed, and may contain layers of ash or tuff, which are usually softer and weather more rapidly. Thus, slope angles have to be reduced accordingly.

Gneiss, quartzite, and hornfels are highly weather resistant and slopes in them may be left almost vertical. Schists vary in character and some of the softer schists may be weathered and tend to slide along their planes of schistosity. Slate generally resists weathering although slips may occur where the cleavage dips into a cut face.

If strata are horizontal, then excavation is relatively straightforward and slopes can be determined with some degree of certainty. Vertical slopes can be excavated in massive limestones and sandstones which are horizontally bedded. In brittle, cemented shales slopes of 60° and 75° are usually safe but increasing fissility and decreasing strength necessitate flatter slopes. Even in weak shales slopes are seldom flatter than 45°. However, excavated slopes may have to be modified in accordance with the dip and strike directions in inclined strata. The most stable excavation in dipping strata is one in which the strata dip into the face. Also, where the face is orientated normal to the strike, there is a low tendency for rocks to slide along their bedding planes. Conversely, the worst situations are likely to occur when the strata dip out of the slope. This is most critical where the rocks dip at angles varying between 30° and 70°. If the dip exceeds 70° and there is no alternative to working against the dip, then the face should be developed parallel to the bedding planes for safety reasons.

Sedimentary sequences in which thin layers of shale or clay are present may have to be treated with caution, especially if the bedding planes are dipping at a critical angle. Weathering may reduce such material to an unstable state within a short period of time which, in turn, can lead to slope failure.

A slope of 1:1.5 is generally used when excavating dry sand, this more or less corresponding to the angle of repose, 30° to 40°. This means that a cutting in a non-cohesive soil will be stable, irrespective of its height, as long as the slope is equal to the lower limit of the angle of internal friction, provided that the slope is suitably drained. In other words the factor of safety (*FS*) with respect to sliding may be obtained from:

$$FS = \frac{\tan \phi}{\tan \beta} \qquad (9.1)$$

where ϕ is the angle of internal friction and β is the slope angle.

Slope failure in frictional soils is a surface phenomenon which is caused by the particles rolling over each other down the slope. The packing density of sands is important, for example, densely packed sands which are very slightly cemented may have excavated faces with high angles which are stable. The water content is of paramount importance in loosely packed sands, for if these are saturated they are likely to flow on excavation.

The most frequently used gradients in many clays vary between 30° and 45°. In certain clays, however, in order to achieve stability the slope angle may have to be less than 20°. The stability of slopes in clay depends not only on its strength and the angle of the slope but also on the depth to which the excavation is taken and on the depth of a firm stratum, if one exists, not far below the base level of the excavation. For example, the critical height (*H*) at which a face of an open excavation in normally

consolidated clay can stand vertically without support can be obtained from:

$$H = \frac{4c}{9.8\gamma} \qquad (9.2)$$

where *c* is the cohesion of the clay and γ its unit weight. Slope failure in a uniform clay takes place along a near 'circular' surface of slippage (Chapter 3).

In stiff, fissured clays the fissures appreciably reduce the strength below that of intact samples, thus, reliable estimation of slope stability is difficult. Generally, steep slopes can be excavated in such clays initially but their excavation means that fissures open due to the relief of residual stress and there is a change from negative to positive pore pressures along the fissures; the former tend to hold the fissures together. This change can occur within a matter of hours or days. Not only does this weaken the clay but it also permits a more significant ingress of water, which means that the clay is softened. Irregular-shaped blocks may begin to fall from the face and slippage may occur along well-defined fissure surfaces which are by no means circular. If there are no risks to property above the crests of slopes in stiff, fissured clays, then they can be excavated at about 35°. Although this will not prevent slips, those which occur are likely to be small.

The stability of the floor of large excavations may be influenced by ground heave, the amount and rate at which it occurs depending on: (i) the degree of reduction in vertical stress during construction operations; (ii) the type and succession of underlying strata; and (iii) the surface and groundwater conditions. Heave is generally greater in the centre of a level excavation in relatively homogeneous ground, as in clays and shales. Long-term swelling involves absorption of water from the ground surface or is due to water migrating from below. Where the excavation is in overconsolidated clays or shales, swelling and softening is quite rapid. In the case of clays with a low degree of saturation, swelling and softening take place very rapidly if surface water gains access to the excavation area.

9.1.3 Methods of excavation

The rock properties which influence drillability include hardness, abrasiveness, grain size, and discontinuities. The harder the rock, the stronger the drilling bit which is required for drilling since higher pressures need to be exerted. Toughness represents the work required to bring about fracture in the rock. Abrasiveness may be regarded as the ability of a rock to wear away drill bits. This property is closely related to hardness and is also influenced by particle shape and texture. The size of the fragments produced during drilling operations influences abrasiveness, for example, large fragments may cause scratching but comparatively little wear whereas the production of dust in tougher but less abrasive rock causes polishing. This may lead to the development of high skin hardness on tungsten carbide bits which in turn may cause them to spall and even diamonds lose their cutting ability upon polishing. Generally, coarse-grained rocks can be drilled more quickly than can fine-grained varieties or those in which the grain size is variable.

The ease of drilling in rocks in which there are many discontinuities is influenced by their orientation in relation to the drillhole. Drilling over an open discontinuity means that part of the energy controlling drill penetration is lost. Where a drillhole crosses discontinuities at a low angle, the bit may stick. It may also lead to excessive wear and to the hole going off line. Drilling across the dip is generally less difficult than drilling with it. If the ground is badly broken, then the drillhole may require casing. Where discontinuities are filled with clay, this may penetrate the flush holes in the bit, causing it to bind or deviate from alignment.

Spacing of the blastholes is determined on the one hand in relation to the strength, density, and fracture pattern within the rock mass, and on the other in relation to the size of the charge. Careful trials are the only certain method of determining the correct burden and blasting pattern in any rock. As a rule spacing will vary between 0.75 and 1.25 times the burden. Generally 1 kg of high explosive will bring down about 8 to 12 tonnes of rock. Good fragmentation reduces or eliminates the amount of secondary blasting while minimizing wear and tear on loading machinery.

Rocks characterized by high specific gravity and high intergranular cohesion with no preferred orientation of mineral grains cause difficulties in blasting. They have high tensile strength and very low brittleness values, the former resisting crack initiation and propagation upon blasting. Examples are gabbros, breccias, and greenstones. A second group which provides difficulties includes certain granites, gneisses, and marbles, which are relatively brittle with a low resistance to dynamic stresses. Blasting in such rocks gives rise to extensive pulverization immediately about the charged holes leaving the area

between almost unfractured. Hence these rocks do not give effective energy transfer from the charge to the rock mass. The third category of rocks giving rise to difficult blasting is those possessing marked preferred orientation (e.g. mica schist). The mechanical anisotropy due to the preferred orientation means that these rocks split easily along the lineation but crack propagation across it is limited.

In many excavations it is important to keep overbreak to a minimum. Apart from the cost of its replacement with concrete, damage to the rock forming the walls or floor may lower its strength and necessitate further excavation. Also, smooth faces allow excavation closer to the payline and are more stable. There are two basic methods which can be used for this purpose, namely, line drilling and presplitting.

Line drilling consists of drilling alternate holes between the main blastholes forming the edge of the excavation. The quantity of explosive placed in each line hole is significantly smaller and indeed if these holes are closely spaced, from 150 to 250 mm, then explosive may be placed only in every second or third hole. The closeness of the holes is controlled by the type of rock being excavated and by the payline. These holes are timed to fire ahead, with or after the nearest normally charged holes of the blasting pattern. The time of firing is also largely dependent on the character of the rock involved.

Presplitting can be defined as the establishment of a free surface or shear plane in the rock mass by the controlled usage of explosives in appropriately aligned and spaced drillholes. A line of trimming holes is charged and fired to produce a shear plane. This acts as a limiting plane for the blast proper and is carried out prior to the drilling and blasting of the main round inside the proposed break lines. The spacing of the trimming holes is governed by the type of rock and the diameter of the hole. Once presplit the rock excavation can be blasted with a normal pattern of holes. In very tight, unfissured rocks difficulty may be experienced in breaking out the main blast to base level.

Damage due to blasting vibrations can be estimated in terms of ground velocity, but it is extremely difficult to determine the limit values of ground velocity for varying degrees of damage. However, a conservative limit of 50 mm/s seems to be commonly accepted as the limit below which no damage will be caused to internal renderings and plasterwork. Nevertheless, low vibration levels may disturb sensitive machinery.

Vibrographs can be placed in locations considered

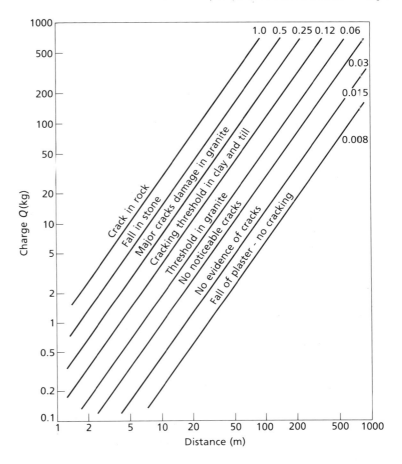

Fig. 9.1 Charge Q as a function of distance for various charge levels.

susceptible to blast damage in order to monitor ground velocity. A record of the blasting effects compared with the size of the charge and distance from the point of detonation is normally sufficient to reduce the possibility of damage to a minimum (Fig. 9.1). The use of multiple-row blasting with short delay ignition reduces the vibration effects.

The major objective of ripping in construction practice is to break the rock just enough to enable economic loading to take place. Rippability depends on intact strength, fracture index, and abrasiveness, that is, strong, massive, and abrasive rocks do not lend themselves to ripping. On the other hand, if sedimentary rocks such as sandstone and limestone are well bedded and jointed or if strong and weak rocks are thinly interbedded, then they can be excavated by ripping rather than by blasting. Some of the weaker sedimentary rocks (less than 1 MPa point-load strength, 15 MPa compressive strength) such as mudstones are not as easily removed by blasting

as their low strength would suggest since they are pulverized in the immediate vicinity of the hole. When blasted, mudstones may lift along bedding planes to fall back when the pressure has been dissipated. Such rocks, particularly if well jointed, are more suited to ripping.

The most common method for determining rippability is by seismic refraction. The seismic velocity of the rock concerned can be compared with a chart of ripper performance based on ripping operations in a wide variety of rocks (Fig. 9.2). Kirsten (1988), however, argued that seismic velocity could only provide a provisional indication of the way in which a rock mass could be excavated. Previously, Weaver (1975) had proposed the use of a modified form of the geomechanics classification (section 9.2.7) as a rating system for the assessment of rock mass rippability (Table 9.1).

The run direction during ripping should be normal to any vertical joint planes, down-dip to any inclined

Table 9.1 Rippability rating chart. (After Weaver 1975.)

	Rock class				
	I	II	III	IV	V
Description	Very good rock	Good rock	Fair rock	Poor rock	Very poor rock
Seismic velocity (m/s)	>2150	2150–1850	1850–1500	1500–1200	1200–450
Rating	26	24	20	12	5
Rock hardness (MPa)	Extremely hard rock (>70)	Very hard rock (20–70)	Hard rock (10–20)	Soft rock (3–10)	Very soft rock (1.7–3.0)
Rating	10	5	2	1	0
Rock weathering	Unweathered	Slightly weathered	Weathered	Highly weathered	Completely weathered
Rating	9	7	5	3	1
Joint spacing (mm)	>3000	3000–1000	1000–300	300–50	<50
Rating	30	25	20	10	5
Joint capacity	Non-continuous	Slightly continuous	Continuous – no gouge	Continuous – some gouge	Continuous – with gouge
Rating	5	6	3	0	0
Joint gouge	No separation	Slight separation	Separation <1 mm	Gouge <5 mm	Gouge >5 mm
Rating	5	5	4	3	1
Strike and dip orientation*	Very unfavourable	Unfavourable	Slightly unfavourable	Favourable	Very favourable
Rating	15	13	10	5	3
Total rating	100–90	90–70†	70–50	50–25	<25
Rippability assessment	Blasting	Extremely hard ripping and blasting	Very hard ripping	Hard ripping	Easy ripping
Tractor selection	—	DD9G/D9G	D9/D8	D8/D7	D7
Horsepower	—	770/385	385/270	270–180	180
Kilowatts	—	575–290	290/200	200–135	135

* Original strike and dip orientation now revised for rippability assessment.

† Ratings in excess of 75 should be regarded as unrippable without preblasting.

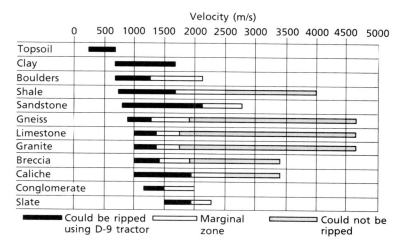

Fig. 9.2 Rippability chart.

strata and, on sloping ground, downhill (Atkinson, 1970). Ripping runs of 70 to 90 m usually give the best results. Where possible the ripping depth should be adjusted so that a forward speed of 3 km/h can be maintained since this is generally found to be the most productive. Adequate breakage depends on the spacing between ripper runs, which is governed by the fracture pattern in the rock.

The diggability of ground is of major importance in the selection of excavating equipment and depends principally upon the intact strength of the ground, its bulk density, bulking factor, and natural water content. The latter influences the adhesion or sticki-ness of soils, especially clay soils.

At the present there is no generally acceptable quantitative measure of diggability, assessment usually being made according to the experience of the operators. However, a fairly reliable indication can be obtained from similar excavations in the area or the behaviour of the ground excavated in trial pits. Attempts have been made to evaluate the performance of excavating equipment in terms of seismic velocity (Fig. 9.3). It would appear that most earth-moving equipment operates best when the seismic velocity of the ground is less than 1000 m/s and will not function above approximately 1800 m/s.

When material is excavated it increases in bulk, this being brought about by the decrease which occurs in density per unit volume. Some examples of typical bulking in soils are given in Table 9.2. The bulking factor is obviously important in relation to loading and removal of material from the working face.

9.1.4 Groundwater and excavation

Groundwater frequently provides one of the most difficult problems during excavation and its removal can prove costly. Not only does water make working conditions difficult, but piping, uplift pressures, and

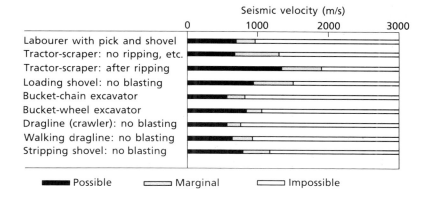

Fig. 9.3 Seismic velocities for determinating diggability. (After Atkinson, 1971.)

Table 9.2 Density, bulking factor, and diggability of some common soils

Soil type	Density (Mg/m^3)	Bulking factor	Diggability
Gravel, dry	1.8	1.25	E
Sand, dry	1.7	1.15	E
Sand and gravel, dry	1.95	1.15	E
Clay, light	1.65	1.3	M
Clay, heavy	2.1	1.35	M-H
Clay, gravel, and sand, dry	1.6	1.3	M

E, easy digging, loose, free-running material such as sand and small gravel; M, medium digging, partially consolidated materials such as clayey gravel and clay; M-H, medium-hard digging, materials such as heavy wet clay, gravels, and large boulders.

flow of water into an excavation can lead to erosion and failure of the sides. Collapsed material has to be removed and damage made good. Subsurface water is normally under pressure, which increases with increasing depth below the water table. Under high-pressure gradients weakly cemented rock can disintegrate. High piezometric pressures may cause the floor of an excavation to heave or, worse still, cause a blowout (Chapter 4). Hence, data relating to the groundwater conditions should be obtained prior to the commencement of operations.

Some of the worst conditions are met in excavations which have to be taken below the water table. In such cases the water level must be lowered by some method of dewatering. The method adopted depends upon the permeability of the ground and its variation within the stratal sequence, the depth of base level below the water table, and the piezometric conditions in underlying horizons. Pumping from sumps within an excavation, bored wells, or wellpoints (Fig. 9.4) are the dewatering methods most frequently used (Bell and Cashman, 1986). Impermeable barriers such as steel-sheet piles, secant piles, diaphragm walls, frozen walls, and

Fig. 9.4 A typical wellpoint layout. (Courtesy of P.M. Cashman.)

grouted walls can be used to keep water out of excavations (Bell and Mitchell, 1986). Ideally these structures should be keyed into an impermeable horizon beneath the excavation.

9.1.5 Methods of slope control and stabilization

It is rarely economical to design a rock slope so that no subsequent rockfalls occur, indeed many roads in rough terrain could not be constructed with the finance available without accepting some such risk. Therefore except where absolute security is essential, slopes should be designed to allow small falls of rock under controlled conditions.

Rock traps in the form of a ditch and/or barrier can be installed at the foot of a slope. Benches may also act as traps to retain rockfall, especially if a barrier is placed at their edge. Wire mesh suspended from the top of the face provides yet another method for controlling rock fall (Fig. 9.5).

Excavation involving the removal of material from the head of an unstable slope, flattening of the slope, benching of the slope, or complete removal of the unstable material helps stabilize a slope. If some form of reinforcement is required to provide support for a rock slope, then it is advisable to install it as quickly as possible after excavation. 'Dentition' refers to masonry or concrete infill placed in fissures or cavities in a rock slope (Fig. 9.6). Thin to medium bedded rocks dipping parallel to the slope can be held in place by steel dowels grouted into drilled holes, which are up to 2 m in length. Rock bolts may be up to 8 m in length with a tensile working load of up to 100 kN. They are put in tension so that the compression induced in the rock mass improves shearing resistance on potential failure plans. Bearing plates, light steel sections, or steel mesh may be used between bolts to support the rock face. Rock anchors are used for major stabilization works, especially in conjunction with retaining structures. They may exceed 30 m in length. For excavated slopes it is more advantageous to improve the properties of the rock slope itself (by anchoring or bolting) than to remove the rock and replace it with concrete.

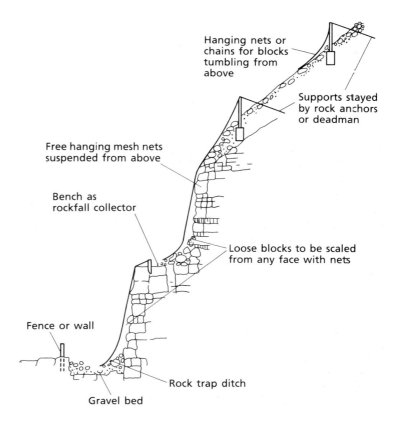

Hanging nets or chains for blocks tumbling from above

Supports stayed by rock anchors or deadman

Free hanging mesh nets suspended from above

Bench as rockfall collector

Loose blocks to be scaled from any face with nets

Fence or wall

Rock trap ditch

Gravel bed

Fig. 9.5 Minimization of rockfall by structural means. (After Fookes and Sweeney, 1976.)

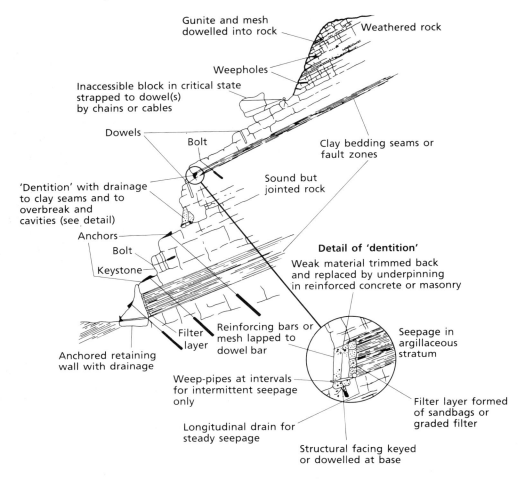

Fig. 9.6 Rockfall control measures. (After Fookes and Sweeney, 1976.)

Gunite or shotcrete are frequently used to preserve the integrity of a rock face by sealing the surface and inhibiting the action of weathering (Fig. 9.7). They are pneumatically applied mortar or concrete respectively. Coatings may be reinforced with wire mesh or used in combination with rock bolts. It is generally considered that such surface treatment offers negligible support to the overall slope structure. Heavily fractured rocks may be grouted in order to stabilize them.

Restraining structures such as retaining walls, cribs, gabions, and buttresses control sliding by increasing the resistance to movement. There are certain limitations which must be considered before retaining walls are used for slope control. These involve the ability of the structure to resist shearing action, overturning and sliding on or below the base of the structure. Retaining walls are often used where there is a lack of space for the full development of a slope, such as along many roads and railways. As retaining walls are subjected to unfavourable loading, a large wall width is necessary to increase slope stability. Reinforced earth can be used for retaining earth slopes. Such a structure is flexible and so can accommodate some settlement. Thus reinforced earth can be used on poor ground where conventional alternatives would require expensive foundations. Reinforced earth walls are constructed by erecting a thin front skin at the face of the wall at the same time as the earth is placed. Strips of steel or geosynthetic are fixed to the facing skin at regular intervals. Cribs may be constructed

Fig. 9.7 Use of shotcrete and rock bolts to protect a road cutting in Natal Group Sandstone, near Mont-aux-Sources, Natal, South Africa.

of precast reinforced concrete or steel units set up in cells which are filled with gravel or stone. Gabions consist of strong wire mesh surrounding placed stones. Concrete buttresses occasionally have been used to support large blocks of rock, usually where they overhang.

Geosynthetic materials, especially geomats and geogrids, are being used increasingly to protect slopes. Geomats are draped over the slope requiring protection and are pegged onto the soil. They are three-dimensional geosynthetics and if filled with soil help establish a vegetative cover.

Drainage is the most generally applicable method for improving the stability of slopes or for the corrective treatment of slides since it reduces the effectiveness of one of the principal causes of instability, namely, excess pore water pressure. The most likely zone of failure must be determined so that the extent of the slope mass which requires drainage treatment can be defined.

Surface runoff should not be allowed to flow unrestrained over a slope. This is usually prevented by the installation of a drainage ditch at the top of an excavated slope to collect drainage from above. The ditch, especially in soils, should be lined to prevent erosion, otherwise it will act as a tension crack. It may be filled with cobble aggregate. Herringbone ditch drainage is usually employed to convey water from the surfaces of slopes. These drainage ditches lead into an interceptor drain at the foot of the slope (Fig. 9.8). Infiltration can be lowered by sealing the cracks in a slope by regrading or filling with cement,

bitumen or clay. A surface covering has a similar purpose and function, for example, the slope may be covered with granular material resting upon filter fabric.

Support and drainage may be afforded by counterfort drains where an excavation is made in sidelong ground likely to undergo shallow, parallel slides. Deep trenches are cut into the slope, lined with filter fabric, and filled with granular material. The granular fill in each trench acts as a supporting buttress or counterfort, as well as providing drainage. However, counterfort drains must extend beneath the potential failure zone, otherwise they merely add unwelcome weight to the slipping mass.

Successful use of subsurface drainage depends on tapping the source of water, the presence of permeable material which aids free drainage, the location of the drain on relatively unyielding material to ensure continuous operation (flexible, PVC drains are now frequently used) and the installation of a filter to minimize silting in the drainage channel. Drainage galleries (Fig. 9.9) are costly to construct and in slipping areas may experience caving. They should be backfilled with stone to ensure their drainage capacity if partially deformed by subsequent movements. Galleries are indispensable in the case of large, slipped masses where drainage has to be carried out over lengths of 200 m or more. Drillholes may be made about the perimeter of a gallery to enhance drainage. Drainage holes with perforated pipes are much cheaper than galleries and are satisfactory over short lengths but it is more difficult to

Fig. 9.8 Surface drainage of a slope in till deposits, near Loch Lomond, Scotland. The aggregate filled ditch drainage leads into an interceptor drain.

Fig. 9.9 Internal drainage gallery in restored slope, near Aberfan, South Wales.

intercept water bearing layers with them. When individual benches are drained by horizontal holes, the latter should lead into a properly graded interceptor trench, which is lined with impermeable material.

9.2 TUNNELS AND TUNNELLING

Geology is the most important factor which determines the nature, form, and cost of a tunnel (Taylor and Conwell, 1981). For example, the route, design, and construction of a tunnel are largely dependent upon geological considerations. Accordingly, tunnelling is an uncertain, often hazardous undertaking because information on ground conditions along the alignment is never complete, no matter how good the site investigation. Estimating the cost of tunnel construction, particularly in areas of geological complexity, is also uncertain.

Prior to tunnel construction the subsurface geology may be explored by means of pits, adits, drilling, and pilot tunnels. Exploration adits before tunnelling proper commences are not usually driven unless a particular section appears to be especially dangerous or a great deal of uncertainty exists. Core drilling aids the interpretation of geological features already identified at the surface.

A pilot tunnel is probably the best method of exploring tunnel locations and should be used if a major sized tunnel is to be constructed in ground that is known to have critical geological conditions. It also drains the rock ahead of the main excavation. If the inflow of water is excessive the rock can be grouted from the pilot tunnel before the main excavation reaches the water bearing zone.

Reliable information about the ground conditions ahead of the advancing face is obviously desirable and can be achieved with a varying degree of success by drilling long, horizontal holes between shafts, or by direct drilling from the tunnel face at regular intervals. In extremely poor ground conditions tunnelling progresses behind an array of probe holes which fan outwards some 10 to 30 m ahead of the tunnel face. Although this slows progress, it ensures completion. Holes drilled upwards from the crown of the tunnel and forwards from the side walls help locate features such as faults, buried channels, weak seams, or solution cavities. Drilling equipment for drilling in a forward direction can be incorporated into a shield or tunnelling machine. The penetration rate of a probe drill must exceed that of the tunnel boring machine; ideally it should be about three times faster. Maintaining the position of the hole, however, presents the major problem during horizontal drilling. In particular, variations in hardness of the ground oblique to the direction of drilling can cause radical deviations. Even in uniform ground rods go off line. The inclination of a hole therefore must be surveyed.

Geophysical investigations can give valuable assistance in determination of subsurface conditions, especially in areas in which the solid geology is poorly exposed. Seismic refraction has been used in measuring depths of overburden in the portal areas of tunnels, in locating faults, weathered zones, or buried channels, and in estimating rock quality (Cratchley et al., 1976; McCann et al., 1990). Seismic testing also can be used to investigate the topography of a river bed and the interface between the alluvium and bedrock. Seismic logging of drillholes can, under favourable circumstances, provide data relating to the engineering properties of rock. Resistivity techniques have proved useful in locating water tables and buried faults, particularly those which are saturated. Resistivity logs of drillholes are used in lateral correlation of layered materials of different resistivities and in the detection of permeable rocks. Ground probing radar offers the possibility of exploring large volumes of rock for anomalies in a short time and at low cost, in advance of major subsurface excavations.

9.2.1 Geological conditions and tunnelling

Large planar surfaces form most of the roof in a rock formation which is not inclined at a high angle and strikes more or less parallel to the axis of a tunnel. In tunnels where jointed strata dip into the side at 30° or more, the updip side may be unstable. Joints which are parallel to the axis of a tunnel and which dip at more than 45° may prove especially treacherous, leading to slabbing of the walls and fallouts from the roof (Fig. 9.10). The effect of joint orientation in relation to the axis of a tunnel is given in Table 9.3.

The presence of flat-lying joints may also lead to blocks becoming dislodged from the roof. When the tunnel alignment is normal to the strike of jointed rocks and the dips are less than 15°, large blocks are again likely to fall from the roof. The sides, however, tend to be reasonably stable. When a tunnel is driven perpendicular to the strike in steeply dipping (Table 9.3) or vertical strata each stratum acts as a beam with a span equal to the width of the cross-section. However, in such a situation blasting operations are generally less efficient. If the axis of a tunnel runs parallel to the strike of vertically dipping

Table 9.3 The effect of joint strike and dip orientations in tunnelling

Strike perpendicular to tunnel axis				Strike parallel to tunnel axis	
Drive with dip		Drive against dip			
Dip 45–90°	Dip 20–45°	Dip 45–90°	Dip 20–45°	Dip 45–90°	Dip 20–45°
Very favourable	Favourable	Fair	Unfavourable	Very unfavourable	Fair
Dip 0–20°: unfavourable, irrespective of strike					

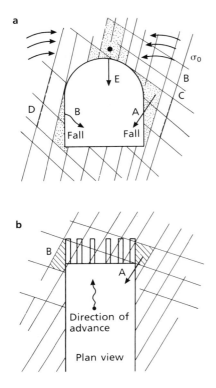

Fig. 9.10 Tunnel in rock with steeply dipping joints: (a) steeply dipping joints (45–90°) which are parallel to the tunnel axis, lead to slabbing of the wall and fallouts from the roof. At point A the slab 'daylights' at the feather-edge bottom and would probably fall with the force of the blast during tunnel advance. The slab, B, may not fall; however, it could be loosened by the original blast and would be susceptible to additional loosening by the shocks of later blasts and by the 'working' of the rock under peak tangential stresses around the tunnel periphery. Unless restrained, the slab, B, might eventually fall. Joints at depth such as C and D may tend to open. Joint blocks at E may be extremely dangerous, appearing stable after the blast but becoming unstable as the tunnel advances and the rock adjusts to the new stress field; (b) block A may be loosened and possibly forcefully ejected by gas pressures during blasting; block B might be loosened but not necessarily removed. Had the tunnel advanced in the opposite direction the relative positions of A and B would be interchanged. (After Robertson, 1974.)

rocks, then the mass of rock above the roof is held by the friction along the bedding planes. In such a situation the upper boundary of loosened rock, according to Terzaghi (1946), does not extend beyond 0.25 times the tunnel width above the crown.

When the joint spacing in horizontally layered rocks is greater than the width of a tunnel, then the beds bridge the tunnel as a solid slab and are only subject to bending under their own weight. Thus, if the bending forces are less than the tensile strength of the rock the roof need not be supported. Where horizontally lying rocks are thickly bedded and contain few joints the roof of the tunnel is flat. Conversely, if the rocks are thinly bedded and are intersected by many joints a peaked roof may form. Nonetheless breakage rarely, if ever, continues beyond a vertical distance equal to half the width of the tunnel above the top of a semicircular payline (Fig. 9.11). This type of stratification is more dangerous where the beds dip at 5 to 10° since this may lead to the roof spalling, as the tunnel is driven forward.

Faults generally mean non-uniform rock pressures on a tunnel and hence at times necessitate special treatment such as the construction of box sections with invert arches. Generally, problems increase as the strike of a fault becomes more parallel to the tunnel opening. However, even if the strike is across the tunnel, faults with low dips can represent a hazard. If the tunnel is driven from the hanging wall, the fault first appears at the invert and it is generally possible to provide adequate support or reinforcement when driving through the rest of the zone. Conversely, when a tunnel is driven from the foot-wall side, the fault first appears in the crown, and there is a possibility that a wedge-shaped block, formed by the fault and the tunnel, will fall from the roof without warning.

Major faults are usually associated with a number of minor faults and the dislocation zone may occur over many metres. What is more, rock material within a faulted zone may be shattered and unstable. Problems tend to increase with increasing width of the fault zone. Sometimes a fault zone is filled with sand-sized crushed rock that has a tendency to flow into the tunnel. If, in addition, the tunnel is located beneath the water table, a sandy suspension may rush into the tunnel. When a fault zone is occupied by clay gouge and a section of a tunnel follows the gouge zone, swelling of this material may occur and cause displacement or breakage of tunnel supports during construction. Large quantities of water in a permeable rock mass are impounded by a fault zone occupied by impervious gouge and are released when tunnelling operations penetrate through the fault zone.

Movements along major active faults in certain parts of the world can disrupt a tunnel lining and

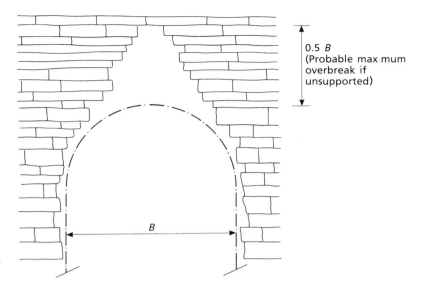

0.5 B
(Probable maximum overbreak if unsupported)

B

Fig. 9.11 Overbreak in thinly bedded horizontal strata with joints. Ultimate overbreak occurs if no support is installed.

even lead to a tunnel being offset. As a consequence it is best to shift the alignment to avoid the fault or, if possible, to use open cut within the active fault.

The earthquake risk to an underground structure is influenced by the material in which it occurs, for example, a tunnel at shallow depth in alluvial deposits will be seriously affected by the large relative displacements of the ground surrounding it. On the other hand, a deep tunnel in solid rock will be subjected to displacements which are considerably less than those which occur at the surface. The main causes of stresses in shallow underground structures arise from the interaction between the structure and the displacements of the ground. If the structure is sufficiently flexible it will follow the displacements and deformations to which the ground is subjected.

Rocks, especially those at depth, are affected by the weight of overburden and the stresses so developed cause the rocks to be strained. In certain areas, particularly orogenic belts, the state of stress is also influenced by tectonic factors. However, because the rocks at depth are confined they suffer partial strain. The stress that does not give rise to strain (i.e. which is not dissipated) remains in the rocks as residual stress. While the rocks remain in a confined condition the stresses accumulate and may reach high values, sometimes in excess of yield point. If the confining condition is removed, as in tunnelling, then the residual stress can cause displacement. The amount of movement depends upon the magnitude of the residual stress. The pressure relief, which

represents a decrease in residual stress, may be instantaneous or slow in character, and is accompanied by movement of the rock mass with variable degrees of violence.

In tunnels driven at great depths, rock may suddenly break from the sides of the excavation, a phenomenon known as rock bursting. In such failures hundreds of tonnes of rock may be released with explosive force. Rock bursts are due to the dissipation of residual stresses which exceed the strength of the ground around the excavation, and their frequency and severity tend to increase with depth — most rock bursts occur at depths in excess of 600 m. The stronger the rock the more likely it is to burst. The most explosive failures occur in rocks which have unconfined compressive strengths and values of Young's modulus greater than 140 MPa and 34 500 MPa, respectively.

Popping is a similar but less violent form of failure. In this case the sides of an excavation bulge before exfoliating. Spalling tends to occur in jointed or cleaved rocks. To a certain extent such a rock mass can bulge as a sheet, collapse occurring when a key block either fails or is detached from the mass.

In fissile rock such as shale, the beds may slowly bend into the tunnel. In this case the rock is not necessarily detached from the main mass, but the deformation may cause fissures and hollows in the rock surrounding the tunnel.

Another pressure relief phenomenon is bumping ground. Bumps are sudden and somewhat violent earth tremors which at times dislodge rock from the sides of the tunnel. They probably are due to rock displacements consequent upon the newly created stress conditions.

9.2.2 Tunnelling in soft ground

All soft ground moves in the course of tunnelling operations (Peck, 1969). In addition some strata change their characteristics on exposure to air. Both factors put a premium on speed of advance and successful tunnelling requires matching the work methods to the stand-up time of the ground.

The difficulties and costs of construction with soft-ground tunnelling depend almost exclusively on the stand-up time of the ground and this in turn is greatly influenced by the position of the water table in relation to the tunnel (Hansmire, 1981). Above the water table the stand-up time principally depends on the shearing and tensile strength of the ground whereas below it, it also is influenced by the permeability of the material involved.

Terzaghi (1950a) distinguished the following types of soft ground:

1 Firm ground. Firm ground has sufficient shearing and tensile strength to allow the tunnel heading to be advanced without support, typical representatives being stiff clays with low plasticity and loess above the water table.

2 Ravelling ground. In ravelling ground, blocks fall from the roof and sides of the tunnel some time after the ground has been exposed. The strength of the ground usually decreases with increasing duration of load. It may also decrease due to dissipation of excess pore water pressures induced by ground movements in clay, or due to evaporation of water with subsequent loss of apparent cohesion in silt and fine sand. If ravelling begins within a few minutes of exposure it is described as fast ravelling, otherwise it is referred to as slow ravelling. Fast ravelling may take place in residual soils and sands with a clay binder below the water table. These materials above the water table are slow ravelling.

3 Running ground. In this type of ground the removal of support from a surface inclined at more than 34° gives rise to a run, the latter occurring until the angle of rest of the material involved is attained. Runs take place in clean, loosely packed gravel, and clean, coarse to medium-grained sand, both above the water table. In clean, fine-grained moist sand a run is usually preceded by ravelling, such behaviour being termed cohesive running.

4 Flowing ground. This type of ground moves like a viscous liquid. It can invade a tunnel from any angle and, if not stopped, ultimately will fill the excavation. Flowing conditions occur in sands and silts below the water table. Such ground above the water table exhibits either ravelling or running behaviour.

5 Squeezing ground. Squeezing ground advances slowly and imperceptibly into a tunnel. There are no signs of fracturing of the sides. Ultimately the roof may give and this can produce a subsidence though at the surface. The two most common reasons why ground squeezes on subsurface excavation are: (i) excessive overburden pressure; (ii) the dissipation of residual stress, both eventually leading to failure. Soft and medium clays display squeezing behaviour. Other materials in which squeezing conditions may obtain include shales and highly weathered granites, gneisses, and schists.

6 Swelling ground. Swelling ground also expands into the excavation but the movement is associated with a considerable volume increase in the ground immediately surrounding the tunnel. Swelling occurs as a result of water migrating into the material of the tunnel perimeter from the surrounding strata. These conditions develop in overconsolidated clays with a plasticity index in excess of about 30% and in certain shales and mudstones, especially those containing montmorillonite. Swelling pressures are of unpredictable magnitude and may be extremely large, for example, the swelling pressure in shallow tunnels may exceed the overburden pressure and in overconsolidated clays it may be as high as 2.0 MPa. The development period may take a few weeks or several months. Immediately after excavation the pressure is insignificant but then it increases at a higher rate. In the final stages the increase slows down.

Boulders within a soft ground matrix may prove difficult to remove whilst if boulders are embedded in a hard cohesive matrix, they may greatly impede the progress of even a hand-mined shield and may render a mechanical excavator of almost any type impotent. Large boulders may be difficult to handle unless they are broken apart by jackhammer or blasting.

9.2.3 Water in tunnels

Construction of a tunnel may alter the groundwater regime of a locality, as a tunnel generally acts as a drain. The amount of water held in a rock mass depends on its reservoir storage properties (Chapter 4), which in turn influence the amount of water which can drain into a tunnel. Isolated heavy flows of water may occur in association with faults, solution pipes, and cavities, or abandoned mine workings, or even from pockets of gravel. Tunnels driven under lakes, rivers, and other surface bodies of water may tap a considerable volume of flow. Flow also may take place from a perched water table to a tunnel beneath.

Generally, the amount of water flowing into a tunnel decreases as construction progresses. This is due to the gradual exhaustion of water at source and to the decrease in hydraulic gradient, and hence in flow velocity. On the other hand, there may be an increase in flow if construction operations cause fissuring, for example, blasting may open new water conduits around a tunnel, shift the direction of flow, and sometimes may even cause partial flooding.

Correct estimation of the water inflow into a projected tunnel is of vital importance, as inflow influences the construction programme (Cripps *et al.*, 1988). One of the principal problems created by water entering a tunnel is that of face stability. Secondary problems include removal of excessively wet muck and the placement of a precision-fitted primary lining or of ribs.

Not only is the value of the maximum inflow required but so is the distribution of inflow along the tunnel section and the changes of flow with time. The greatest groundwater hazard in underground work is the presence of unexpected water bearing zones, and therefore whenever possible the position of hydrogeological boundaries should be located. Obviously, the location of the water table, and its possible fluctuations are of major consequence.

Water pressures are more predictable than water flows as they are nearly always a function of the head of water above the tunnel location. They can be very large, especially in confined aquifers. Hydraulic pressures should be taken into account when considering the thickness of rock that will separate an aquifer from the tunnel. Unfortunately, however, the hydrogeological situation is rarely so easily interpreted as to make accurate quantitative estimates possible.

Sulphate bearing solutions attack concrete; thus water quality must be investigated. Particular attention should be given to water flowing from sequences containing gypsum and anhydrite. Rocks containing iron pyrite may also give rise to waters carrying sulphates.

Most of the serious difficulties encountered during tunnelling operations are directly or indirectly caused by the percolation of water towards the tunnel and most of the techniques for improving ground conditions are directed towards its control. This may be achieved by using drainage, compressed air, grouting, or freezing techniques.

9.2.4 Gases in tunnels

Naturally occurring gas can occupy the pore spaces and voids in rock. This gas may be under pressure and there have been occasions where gas under pressure has burst into underground workings causing the rock to fail with explosive force. Wherever possible the likelihood of gas hazards should be noted during the geological survey, but this is one of the most difficult tunnel hazards to predict. If the flow of gas appears to be fairly continuous, the entrance of the flow may be sealed with concrete. Often the supply of gas is quickly exhausted, but cases have been reported where it continued for up to three weeks.

Many gases are dangerous; methane, which may be encountered in Coal Measures, is lighter than air and can readily migrate from its point of origin. Not only is it toxic, it is also combustible and highly explosive when mixed with air. Carbon dioxide and carbon monoxide are both toxic. The former is heavier than air and hangs about the floor of an excavation. Carbon monoxide is slightly lighter than air and like carbon dioxide and methane it is found in Coal Measures strata. Carbon dioxide also may be associated with volcanic deposits and limestones. Hydrogen sulphide is heavier than air and is highly toxic. It is also explosive when mixed with air. The gas may be generated by the decay of organic substances or by volcanic activity. Hydrogen sulphide may be absorbed by water which then becomes injurious as far as concrete is concerned. Sulphur dioxide is a colourless, pungent, asphyxiating gas which dissolves readily in water to form sulphuric acid. It is usually associated with volcanic emanations or it may be formed by the breakdown of pyrite.

9.2.5 Temperatures in tunnels

Temperatures in tunnels are not usually of concern unless the tunnel is more than 170 m below the surface. When rock is exposed by excavation the amount of heat liberated depends on (i) the virgin rock temperature (VRT); (ii) the thermal properties of the rock; (iii) the length of time of exposure; (iv) the area, size, and shape of exposed rock; (v) the wetness of rock; (vi) the air-flow rate; (vii) the dry-bulb temperature; and (viii) the humidity of the air.

In deep tunnels high temperatures can make work more difficult, and along with rock pressures limit the depth of tunnelling. The moisture content of the air in tunnels is always high and in saturated air the efficiency of labour declines when the temperature exceeds 25°C, dropping to almost zero when the temperature reaches 35°C. Conditions can be improved by increased ventilation, by water spraying, or by using refrigerated air. Air refrigeration is essential when the VRT exceeds 40°C.

The rate of increase in rock temperature with depth depends on the geothermal gradient, which in turn is inversely proportional to the thermal conductivity of the material involved:

$$\text{Geothermal gradient} = \frac{0.05}{k} \text{ (approximately) } °C/m$$

$$(9.3)$$

where k is the thermal conductivity. Although the geothermal gradient varies with locality, and according to rock type and structure, on average it increases at a rate of 1°C per 30 to 35 m depth. In geologically stable areas the mean gradient is 1°C for every 60 to 80 m whereas in volcanic districts it may be as much as 1°C for every 10 to 15 m depth. The geothermal gradient under mountains is larger than under plains; in the case of valleys, the situation is reversed.

The temperature of the rocks influences the temperature of any water they may contain. Fissure water that flows into workings acts as an efficient carrier of heat. This may be locally more significant than the heat conducted through the rocks themselves, for example, for every litre of water which enters the workings at a VRT of 40°C, if the water cools to 25°C before it reaches the pumps, then the heat added to the ventilating air stream will be 62.8 kW.

Earth temperatures can be measured by placing thermometers in drillholes, measurement being taken when a constant temperature is attained. The results, in the form of geoisotherms, can be plotted on the longitudinal section of the tunnel.

9.2.6 Excavation of tunnels

In soft ground, support is vital and so tunnelling is carried out by using shields. A shield is a cylindrical drum with a cutting edge around the circumference, the cut material being delivered onto a conveyor for removal. Shields without cutting heads may be used in sands, silts, or clays. The limits of these machines are usually given as an unconfined compressive strength of 20 MPa.

The use of bentonite slurry in a pressure bulkhead machine to support the face in soft ground was introduced in the late 1960s. It represented a major innovation in mechanized tunnelling, particularly in granular sediments not suited to compressed air. The bentonite slurry counterbalances the hydrostatic head of groundwater in the soil and stability is further increased as the bentonite is forced into the pores of the soil, gelling once penetration occurs. The bentonite forms a seal on the surface. However, boulders in soils, such as till, create an almost impossible problem for slurry-face machines. A mixed face of hard rock and cohesionless soil below the water table presents a similar dilemma.

Mechanical cutting using moles or blasting is used in hard rock tunnelling. The performance of tunnel boring machines (TBMs) is more sensitive to changes in rock properties than conventional drilling and blasting techniques. Consequently, their use in rock masses which have not been thoroughly investigated involves a high risk. Giant TBMs are capable of drilling holes several metres in diameter through rock formations.

Apart from ground stability and support, the most important economic factors in machine tunnelling in hard rock are cutter costs and penetration rate. The rate of wear is basically a function of the abrasive characteristics of the rock mass involved. Penetration rate is a function of cutter geometry, thrust of the machine, and the rock strength.

Tunnel boring machines have successfully cut most sedimentary rock types. In fact these machines have achieved a faster rate of drivage than conventional tunnelling methods in rocks with unconfined compressive strengths of up to 150 MPa. For example, tunnel boring machines have occasionally excavated 100 m per day in moderately strong rocks (70 to 140 MPa unconfined compressive strength). They are normally much slower in very soft rock because of time spent in ground control and in very hard rock because of slow penetration rate.

The stresses imposed on the surrounding rock by machine tunnelling are much less than those produced during blasting and therefore damage to the perimeter is minimized and a sensibly smooth base is usually achieved. Overbreak is also less, on average 5% as compared with up to 25% for conventional methods. This means that less support is required.

As a result, machine tunnelling is generally less expensive than the conventional method. Nevertheless, the decision to use a machine must be based upon a particularly thorough knowledge of the anticipated geological conditions.

Overbreak refers to the removal of rock material beyond the payline, the cost of which has to be met by the contractor. Obviously, every effort must be made to keep overbreak to a minimum. The amount of overbreak is influenced by rock type and discontinuities as well as the type of excavation.

The conventional method of advancing a tunnel in hard rock is by full-face driving in which the complete face is drilled and blasted as a unit. However, full-face driving should be used with caution where the rocks are variable. The usual alternatives are the top-heading and bench method or the top-heading method whereby the tunnel is worked on an upper and lower section or heading. The sequence of operations in these three methods is illustrated in Figure 9.12.

In tunnel blasting a cut is opened up approximately in the centre of the face in order to provide a cavity into which subsequent shots can blast. Delay detonation allows a face to be charged, stemmed, and fired, the shots being detonated in a predetermined sequence. The first shots in the round blast out the cut and subsequent shots blast in sequence to the free face so formed.

Drilling and blasting can damage the rock structure depending on the properties of the rock mass and the blasting technique. As far as technique is concerned, attention should be given to the need to maintain adequate depths of pull, to minimize overbreak and to maintain blasting vibrations below acceptable levels. The stability of a tunnel roof in fissured rocks depends upon the formation of a natural arch and this is influenced by the extent of the disturbance, the irregularities of the profile, and the relationship between tunnel size and fracture pattern. The amount of overbreak tends to increase with increased depths of pull since drilling inaccuracies are magnified. In such situations not only does the degree of overbreak become very expensive in terms of grout and concrete backfill but it may give rise to support problems and subsidence over the crown of the tunnel. However, overbreak can be reduced by accurate drilling and a carefully controlled scale of blasting. Controlled blasting may be achieved either by presplitting the face to the desired contour or by smooth blasting.

In the presplitting method a series of holes is drilled around the perimeter of the tunnel, loaded with explosives which have a low charging density,

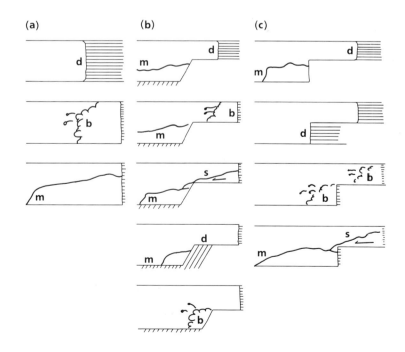

Fig. 9.12 Tunnelling methods: (a) full face; (b) top heading and bench; (c) top heading. Bench drilled horizontally. Phases: d, drilling; b, blasting; m, mucking; s, scraping.

and detonated before the main blast. The initial blast develops a fracture which spreads between the holes. Hence, the main blast leaves an accurate profile. The technique is not particularly suited to slates and schists because of their respective cleavage and schistosity. Indeed, slates tend to split along, rather than across their cleavage. Although it is possible to presplit jointed rock masses adequately, the tunnel profile is still influenced by the pattern of the jointing.

Smooth blasting has proved a more successful technique than presplitting. Here again explosives with a low charging density are used in closely spaced perimeter holes, for example, the ratio between burden and hole spacing is usually 1:0.8 which means that crack formation is controlled between the drillholes and hence is concentrated within the final contour. The holes are fired after the main blast, their purpose being to break away the last fillet of rock between the main blast and the perimeter. Smooth blasting cannot be carried out without good drilling precision. Normally, smooth blasting is restricted to the roof and walls of a tunnel but occasionally it is used in the excavation of the floor. Because fewer cracks are produced in the surrounding rock it is stronger and so the desired roof curvature can be maintained to the greatest possible extent. Hence, the load-carrying capacity of the rock is properly utilized.

9.2.7 Analysis of tunnel support

The time a rock mass may remain unsupported in a tunnel is called its stand-up time or bridging capacity. This mainly depends on the magnitude of the stresses within the unsupported rock mass, which in their turn depend on the span, strength, and discontinuity pattern. If the bridging capacity of the rock is high, the rock material next to the heading will stay in place for a considerable time. By contrast, if the bridging capacity is low, the rock will immediately start to fall at the heading so that supports have to be positioned as soon as possible.

The primary support for a tunnel in rock may be provided by rock bolts, shotcrete, or steel arches (Clough, 1981). Rock bolts maintain the stability of an opening: (i) by suspending the dead weight of a slab from the rock behind; (ii) by providing a normal stress on the rock surface to clamp discontinuities together and develop beam action; (iii) by providing a confining pressure to increase shearing resistance and develop arch action (the ability of the rock mass to transfer load to the tunnel sides); and (iv) by

preventing key blocks becoming loosened so that the strength and integrity of the rock mass is maintained. Shotcrete can be used for lining tunnels, e.g. a layer of 150 mm thick around a tunnel 10 m in diameter can safely carry a load of 500 kPa corresponding to a burden of approximately 23 m of rock, more than has ever been observed with rock falls. When combined with rock bolting, shotcrete has proved an excellent temporary support for all qualities of rock. In very bad cases steel arches can be used for reinforcement of weaker tunnel sections.

A classification of rock masses is of primary importance in relation to the design of the type of tunnel support. Lauffer's (1958) classification represented an appreciable advance in the art of tunnelling since it introduced the concept of an active unsupported rock span and the corresponding stand-up time, both of which are very relevant parameters for determination of the type and amount of primary support in tunnels. The active span is the width of the tunnel or the distance from support to the face in cases where this is less than the width of the tunnel. The relationships found by Lauffer are given in Figure 9.13.

Bieniawski (1974, 1989) maintained that (i) the uniaxial compressive strength of rock material; (ii) the rock quality designation; (iii) the spacing, orientation, and condition of the discontinuities; and (iv) groundwater inflow were the factors which should be considered in any engineering classification of rock. His classification of rock masses based on these parameters is given in Table 9.4. Each parameter is grouped into five categories and the categories are given a rating. Once determined, the ratings of the individual parameters are summed to give the total rating or class of the rock mass. The higher the total rating, the better the rock mass conditions. However, the accuracy of the rock mass rating in certain situations may be open to question, for example, it does not take into account the effects of blasting on rock masses. Neither does it consider the influence of *in situ* stress on stand-up time nor the durability of the rock (Varley, 1990). The latter can be assessed in terms of the geodurability classification (Chapter 3).

Suitable support measures at times must be adopted to attain a stand-up time longer than that indicated by the total rating or class of the rock mass. These measures constitute the primary or temporary support. Their purpose is to ensure tunnel stability until the secondary or permanent support system, for example, a concrete lining, is installed. The form of primary support depends on: (i) depth

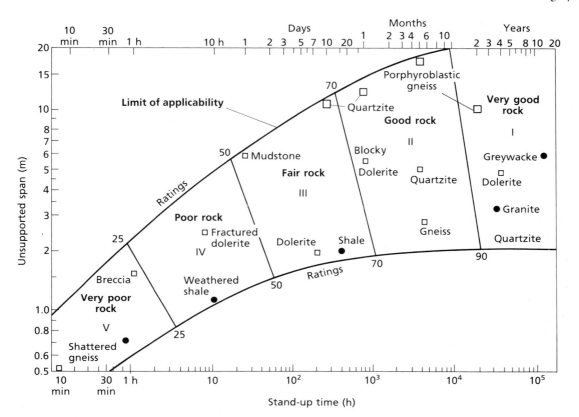

Fig. 9.13 Geomechanics classification of rock masses for tunnelling. South African case studies are indicated by squares while those from Alpine countries are shown by dots. (After Lauffer, 1958.)

below the surface; (ii) tunnel size and shape; and (iii) method of excavation. Table 9.5 indicates the primary support measures for shallow tunnels 5 m to 12 m in diameter driven by drilling and blasting.

Barton *et al.* (1975) pointed out that Bieniawski (1974) in his analysis of tunnel support more or less ignored the roughness of joints, the frictional strength of the joint fillings, and the rock load. They therefore proposed the concept of rock mass quality (Q) which could be used as a means of rock classification for tunnel support. They defined the rock mass quality in terms of six parameters:

1 the RQD or an equivalent estimate of joint density;

2 the number of joint sets (J_n), which is an important indication of the degree of freedom of a rock mass. The RQD and the number of joint sets provide a crude measure of relative block size;

3 the roughness of the most unfavourable joint set (J_r). The joint roughness and the number of joint

sets determine the dilatancy of the rock mass;

4 the degree or alteration or filling of the most unfavourable joint set (J_a). The roughness and degree of alteration of the joint walls or filling materials provides an approximation of the shear strength of the rock mass;

5 the degree of water seepage (J_w);

6 the stress reduction factor (SRF) which accounts for the loading on a tunnel caused either by loosening loads in the case of clay-bearing rock masses, or unfavourable stress–strength ratios in the case of massive rock. Squeezing and swelling is also taken account of in the SRF.

They provided a rock mass description and ratings for each of the six parameters (Table 9.6) which enabled the rock mass quality (Q) to be derived from:

$$Q = \frac{\text{RQD}}{J_n} \times \frac{J_r}{J_a} \times \frac{J_w}{\text{SRF}} \tag{9.4}$$

Table 9.4 The rock mass rating system (geomechanics classification of rock masses). (After Bieniawski, 1989.)

(a) Classification parameters and their ratings

Parameter		Ranges of values						
1	Strength of intact rock material	Point-load strength index (MPa)	>10	4–10	2–4	1–2	For this low range, uniaxial compressive test is preferred	
		Uniaxial compressive strength (MPa)	>250	100–250	50–100	25–50	5–25	1–5	<1
	Rating		15	12	7	4	2	1	0
2	Drill core quality RQD (%)		90–100	75–90	50–75	25–50	<25	
	Rating		20	17	13	8	3	
3	Spacing of discontinuities		>2 m	0.6–2 m	200–600 mm	60–200 mm	<60 mm	
	Rating		20	15	10	8	5	
4	Condition of discontinuities		Very rough surfaces Not continuous No separation Unweathered wall rock	Slightly rough surfaces Separation <1 mm Slightly weathered walls	Slightly rough surfaces Separation <1 mm Highly weathered wall	Slickensided surfaces or Gouge <5 mm thick or Separation 1–5 mm Continuous	Soft gouge >5 mm thick or Separation >5 mm Continuous	
	Rating		30	25	20	10	0	
5	Groundwater	Inflow per 10 m tunnel length (l/min)	None	<10	10–25	25–125	>125	
		Ratio $\dfrac{\text{Joint water pressure}}{\text{Major principal stress}}$	0	<0.1	0.1–0.2	0.2–0.5	>0.5	
			or	or	or	or	or	
		General conditions	Completely dry	Damp	Wet	Dripping	Flowing	
	Rating		15	10	7	4	0	

Table 9.4 *Continued*

(b) Rating adjustment for discontinuity orientations

Parameter	Ranges of values				
	Very favourable	Favourable	Fair	Unfavourable	Very unfavourable
Strike and dip orientations of discontinuities					
Ratings Tunnels and mines	0	−2	−5	−10	−12
Foundations	0	−2	−7	−15	−25
Slopes	0	−5	−25	−50	−60

(c) Rock mass classes determined from total ratings

Rating	100←81	80←61	60←41	40←21	<20
Class no.	I	II	III	IV	V
Description	Very good rock	Good rock	Fair rock	Poor rock	Very poor rock

(d) Meaning of rock mass classes

Class no.	I	II	III	IV	V
Average stand-up time	20 yr for 15 m span	1 yr for 10 m span	1 wk for 5 m span	10h for 2.5 m span	30 min for 1 m span
Cohesion of the rock mass (kPa)	>400	300−400	200−300	100−200	<100
Friction angle of the rock mass (deg.)	>45	35−45	25−35	15−25	<15

Table 9.5 Guide for selection of primary support in tunnels at shallow depth size: 5 m to 15 m; construction by drilling and blasting. (After Bieniawski, 1974.)

Rock mass class	Alternative support systems		
	Mainly rockbolts (20 mm dia., length half tunnel width, resin bonded)	Mainly shotcrete	Mainly steel ribs
I	Generally no support is required		
II	Rockbolts spaced 1.5 to 2.0 m plus occasional wire mesh in crown	Shotcrete 50 mm in crown	Uneconomic
III	Rockbolts spaced 1.0 to 1.5 m plus wire mesh and 30 mm shotcrete in crown where required	Shotcrete 100 mm in crown and 50 mm in sides plus occasional wire mesh and rockbolts where required	Light sets spaced 1.5 to 2 m
IV	Rockbolts spaced 0.5 to 1.0 m plus wire mesh and 30−50 mm shotcrete in crown and sides	Shotcrete 150 mm in crown and 100 mm in sides plus wire mesh and rockbolts, 3 m long spaced 1.5 m	Medium sets spaced 0.7 to 1.5 m plus 50 mm shotcrete in crown and sides
V	Not recommended	Shotcrete 200 mm in crown and 150 mm in sides plus wire mesh, rockbolts and light steel sets. Seal face. Close invert	Heavy sets spaced 0.7 m with lagging. Shotcrete 80 mm thick to be applied immediately after blasting

The numerical value of Q ranges from 0.001 for exceptionally poor quality squeezing ground, to 1000 for exceptionally good quality rock which is practically unjointed. Rock mass quality, together with the support pressure and the dimensions, and purpose of the underground excavation, are used to estimate the type of suitable permanent support. A fourfold change in Q value indicates the need for a different support system. Zones of different Q value are mapped and classified separately. However, in variable conditions where different zones occur within a tunnel, each for only a few metres, it is more economic to map the overall quality and to estimate an average value of Q, from which a design of a compromise support system can be made (Barton, 1988).

The Q value is related to the type and amount of support by deriving the equivalent dimension of the excavation. The latter is related to the size and purpose of the excavation and is obtained from:

$$\text{equivalent dimension} = \frac{\text{span or height of wall}}{\text{ESR}} \quad (9.5)$$

where ESR is the excavation support ratio related to the use of the excavation and the degree of safety required. Some values of ESR are shown in Table 9.7.

Stacey and Page (1986) made use of the Q system to develop design charts to determine the factor of safety for unsupported excavations, the spacing of rockbolts over the face of an excavation, and the thickness of shotcrete on an excavation (Figs 9.14a, b, and c, respectively). For civil engineering applications a factor of safety exceeding 1.2 is required if the omission of support is to be considered. The support values suggested in the charts are for primary support. The values should be doubled for long-term support.

9.3 UNDERGROUND CAVERNS

The site investigation for an underground cavern has to locate a sufficiently large mass of sound rock in which the cavern can be excavated. Because caverns usually are located at appreciable depth below ground surface, the rock mass often is beneath the

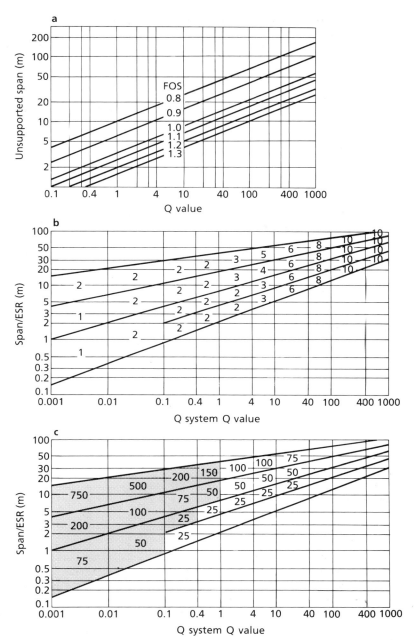

Fig. 9.14 (a) Relationship between unsupported span and Q value; (b) bolt-support estimation using the Q system, bolt spacing — m^2 of excavation per bolt, where the area per bolt is greater than 6 m^2, spot bolting is implied; (c) shotcrete and wire-mesh support estimation using the Q system. Thickness of shotcrete in mm (mesh reinforcement in the shaded areas). Note that the very thick applications of shotcrete are not practical but the values are included for completeness. (After Stacey and Page, 1986.) The support intensity given in design charts b and c is appropriate for primary support; where long-term support is required the design chart values should be modified as follows: (i) divide area per bolt by 2; (ii) multiply shotcrete thickness by 2. FOS, factor of safety.

influence of weathering and consequently, the chief considerations are rock quality, geological structure, and groundwater conditions. The orientation of an underground cavern is usually based on an analysis of the joint pattern, including the character of the different joint systems in the area and, where re-levant, also on the basis of the stress distribution. Usually it is considered necessary to avoid an orientation whereby the long axis is parallel to steeply inclined major joint sets. Wherever possible caverns should be orientated so that fault zones are avoided.

Displacement data provide a direct means of

Table 9.6 Classification of individual parameters used in the Norwegian Geotechnical Institute (NGI) tunnelling quality index or "Q" system. (After Barton et al., 1975.)

Description	Value	Notes
1 Rock quality designation	RQD	1 Where RQD is reported or measured as ≤ 10 (including 0), a nominal value of 10 is used to evaluate Q
A Very poor	0–25	
B Poor	25–50	2 RQD intervals of 5, i.e. 100, 95, 90, etc. are sufficiently accurate
C Fair	50–75	
D Good	75–90	
E Excellent	90–100	
2 Joint set number	J_n	1 For intersections use $(3.0 \times J_n)$
A Massive, no or few joints	0.5–1.0	2 For portals use $(2.0 \times J_n)$
B One joint set	2	
C One joint set plus random	3	
D Two joint sets	4	
E Two joint sets plus random	6	
F Three joint sets	9	
G Three joint sets plus random	12	
H Four or more joint sets, random, heavily jointed 'sugar cube', etc.	15	
J Crushed rock, earthlike	20	
3 Joint roughness number	J_r	1 Add 1.0 if the mean spacing of the relevant joint set is greater than 3 m
(a) Rock wall contact and		
(b) Rock wall contact before 10 cm shear		2 $J_r = 0.5$ can be used for planar, slickensided joints having lineations, provided the lineations are orientated for minimum strength
A Discontinuous joints	4	
B Rough or irregular, undulating	3	
C Smooth, undulating	2	
D Slickensided, undulating	1.5	
E Rough or irregular, planar	1.5	
F Smooth, planar	1.0	
G Slickensided, planar	0.5	
(c) No rock wall contact when sheared		
H Zone containing clay minerals thick enough to prevent rock wall contact	1.0	
J Sandy, gravelly, or crushed zone thick enough to prevent rock wall contact	1.0	

4 Joint alteration number	J_a	ϕ_r (approx.)
(a) Rock wall contact		
A Tightly healed, hard, non-softening, impermeable filling	0.75	—
B Unaltered joint walls, surface staining only	1.0	(25–35°)
C Slightly altered joint walls, non-softening mineral coatings, sandy particles, clay-free disintegrated rock, etc.	2.0	(25–30°)
D Silty, or sandy clay coatings, small clay-fraction (non-softening)	3.0	(20–25°)
E Softening or low-friction clay mineral coatings, i.e. kaolinite, mica. Also chlorite, talc, gypsum, and graphite etc., and small quantities of swelling clays. (Discontinuous coatings, 1–2 mm or less in thickness)	4.0	(8–16°)
(b) Rock wall contact before 10cm shear		
F Sandy particles, clay-free disintegrated rock, etc.	4.0	(25–30°)
G Strongly overconsolidated, non-softening clay mineral fillings (continuous, <5mm thick)	6.0	(16–24°)
H Medium or low overconsolidation, softening, clay mineral fillings (continuous, <5mm thick)	8.0	(12–16°)
J Swelling clay fillings, i.e. montmorillonite (continuous, <5mm thick). Values of J_a depend on percentage of swelling clay-size particles, and access to water	8.0–12.0	(6–12°)
(c) No rock wall contact when sheared		
K Zones or bands of disintegrated or crushed rock and clay	6.0	
L (see G, H, and J for clay conditions)	8.0	
M	8.0–12.0	(6–24°)
N Zones or bands of silty or sandy clay, small clay fraction (non-softening)	5.0	
Q Thick, continuous zones or bands of clay (see G, H and J	10.0–13.0	
P for clay conditions)	13.0–20.0	(6–24°)
R		

1 Values of ϕ_r, the residual friction angle, are intended as an approximate guide to the mineralogical properties of the alteration products, if present

Continued on p. 314

Table 9.6 *Continued*

Description	Value	Approx. water pressure (kgf/cm²)	Notes
5 *Joint water reduction factor*	J_w		1 Factors C to F are crude estimates. Increase J_w if drainage measures are installed
A Dry excavations or minor inflow, i.e. <5l/min locally	1.0	<1.0	2 Special problems caused by ice formation are not considered
B Medium inflow or pressure, occasional outwash of joint fillings	0.66	1.0–2.5	
C Large inflow or high pressure in competent rock with unfilled joints	0.5	2.5–10.0	
D Large inflow or high pressure, considerable outwash of fillings	0.33	2.5–10.0	
E Exceptionally high inflow or pressure at blasting, decaying with time	0.2–0.1	>10	
F Exceptionally high inflow or pressure continuing without decay	0.1–0.05	>10	
6 *Stress reduction factor*	SRF		1 Reduce these values of SRF by 25–50% if the relevant shear zones only influence but do not intersect the excavation
(a) Weakness zones intersecting excavation, which may cause loosening of rock mass when tunnel is excavated			2 For strongly anisotropic virgin stress field (if measured): when $5 \leq \sigma_1/\sigma_3 \leq 10$, reduce σ_c to $0.8\sigma_c$ and σ_t to $0.8\sigma_t$. When $\sigma_1/\sigma_3 > 10$, reduce σ_c to $0.6\sigma_c$ and σ_t to $0.6\sigma_t$, where $\sigma_c =$ unconfined compressive strength, and $\sigma_t =$ tensile strength (point load) and σ_1 and σ_3 are the major and minor principal stresses
A Multiple occurrences of weakness zones containing clay or chemically disintegrated rock, very loose surrounding rock (any depth)	10.0		3 Few case records available where depth of crown below surface is less than span width. Suggest SRF increase from 2.5 to 5 for such cases (see H).
B Single weakness zones containing clay, or chemically disintegrated rock (excavation depth <50m)	5.0		
C Single weakness zones containing clay, or chemically disintegrated rock (excavation depth >50m)	2.5		
D Multiple shear zones in competent rock (clay-free), loose surrounding rock (any depth)	7.5		
E Single shear zones in competent rock (clay-free) (depth of excavation <50m)	5.0		
F Single shear zones in competent rock (clay-free) (depth of excavation >50m)	2.5		
G Loose open joints, heavily jointed or 'sugar cube' (any depth)	5.0		

	σ_c/σ_1	σ_t/σ_1	SRF
(b) Competent rock, rock-stress problems			
H Low stress, near surface	>200	>13	2.5
J Medium stress	200–10	13–0.66	1.0
K High stress, very tight structure (usually favourable to stability, may be unfavourable for wall stability)	10–5	0.66–0.33	0.5–2
L Mild rock burst (massive rock)	5–2.5	0.33–0.16	5–10
M Heavy rock burst (massive rock)	<2.5	<0.16	10–20
(c) Squeezing rock, plastic flow of incompetent rock under the influence of high rock pressure			
N Mild squeezing-rock pressure			5–10
O Heavy squeezing-rock pressure			10–20
(d) Swelling rock, chemical swelling activity depending upon presence of water			
P Mild swelling rock pressure			5–10
R Heavy swelling rock pressure			10–20

Additional notes on the use of these tables

When making estimates of the rock mass quality (Q) the following guidelines should be followed, in addition to the notes listed in the tables:

1 When drillhole core is unavailable, RQD can be estimated from the number of joints per unit volume, in which the number of joints per metre for each joint set are added. A simple relation can be used to convert this number to RQD for the case of clay-free rock masses: $RQD = 115 - 3.3J_v$ (approx.) where J_v = total number of joints per m^3 ($RQD = 100$ for $J_v < 4.5$).

2 The parameter J_n representing the number of joint sets will often be affected by foliation, schistosity, slaty cleavage, or bedding, etc. If strongly developed these parallel 'joints' should obviously be counted as a complete joint set. However, if there are few 'joints' visible, or only occasional breaks in the core due to these features, then it will be more appropriate to count them as 'random' when evaluating J_n.

3 The parameters J_r and J_a (representing shear strength) should be relevant to the weakest significant joint set or clay-filled discontinuity in the given zone. However, if the joint set or discontinuity with the minimum value of (J_r/J_a) is favourably oriented for stability, then a second, less favourably oriented joint set or discontinuity may sometimes be more significant, and its higher value of J_r/J_a should be used when evaluating Q. The value of J_r/J_a should in fact relate to the surface most likely to allow failure to initiate.

4 When a rock mass contains clay, the factor SRF appropriate to loosening loads should be evaluated. In such cases the strength of the intact rock is of little interest. However, when jointing is minimal and clay is completely absent the strength of the intact rock may become the weakest link, and the stability will then depend on the ratio rock-stress/rock-strength. A strongly anisotropic stress field is unfavourable for stability and is roughly accounted for as in note 2 in the table for stress reduction factor evaluation.

5 The compressive and tensile strengths (σ_c and σ_t) of the intact rock should be evaluated in the saturated condition if this is appropriate to present or future in situ conditions. A very conservative estimate of strength should be made for those rocks that deteriorate when exposed to moist or saturated conditions.

Table 9.7 Equivalent support ratio for different excavations

Excavation category	ESR
1 Temporary mine openings	3–5

2 Vertical shafts:
  Circular section
  Rectangular/square section
	
	2.5
	2.0

3 Permanent mine openings, water tunnels for hydropower (excluding high-pressure penstocks), pilot tunnels, drifts, and headings for large excavations	1.6
4 Storage caverns, water treatment plants, minor highway and railroad tunnels, surge chambers, access tunnels	1.4
5 Power stations, major highway or railroad tunnels, civil defence chambers, portals, intersections	1.0
6 Underground nuclear power stations, railroad stations, factories	0.8

evaluating cavern stability. Displacements which have exceeded the predicted elastic displacements by a factor of 5 or 10 have usually resulted in decisions to modify support and excavation methods. In a creep-sensitive material, such as may occur in a major shear zone or zone of soft, altered rock, the natural stresses concentrated around an opening cause time-dependent displacement which, if restrained by a support, result in a build-up of stress on the support. Conversely if a rock mass is not sensitive to creep, stresses around an opening are normally relieved as blocks displace towards the opening. However, initial movements may be very much influenced by the natural stresses concentrated around an opening and under certain boundary conditions may continue to act even after large displacements have occurred.

The angle of friction for tight irregular joint surfaces is commonly greater than 45° and as a consequence the included angle of any wedge opening into the roof of a cavern has to be 90° or more, if the wedge is to move into the cavern. A tight, rough joint system therefore only presents a problem when it intersects the surface of a cavern at relatively small angles or is parallel to the surface of the cavern. However, if the material occupying a thick shear zone has been reduced to its residual strength, then the angle of friction could be as low as 15° and in

such an instance the included angle of the wedge would be 30°. Such a situation would give rise to a very deep wedge which could move into a cavern. Displacement of wedges into a cavern is enhanced if the ratio of the intact unconfined compressive strength to the natural stresses concentrated around the cavern is low. Values of less than 5 are indicative of stress conditions in which new extension fractures develop about the cavern during its excavation. Wedge failures are facilitated by shearing and crushing of the asperities along discontinuities as wedges are displaced.

The walls of a cavern may be greatly influenced by the prevailing state of stress, especially if the tangential stresses concentrated around the chamber approach the intact compressive strength of the rock. In such cases extension fractures develop near the surface of the cavern as it is excavated and cracks produced by blast damage become more pronounced. The problem is accentuated if any lineation structures or discontinuities run parallel with the walls of the cavern. Indeed, popping of slabs of rock may take place from cavern walls.

Rock bursts have occurred in underground caverns at rather shallow depths, particularly where they were excavated in the sides of valleys and on the inside of faults when the individual fault passed through a cavern and dipped towards an adjacent valley. Bursting can take place at depths of 200 to 300 m where the tensile strength of the rocks varies between 3 and 4 MPa.

Three methods of blasting are normally used to excavate underground caverns (Fig. 9.15):
1 in the overhead tunnel the entire profile is drilled and blasted together or in parts by horizontal holes;
2 benching with horizontal drilling is commonly used to excavate the central parts of a large cavern;
3 the bottom of a cavern may be excavated by benching, the blastholes being drilled vertically.
The central part of a cavern can also be excavated by vertical benching provided that the upper part has been excavated to a sufficient height or that the walls of the cavern are inclined. Smooth blasting is used to minimize fragmentation in the surrounding rock. Once the crown of the cavern has been excavated economical excavation of the walls calls for large, deep bench cuts exposing substantial areas of wall in a single blast. Under these conditions, however, an unstable wedge can be exposed and fail before it is supported.

The support pressures required to maintain the stability of a cavern increase as its span increases so

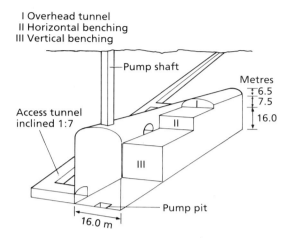

I Overhead tunnel
II Horizontal benching
III Vertical benching

Pump shaft

Metres
6.5
7.5
16.0

Access tunnel
inclined 1:7

Pump pit

16.0 m

Fig. 9.15 Main stages in excavation of underground caverns.

that for the larger caverns standard-sized bolts arranged in normal patterns may not be sufficient to hold the rock in place. Most caverns have arched crowns with span to rise ratios (B/R) of 2.5 to 5.0. In general, higher support pressures are required for flatter roofs. Frequently, the upper parts of the walls are more heavily bolted in order to help support the haunches and the roof arch while the lower walls may be either slightly bolted or even unbolted.

9.4 SHAFTS AND RAISES

A shaft is driven vertically downwards whereas a raise is driven either vertically or at a steep angle upwards. In shaft sinking, drilling is usually easier than in tunnelling but blasting is against gravity and mucking is slow and therefore expensive. By contrast, mucking and blasting are simpler in raising operations but drilling is difficult. As the cost of drilling is generally exceeded by that of blasting and mucking, raising is more economic than shaft sinking where, of course, both are practical. A raise is of small cross-sectional area. If the excavation is to have a large diameter, then it is enlarged from above, the primary raise excavation being used as a muck shute. Where a raise emerges at the surface through unconsolidated material, then this section is excavated from the surface. Shafts and raises are also excavated by boring machines.

A shaft will usually sink through a series of different rock types and the two principal problems likely

to be encountered are: (i) varying stability of the walls; and (ii) ingress of water. These two problems frequently occur together and are likely to be met with in rock with a high fracture index — weak zones being particularly hazardous. They are most serious, however, in unconsolidated deposits, especially loosely packed gravels, sands, and silts. The position of the water table is highly significant. The geological investigation prior to shaft sinking should therefore provide detailed information relating to the character of the soils and rocks involved, noting where appropriate, their fracture index, strength, porosity, permeability, the position of the water table, and the pore water pressure.

A simple and effective method of dealing with groundwater in shaft sinking is to pump from a sump within the shaft. However, problems arise when the quantity pumped is so large that the rate of inflow under high head causes instability in the sides of the shaft or prevents the fixing and back-grouting of the shaft lining. Although the idea of surrounding a shaft by a ring of bored wells is at first sight attractive, there are practical difficulties in achieving effective lowering of the water table. In fissured rocks or variable water bearing soils there is a tendency for the water flow to by-pass the wells and take preferential paths directly into the excavation.

Where the stability of the wall and/or the ingress of water are likely to present problems in shaft sinking, one of the most frequent techniques resorted to is ground freezing. Freezing transforms weak, waterlogged materials into ones which are self-supporting and impervious. It therefore affords temporary support to an excavation as well as being a means of excluding groundwater. Normally the freeze probes are laid out in linear fashion so that an adequate boundary wall encloses the future excavation when the radial development of ice about each probe unites to form a continuous section of frozen ground. The usual coolant is brine although liquid nitrogen is used in special circumstances.

Ground freezing means heat transfer and, in the absence of moving groundwater, this is brought about by conduction. Thus, the thermal conductivities of the materials involved govern the rate at which freezing proceeds. These values fall within quite narrow limits for all types of frozen ground. This is why ground freezing is a versatile technique and can deal with a variety of soil and rock types in a stratal sequence. But a limitation is placed upon the freezing process by unidirectional flow of groundwater. For a brine freezing project a velocity exceeding 2 m per day will seriously affect and distort the growth of an ice wall. The tolerance is much wider when liquid nitrogen is used.

Grouting can be an economical method of eliminating or reducing the flow of groundwater into shafts if the soil or rock conditions are suitable for accepting cement or chemical grouts. This is because the perimeter of the grouted zone is relatively small in relation to the depth of the excavation.

9.5 RESERVOIRS

Although most reservoirs today are multipurpose, their principal function, no matter what their size, is to stabilize the flow of water in order to satisfy a varying demand from consumers or to regulate water supplied to a river course. In other words, water is stored at times of excess flow to conserve it for later release at times of low flow, or to reduce flood damage downstream.

The most important physical characteristic of a reservoir is its storage capacity. Probably the most important aspect of storage in reservoir design is the relationship between capacity and yield. The yield is the quantity of water which a reservoir can supply at any given time. The maximum possible yield equals the mean inflow less evaporation and seepage loss. In any consideration of yield the maximum quantity of water which can be supplied during a critical dry period (i.e. during the lowest natural flow on record) is of prime importance and is defined as the safe yield.

The maximum elevation to which the water in a reservoir basin will rise during ordinary operating conditions is referred to as the top water or normal pool level. For most reservoirs this is fixed by the top of the spillway. Conversely minimum pool level is the lowest elevation to which the water is drawn under normal conditions, this being determined by the lowest outlet. Between these two levels the storage volume is termed the useful storage, whilst the water below the minimum pool level, because it cannot be drawn upon, is the dead storage. During floods the water level may rise above top water level but this surcharge cannot be retained since it is above the elevation of the spillway.

In any adjustment of a river regime to the new conditions imposed by a reservoir, problems may emerge both up- and downstream. Deposition around the head of a reservoir may cause serious aggradation upstream resulting in a reduced capacity of the stream channels to contain flow. Hence flooding becomes more frequent and the water table rises. Removal of sediment from the outflow of a reservoir can lead to erosion in the river regime downstream of the dam, with consequent acceleration of headward erosion in tributaries and lowering of the water table.

9.5.1 Investigation of reservoir sites

In an investigation of a potential reservoir site, consideration must be given to the amount of rainfall, runoff, infiltration, and evapotranspiration which occurs in the catchment area. The climatic, topographical, and geological conditions are therefore important, as is the type of vegetative cover. Accordingly, the two essential types of basic data needed for reservoir design studies are adequate topographical maps and hydrological records. Indeed, the location of a large, impounding direct-supply reservoir is very much influenced by topography since this governs its storage capacity. Initial estimates of storage capacity can be made from topographic maps or aerial photographs, more accurate information being obtained, where necessary, from subsequent surveying. Catchment areas and drainage densities also can be determined from maps and airphotos.

Reservoir volume can be estimated by planimetering areas upstream of the dam site for successive contours up to proposed top water level. Then the mean of two successive contour areas is multiplied by the contour interval to give the interval volume, the summation of the interval volumes providing the total volume of the reservoir site.

Records of stream flow are required for determining the amount of water available for conservation purposes. Such records contain flood peaks and volumes which are used to determine the amount of storage needed to control floods and to design spillways and other outlets. Records of rainfall are used to supplement stream flow records or as a basis for computing stream flow where there are no flow records obtainable. Losses due to seepage and evaporation also must be taken into account.

The field reconnaissance provides indications of the areas where detailed geological mapping may be required and where to locate drillholes, such as low, narrow saddles or other seemingly critical areas in the reservoir rim. Drillholes on the flanks of reservoirs should be drilled at least to floor level. Permeability and pore water pressure tests can be carried out in these drillholes.

9.5.2 Leakage from reservoirs

The most attractive site for a large impounding reservoir is a valley constricted by a gorge at its outfall with steep banks upstream so that a small dam can impound a large volume of water with a minimum extent of water spread. However, two other factors have to be taken into consideration: (i) the watertightness of the basin; and (ii) bank stability. The question of whether or not significant water loss will take place is chiefly determined by the groundwater conditions, more specifically by the hydraulic gradient. Accordingly, once the groundwater conditions have been investigated an assessment can be made of watertightness and possible groundwater control measures.

Leakage from a reservoir takes the form of sudden increases in stream flow downstream of the dam site with boils in the river and the appearance of springs on the valley sides. It may be associated with major defects in the geological structure such as solution channels, fault zones, or buried channels through which large and essentially localized flows take place. Seepage is a more discreet flow, spread out over a larger area, but may be no less in total amount.

The economics of reservoir leakage vary. Although a highly leaky reservoir may be acceptable in an area where runoff is evenly distributed throughout the year, a reservoir basin with the same rate of water loss may be of little value in an area where runoff is seasonally deficient. A river regulating scheme can operate satisfactorily despite some leakage from the reservoir and reservoirs used solely for flood control may be effective even if they are very leaky. By contrast, leakage from a pumped storage reservoir must be assessed against pumping costs.

Serious water loss has in some instances led to the abandonment of the reservoir scheme, examples being the Jerome reservoir in Idaho, the Cedar reservoir in Washington, the Monte Jacques reservoir in Spain, the Hales Bar reservoir in Tennessee, and the Hondo reservoir in New York.

Apart from the conditions in the immediate vicinity of the dam, the two factors which determine the retention of water in reservoir basins are the piezometric conditions in, and the natural permeability of, the floor and flanks of the basin. Knill (1971) pointed out that four groundwater conditions existed on the flanks of a reservoir, namely:

1 The groundwater divide and piezometric level are at a higher elevation than that of the proposed top water level. In this situation no significant water loss takes place.

2 The groundwater divide, but not the piezometric level, is above the top water level of the reservoir. In these circumstances seepage can take place through the separating ridge into the adjoining valley. Deep seepage can take place but the rate of flow is determined by the *in situ* permeability.

3 Both the groundwater divide and piezometric conditions are at a lower elevation than the top water level but higher than that of the reservoir floor. In this case the increase in groundwater head is low and the flow from the reservoir may be initiated under conditions of low piezometric pressure in the reservoir flanks.

4 The water table is depressed below the base of the reservoir floor. This indicates deep drainage of the rock mass or very limited recharge. A depressed water table does not necessarily mean that reservoir construction is out of the question but groundwater recharge will take place on filling which will give rise to a changed hydrogeological environment as the water table rises. In such instances the impermeability of the reservoir floor is important. When permeable beds are more or less saturated, particularly when they have no outlet, seepage is appreciably decreased. At the same time the accumulation of silt on the floor of the reservoir tends to reduce seepage. If, however, the permeable beds contain large pore spaces or discontinuities and they drain from the reservoir, then seepage continues.

Troubles from seepage can usually be controlled by exclusion or drainage techniques. Cutoff trenches, carried into bedrock, may be constructed across cols occupied by permeable deposits. Grouting may be effective where localized fissuring is the cause of leakage. Impervious linings (Fig. 9.16) consume large amounts of head near the source of water thereby reducing hydraulic gradients and saturation at the points of the exit and increasing resistance to seepage loss. Clay blankets or layers of silt have been used to seal exits from reservoirs.

Because of the occurrence of permeable contacts, close jointing, pipes, and vesicles, and the possible presence of tunnels and cavities, recent accumulations of basaltic lava flows can prove highly leaky and treacherous rocks with respect to watertightness. Lava flows are frequently interbedded, often in an irregular fashion, with pyroclastic deposits. Deposits of ash and cinders tend to be highly permeable.

Reservoir sites in limestone terrains vary considerably in their suitability. Massive, horizontally bedded

Fig. 9.16 An asphalt membrane at the side of Mornos reservoir, near Lidoriki, north of Athens, Greece, to prevent leakage into folded and broken strata which includes limestones.

limestones, relatively free from solution features, form excellent sites. On the other hand, well-jointed, cavernous, and deformed limestones are likely to present problems in terms of stability and watertightness. Serious leakage has usually taken place as a result of cavernous conditions which are not fully revealed or appreciated at the site investigation stage. Sites are best abandoned where large, numerous solution cavities extend to considerable depths. Where the problem is not so severe solution cavities can be cleaned and grouted.

Sinkholes and caverns can develop in thick beds of gypsum more rapidly than they can in limestone. In the United States they have been known to form within a few years where beds of gypsum are located below reservoirs. Extensive surface cracking and subsidence has occurred in Oklahoma and New Mexico due to the collapse of cavernous gypsum. The problem is accentuated by the fact that gypsum is weaker than limestone and therefore collapses more readily. Uplift is a problem which has been associated with the hydration of anhydrite beneath reservoirs.

Buried channels may be filled with coarse, granular stream deposits or deposits of glacial origin and if they occur near the perimeter of a reservoir they almost invariably pose leakage problems. Indeed leakage through buried channels, via the perimeter of a reservoir, is usually more significant than through the main valley. Hence the bedrock profile, the type of deposits, and groundwater conditions should be determined.

A thin layer of relatively impermeable superficial deposits does not necessarily provide an adequate seal against seepage. A controlling factor in such a situation is the groundwater pressure immediately below the deposits. Where artesian conditions exist, springs may break the thinner parts of the superficial cover. If the water table below the deposits is depressed, then there is a risk that the weight of water in the reservoir may puncture them. What is more, on filling a reservoir there is a possibility that the superficial material may be ruptured or partially removed to expose the underlying rocks. This happened at the Monte Jacques reservoir in northern Spain where alluvial deposits covered cavernous limestone. The alluvium was washed away to expose a large sinkhole down which the water escaped.

Leakage along faults generally is not a serious problem as far as reservoirs are concerned since the length of the flow path is usually too long. However, fault zones occupied by permeable fault breccia running beneath the dam must be given special consideration. When the reservoir basin is filled the hydrostatic pressure may cause removal of loose material from such fault zones and thereby accentuate leakage. Permeable fault zones can be grouted, or if a metre or so wide, excavated and filled with rolled clay or concrete.

9.5.3 Stability of the sides of reservoirs

The formation of a reservoir upsets the groundwater regime and represents an obstruction to water flowing downhill. The greatest change involves the raising of the water table. Some soils or rocks, which are brought within the zone of saturation, may then become unstable and fail, as saturated material is weaker than unsaturated. This can lead to slumping and sliding on the flanks of a reservoir. In glaciated valleys morainic material generally rests on a rock slope smoothed by glacial erosion, which accentuates the problem of slip. Landslides which occur after a reservoir is filled reduce its capacity. Also ancient landslipped areas which occur on the rims of a reservoir may present a leakage hazard and could be reactivated.

The worst man-induced landslide on record took place into the Vajont reservoir in northern Italy on 9 October 1963 (Jaeger, 1965). More than $300 \times 10^6 \, m^3$ moved downhill with such momentum that it crossed the 99 m wide gorge and rode 135 m up the opposite side. It filled the reservoir for a distance of 2 km with slide material, which in places reached heights of 175 m above top water level and displaced water in the reservoir, thereby generating a huge wave which overtopped the dam to a height of some 100 m above its crest.

9.5.4 Sedimentation in reservoirs

Although it is seldom a decisive factor in determining location, sedimentation in reservoirs is an important problem in certain countries, for example, investigations in the United States suggest that sedimentation will limit the useful life of most reservoirs to less than 200 years. Sedimentation in a reservoir may lead to one or more of its major functions being seriously curtailed or even to it becoming inoperative. In a small reservoir sedimentation may seriously affect the available carry-over water supply and ultimately necessitate abandonment.

In those areas where streams carry heavy sediment loads the rates of sedimentation must be estimated accurately in order that the useful life of any proposed reservoir may be determined. The volume of sediment carried varies with stream flow, but usually the peak sediment load occurs prior to the peak stream flow discharge. Frequent sampling accordingly must be made to ascertain changes in sediment transport. Volumetric measurements of sediment in reservoirs are made by soundings taken to develop the configuration of the reservoir sides and bottom below the water surface.

Size of a drainage basin is the most important consideration as far as sediment yield is concerned, the rock types, drainage density, and gradient of slope also being important. The sediment yield is also influenced by the amount and seasonal distribution of precipitation and the vegetative cover. Poor cultivation practices, overgrazing, improper disposal of mine waste, and other human activities may accelerate erosion or contribute directly to stream loads.

The ability of a reservoir to trap and retain sediment is known as its trap efficiency and is expressed as the percentage of incoming sediment which is retained. Trap efficiency depends on total inflow, rate of flow, sediment characteristics, and the size of the reservoir.

9.6 DAMS AND DAM SITES

The type and size of dam constructed depend upon the need for and the amount of water available, the topography and geology of the site, and the construction materials which are readily obtainable. Dams can be divided into two major categories according to the type of material with which they are constructed, namely, concrete dams and earth dams. Concrete dams can be subdivided into gravity, arch, and buttress dams. Earth dams comprise rolled-fill and rockfill embankments. In dam construction the prime concern is safety (this coming before cost), that is, the foundations and abutments must be adequate for the type of dam selected.

A gravity dam is a rigid, monolithic structure which is usually straight in plan although sometimes it may be slightly curved. Its cross-section is roughly trapezoidal. Generally, gravity dams can tolerate only the smallest differential movements and their resistance to dislocation by the hydrostatic pressure of the reservoir water is due to their own weight. A favourable site is usually one in a constricted area of a valley where sound bedrock is reasonably close to the surface, both in the floor and abutments.

An arch dam consists of a concrete wall, of high-strength concrete, curved in plan, with its convex face pointing upstream (Fig. 9.17). Arch dams are relatively thin walled and lighter in weight than gravity dams. They will stand up to large deflections in the foundation rock provided that the deflections are uniformly distributed. They transmit most of the horizontal thrust of the reservoir water to the abutments by arch action and this, together with their relative thinness, means that they impose high stresses upon narrow zones at the base, as well as the abutments. Therefore the strength of the rock

Fig. 9.17 Kariba dam on the Zambesi between Zambia and Zimbabwe.

mass at the abutments and immediately downvalley of the dam must be unquestionable, and its modulus of elasticity must be high enough to ensure that its deformation under thrust from the arch is not so great as to induce excessive stresses in the arch. Ideal locations for arch dams are provided by narrow gorges where the walls are capable of withstanding the thrust produced by the arch action.

In locations where the foundation rocks are competent, buttress dams provide an alternative to other concrete dams. A buttress dam consists principally of a slab of reinforced concrete which slopes upstream and is supported by a number of buttresses whose axes are normal to the slab (Fig. 9.18). The buttresses support the slab and transmit the water load to the foundation. They are rather narrow and act as heavily loaded walls thus exerting tremendous unit pressures on the foundations. In weak rocks buttresses may punch into the ground causing upheaval of material between them.

Earth dams are embankments of earth or rock with an impermeable core to control seepage (Fig. 9.19). This usually consists of clayey material or if sufficient quantities are not available, then concrete or asphaltic concrete membranes are used. The core is normally extended as a cutoff or grout curtain below ground level when seepage beneath the dam has to be controlled. Drains of sand and/or gravel installed within and beneath the dam also afford seepage control. Because of their broad base, earth dams impose much lower stresses on the foundations than concrete dams. Furthermore, they can more readily accommodate deformation such as that due to settlement. Thus, earth dams have been con-

Fig. 9.18 Errochty dam — an example of a buttress dam. (Courtesy of North of Scotland Hydro-Electric Board.)

Fig. 9.19 Hardap dam, near Mariental, Namibia.

structed on a great variety of foundations ranging from weak, unconsolidated stream or glacial deposits to high strength rocks.

An earth dam may be zoned or homogeneous, the former type being more common. A zoned dam is a rolled-fill dam composed of several zones that increase in permeability from the core towards the outer slopes (Fig. 9.20). The number of zones depends on the availability and type of borrow material. Stability of a zoned dam is mostly due to the weight of the heavy outer zones.

If there is only one type of borrow material readily available, a homogeneous embankment is constructed, that is, it is constructed entirely (or almost entirely) of one type of material, usually fine-grained, although sand and sand–gravel mixtures have been used.

Rockfill dams usually consist of three basic elements: (i) a loose rockfill dump, which forms the bulk of the dam and resists the thrust of the reser-

voir; (ii) an impermeable facing on the upstream side or an impermeable core; and (iii) rubble masonry between to act as a cushion for the membrane and to resist destructive deflections. Consolidation of the main rock body may leave the face unsupported with the result that cracks form through which seepage can occur. Flexible asphalt membranes overcome this problem.

Some sites which are geologically unsuitable for a specific type of dam design may support one of composite design. For instance, a broad valley which has strong rocks on one side and weaker ones on the other can possibly be spanned by a combined gravity and embankment dam (Fig. 9.21).

The construction of a dam and the filling of a reservoir behind it impose a load on the sides and floor of a valley creating new stress conditions. These stresses must be analysed so that there is ample assurance that there will be no possibility of failure. A concrete dam behaves as a rigid, mono-

Fig. 9.20 Cross-section of Cedar Springs dam. Zone 1, clay core from lake bed deposit; zone 2, silty sand from Harold formation; zone 3, processed sand–gravel transition from river alluvium or crushed rock; zones 4 & 4a, rolled, processed rockfill (75 mm minimum and 750 mm maximum); zone 5, dumped rockfill, processed (450 mm minimum). (After Sherard *et al.*, 1974.)

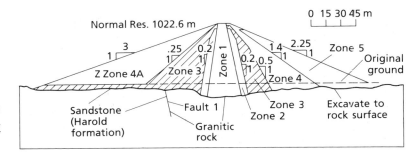

Fig. 9.21 Inanda dam, a composite dam north of Durban, South Africa.

lithic structure, the stress acting on the foundation being a function of the weight of the dam as distributed over the total area of the foundation. By contrast, earthfill dams exhibit semiplastic behaviour and the pressure on the foundation at any point depends on the thickness of the dam above that point. Vertical static forces act downwards and include the weight of both the structure and the water, although a large part of the dam is submerged and therefore the buoyancy effect reduces the influence of these two forces. The most important dynamic forces acting on a dam are wave action, overflow of water, and seismic shocks.

Horizontal forces are exerted on a dam by the lateral pressure of water behind it. These, if excessive, may cause concrete dams to slide. The tendency towards sliding at the base of such dams is of particular significance in fissile rocks such as shales, slates, and phyllites. Weak zones, such as interbedded ashes in a sequence of basalt lava flows, can prove troublesome. The presence of flat-lying joints may destroy much of the inherent shear strength of a rock mass and reduce the problem of resistance of a foundation to horizontal forces to one of sliding friction so that the roughness of joint surfaces becomes a critical factor. The rock surface should be roughened to prevent sliding, and keying the dam some distance into the foundation is advisable. Another method of reducing sliding is to give a downward slope to the base of the dam in the upstream direction of the valley.

Variations in pore water pressure cause changes in the state of stress in rock masses. Increasing pore water pressure may lift beds and the dam itself, and so decrease the shearing strength and resistance to sliding within the rock mass. Pore water reduces the compressive strength of rocks and causes an increase in the amount of deformation they undergo. It also may be responsible for swelling in certain rocks and for an acceleration in their rate of alteration. Pore water in the stratified rocks of a dam foundation reduces the coefficient of friction between the individual beds, and between the foundation and the dam.

Percolation of water through the foundations of concrete dams, even when the rock masses concerned are of good quality and of minimum permeability, is always a decisive factor in the safety and performance of dams. Such percolation can remove filler material which may be occupying joints, which in turn can lead to differential settlement of the foundations. It may also open joints which decreases the strength of the rock mass.

In highly permeable rock masses excessive seepage beneath a dam may damage the foundation. Seepage rates can be lowered by reducing the hydraulic gradient beneath the dam by incorporating a cutoff into the design. A cutoff lengthens the flow path so reducing the hydraulic gradient. It extends to an impermeable horizon or some specified depth and is usually located below the upstream face of the dam. The rate of seepage can also be effectively reduced by placing an impervious earthfill against the lower part of the upstream face of a dam.

Uplift pressure acts against the base of a dam and is caused by water seeping beneath it which is under hydrostatic head from the reservoir. Uplift pressure should be distinguished from the pore water pressure in the material beneath the dam. The uplift pressure on the heel of a dam is approximately equal to the depth of the foundation below water level multiplied by the unit weight of the water. In the simplest case it is assumed that the difference in hydraulic heads between the heel and the toe of the dam is dissipated uniformly between them. The uplift pressure can be reduced by allowing water to be conducted downstream by drains incorporated into the foundation and base of the dam.

When load is removed from a rock mass on excavation it is subject to rebound. The amount of rebound depends on the modulus of elasticity of the rocks concerned, the larger the modulus of elasticity the smaller the rebound. The rebound process in rocks generally takes a considerable time to achieve completion and will continue after a dam has been constructed if the rebound pressure or heave developed by the foundation material exceeds the effective weight of the dam. Hence, if heave is to be counteracted a dam should impose a load on the foundation equal to or slightly in excess of the load removed.

All foundation and abutment rocks yield elastically to some degree. In particular, the modulus of elasticity of the rock is of primary importance as far as the distribution of stresses at the base of a concrete dam is concerned. Also, tensile stresses may develop in concrete dams when the foundations undergo significant deformation. The modulus of elasticity is used in the design of gravity dams for comparing the different types of foundation rocks with each other and with the concrete of the dam. In the design of arch dams, if Young's modulus of the foundation has a lower value than that of the concrete or varies widely in the rocks against which the dam abuts, dangerous stress conditions may develop in the dam. The elastic properties of a rock and existing strain conditions assume importance in proportion to the height of a dam since this influences the magnitude of the stresses imparted to the foundation and abutments. The influence of geological structures in lowering Young's modulus must be accounted for by the provision of adequate safety factors. It should also be borne in mind that blasting during excavation of foundations can open up fissures and joints which leads to greater deformability of the rock mass. The deformability of the rock mass, any possible settlements, and the amount of increase of deformation with time can be taken into consideration by assuming lower modulus of elasticity in the foundation or by making provisions for prestressing.

9.6.1 Geology and dam sites

Of the various natural factors which directly influence the design of dams none are more important than geological ones. Not only do they control the character of the foundation but they also govern the materials available for construction (Anderson and McNicol, 1989). The major questions which need answering include the depth at which adequate foundations exist, the strengths of the rocks involved, the likelihood of water loss, and any special features which have a bearing on excavation. The character of the foundations upon which dams are built and their reaction to the new conditions of stress and strain, of hydrostatic pressure, and of exposure to weathering must be ascertained so that the proper factors of safety may be adopted to ensure against subsequent failure. Excluding the weaker types of compaction shales, mudstones, marls, pyroclasts, and certain very friable types of sandstone, there are few foundation materials deserving the name rock that are incapable of resisting the bearing loads even of high dams.

In their unaltered state plutonic rocks are essentially sound and durable with adequate strength for any engineering requirement. In some instances, however, intrusives may be highly altered by weathering or hydrothermal attack. Generally, the weathered product of plutonic rocks has a large clay content, although that of granitic rocks is sometimes porous with a permeability comparable to that of medium-grained sand, so that it requires some type of cutoff or special treatment of the upstream surface.

Thick, massive basalts make satisfactory dam sites but many basalts of comparatively young geological age are highly permeable, transmitting water via their open joints, pipes, cavities, tunnels, and contact zones. Foundation problems in young volcanic sequences are twofold: (i) weak beds of ash and tuff may occur between the basalt flows which give rise to problems of differential settlement or sliding; (ii) weathering during periods of volcanic inactivity may have produced fossil soils, these being of much lower strength.

Rhyolites, and frequently andesites, do not present the same leakage problems as basalts. They frequently offer good foundations for concrete dams although at some sites chemical weathering may mean that embankment designs have to be adopted.

Pyroclastics usually give rise to extremely variable foundation conditions due to wide variations in

strength, durability, and permeability. Their behaviour very much depends upon their degree of
induration, for example, many agglomerates have a
high enough strength to support a concrete dam and
also have a low permeability. By contrast, ashes are
invariably weak and often highly permeable. One
particular hazard concerns ash which is metastable
and has not been previously wetted, it undergoing a
significant reduction in void ratio on saturation.
Clay-cement grouting at high pressures may turn ash
into a satisfactory foundation. Tuffs and ashes are
frequently prone to sliding. Montmorillonite is not
an uncommon constituent in these rocks when they
are weathered.

Fresh metamorphosed rocks such as quartzite and
hornfels are very strong and afford excellent dam
sites. Marble has the same advantages and disadvantages as other carbonate rocks. Generally
speaking, gneiss has proved a good foundation rock
for dams.

Cleavage, schistosity, and foliation in regional
metamorphic rocks may adversely affect their
strength and make them more susceptible to decay.
Moreover, areas of regional metamorphism have
usually suffered extensive folding so that rocks may
be fractured and deformed. Certain schists, slates,
and phyllites are variable in quality, some being
excellent for dam site purposes, others, regardless
of the degree of their deformation or weathering,
are so poor as to be wholly undesirable in foundations and abutments, for example, talc, chlorite,
and sericite schists are weak rocks containing closely
spaced planes of schistosity.

Some schists become slippery upon weathering
and therefore fail under a moderately light load,
whereas slates and phyllites tend to be durable.
Although slates and phyllites are suitable for concrete dams where good load-bearing strata occur at
a relatively shallow depth, problems may arise in
excavating broad foundations. Particular care is
required in blasting slates, phyllites, and schists,
otherwise considerable overbreak or shattering may
result. It may be advantageous to use smooth blasting for final trimming purposes. When compacted in
lifts using a vibratory roller, these rocks break down
to give a well-graded permeable fill. Consequently,
rockfill embankments are being increasingly adopted
at such sites.

Joints and shear zones are responsible for the
unsound rock encountered at dam sites on plutonic
and metamorphic rocks. Unless they are sealed they
may permit leakage through foundations and abut-

ments. Slight opening of joints on excavation leads
to imperceptible rotations and sliding of rock blocks,
large enough to appreciably reduce the strength and
stiffness of the rock mass. Sheet or flat-lying joints
tend to be approximately parallel to the topographic
surface and introduce a dangerous element of weakness into valley slopes. Their width varies and if they
remain untreated large quantities of water may escape through them from the reservoir. Moreover,
joints may transmit hydrostatic pressures into the
rock masses downstream from the abutments which
are high enough to dislodge sheets of rock. If a joint
is very wide and located close to the rock surface it
may close up under the weight or lateral pressure
exerted by the dam and cause important differential
settlement.

Sandstones have a wide range of strength depending largely upon the amount and type of cement-
matrix material occupying the voids. With the exception of shaly sandstone, sandstone is not subject
to rapid surface deterioration on exposure. As a
foundation rock even poorly cemented sandstone is
not susceptible to plastic deformation. However,
friable sandstones introduce problems of scour within the foundation. Moreover, sandstones are highly
vulnerable to the scouring and plucking action of the
overflow from dams and have to be adequately
protected by suitable hydraulic structures. A major
problem of dam sites located in sandstones results
from the fact that they are generally transected by
joints, which reduce resistance to sliding. Generally,
however, sandstones have high coefficients of
internal friction which give them high shearing
strengths, when restrained under load.

Sandstones are frequently interbedded with shale.
These layers of shale may constitute potential sliding
surfaces. Sometimes such interbedding accentuates
the undesirable properties of the shale by permitting
access of water to the shale–sandstone contacts.
Contact seepage may weaken shale surfaces and
cause slides in formations which dip away from
abutments and spillway cuts. Severe uplift pressures
may also develop beneath beds of shale in a dam
foundation and appreciably reduce its resistance to
sliding. Foundations and abutments composed
of interbedded sandstones and shales also present
problems of settlement and rebound, the magnitude
of these factors depending upon the character of the
shales.

The permeability of sandstone depends upon the
amount of cement in the voids and especially on the
incidence of discontinuities. The porosity of sand-

stones generally does not introduce leakage problems of moment, though there are exceptions. The sandstones in a valley floor may contain many open joints that wedge out with depth and these are often caused by rebound of interbedded shales. Conditions of this kind in the abutments and foundations of dams greatly increase the construction costs for several reasons. They have a marked influence on the depth of stripping, especially in the abutments. They must be cut-off by an elaborate programme of pressure grouting and drainage for the combined purposes of preventing excessive leakage and reducing the undesirable uplift effects of hydrostatic pressure of reservoir water on the base of the dam or on the base of some bedding contact within the dam foundation.

Limestone dam sites vary widely in their suitability. Thick bedded horizontally lying limestones relatively free from solution cavities afford excellent dam sites. Limestone requires no special treatment to ensure a good bond with concrete. On the other hand, thin bedded, highly folded, or cavernous limestones are likely to present serious foundation or abutment problems involving bearing capacity or watertightness, or both (Soderburg, 1979). Resistance to sliding involves the shearing strength of limestone. If the rock mass is thin bedded, a possibility of sliding may exist. This should be guarded against by suitably keying the structure into the foundation rock. Beds separated by layers of clay or shale, especially those inclined downstream, may, under certain conditions, serve as sliding planes and give rise to failure.

Some solution features will always be present in limestone. The size, form, abundance, and downward extent of these features depends upon the geological structure and the presence of interbedded impervious layers. Individual cavities may be open, they may be partially or completely filled with clay, silt, sand, or gravel mixtures, or they may be water-filled conduits. Solution cavities present numerous problems in the construction of large dams, among which bearing capacity and watertightness are paramount. Few dam sites are so bad that it is impossible to construct safe and successful structures upon them but the cost of the necessary remedial treatment may be prohibitive (Vick and Bromwell, 1989). Dam sites should be abandoned where the cavities are large and numerous, and extend to considerable depths. Sufficient bearing strength generally may be obtained in cavernous rock by deeper excavation than otherwise would be necessary. Watertightness

may be attained by removing the material from cavities, and refilling with concrete. The small filled cavities may be sealed effectively by washing out and then grouting with cement. The establishment of a watertight cutoff through cavernous limestone presents difficulties in proportion to the size and extent of the solution openings.

The removal of evaporites by solution can result in subsidence and collapse of overlying strata. Indeed cavities have been known to form in the USA within a matter of a few years where thick beds of gypsum have occurred beneath dams. Investigations have proved that when anhydrite and gypsum are interbedded with marl (mudstone) they are generally sound.

Well cemented shales, under structurally sound conditions, present few dam site problems, though their strength limitations and elastic properties may be factors of importance in the design of concrete dams of appreciable height. They, however, have lower moduli of elasticity and lower shear strength values than concrete and therefore are unsatisfactory foundation materials for arch dams. Moreover, if the lamination is horizontal and well developed, then the foundations may offer little shear resistance to the horizontal forces exerted by a dam. A structure keying the dam into such a foundation then is required.

Severe settlements may take place in low grade compaction shales. Thus, such sites are generally developed with earth dams, but associated concrete structures such as spillways will involve these problems. Rebound in deep spillway cuts may cause buckling of spillway linings and differential rebound movements in the foundations may require special design provisions.

The stability of slopes in cuts is one of the major problems in shale both during and after construction. If a spillway is to be deeply cut into shale, no major slides must occur since blocking the channel might cause overtopping and failure of the dam. Similarly, cuttings in shale above other structures must be made stable. This problem becomes particularly acute in dipping formations and in formations containing montmorillonite.

The opening of joints and the development of shear planes in shales for considerable distances behind the normal zones of creep on valley sides result from a combination of elastic rebound and oversteepening of slopes. These deep-seated disturbances may give rise to dangerous hydrostatic pressures on the abutment rocks downstream from the dam, leakage around the ends of the dam, and reduced resistance of the rock to the horizontal forces. The situation may be complicated by the fact that most of the open joints are filled with clay and so grouting may not be feasible.

Earth dams are usually constructed on clays as they lack the load bearing properties necessary to support concrete dams. Beneath valley floors, clays are frequently contorted, fractured, and softened due to valley creep so that the load of an earth dam may have to spread over wider areas than is the case with shales and mudstones. Rigid ancillary structures necessitate spread footings or raft foundations. Deep cuts involve problems of rebound if the weight of removed material exceeds that of the structure. Slope stability problems also arise, with rotational slides a hazard.

Among the many manifestations of glaciation are the presence of buried channels, disrupted drainage systems, deeply filled valleys, sand−gravel terraces, narrow overflow channels connecting open valleys, and extensive deposits of lacustrine silts and clays, till, and outwash sands and gravels. Deposits of peat and head (solifluction debris) may be interbedded with these glacial deposits. Consequently, glacial deposits may be notoriously variable in composition, both laterally and vertically. As a result dam sites in glaciated areas are among the most difficult to appraise on the basis of surface evidence. Knowledge of the preglacial, glacial, and postglacial history of a locality is of vital importance in the search for the most practicable sites. A primary consideration in glacial terrains is the discovery of sites where rock foundations are available for spillway, outlet, and powerhouse structures. Generally, earth dams are constructed in areas of glacial deposits. Concrete dams, however, are feasible in postglacial, rock-cut valleys, or composite dams are practicable in valleys containing rock benches.

The major problems associated with foundations on alluvial deposits generally result from the fact that the deposits are poorly consolidated. Silts and clays are subject to plastic deformation or shear failure under relatively light loads and undergo consolidation for long periods of time when subjected to appreciable loads. Many large earth dams have been built upon such materials but this demands a thorough exploration and testing programme in order to design safe structures. Soft alluvial clays at ground level have generally been removed if economically feasible. The slopes of an embankment dam may be flattened in order to mobilize greater foundation shear strength, or berms may be introduced. Where soft alluvial clays are not more than 2.3 m thick they should consolidate during construction if covered with a drainage blanket, especially if resting on sand and gravel. With thicker deposits it may be necessary to incorporate vertical sand drains within the clays. However, coarser sands and gravels undergo comparatively little consolidation under load and therefore afford excellent foundations for earth dams. Their primary problems result from their permeability. Alluvial sands and gravels form natural drainage blankets under the higher parts of an earth or rockfill dam, so that seepage through them beneath the dam must be cut off. Problems relating to underseepage through pervious strata may be met by a grout curtain. Alternatively, underseepage may be checked by the construction of an impervious upstream blanket to lengthen the path of percolation and the installation on the downstream side of suitable drainage facilities to collect the seepage.

Talus or scree may clothe the lower slopes in mountainous areas and because of its high permeability must be avoided in locating a dam site, unless it is sufficiently shallow to be economically removed from under the dam.

Landslips are a common feature of valleys in mountainous areas and large slips often cause narrowing of a valley which thus looks topographically suitable for a dam. Unless they are shallow seated and can be removed or effectively drained, it is prudent to avoid landslipped areas in dam location, because their unstable nature may result in movement during construction or subsequently on drawdown.

Fault zones may be occupied by shattered or crushed material and so represent zones of weakness which may give rise to landsliding upon excavation for a dam. The occurrence of faults in a river is not unusual and this generally means that the material along the fault zone is highly altered, thus necessitating a deep cutoff.

In most known instances of historic fault breaks the fracturing has occurred along a pre-existing fault. Movement along faults occurs not only in association with large and infrequent earthquakes but also in association with small shocks and continuous slippage known as fault creep. Earthquakes resulting from displacement and energy release on one fault can sometimes trigger small displacements on other unrelated faults many kilometres distant. Breaks on subsidiary faults have occurred at distances as great as 25 km from the main fault. Obviously, with increasing distance from the main fault the amount of displacement decreases.

Individual fault breaks during earthquakes have ranged in length from less than a kilometre to several hundred kilometres. However, the length of the fault break during a particular earthquake is generally only a fraction of the true length of the fault. The longer fault breaks have greater displacements and generate larger earthquakes. The maximum displacement is less than 6 m for the great majority of fault breaks and the average displacement along the length of the fault is less than half the maximum. These figures suggest that zoned embankment dams can be built with safety at sites with active faults.

All major faults located in regions where strong earthquakes have occurred should be regarded as potentially active unless convincing evidence exists to the contrary. In stable areas of the world little evidence exists of fault displacements in the recent past. Nevertheless, an investigation should be carried out to confirm the absence of active faults at or near any proposed major dam in any part of the world.

9.6.2 Construction materials for earth dams

Wherever possible, construction materials for an earth dam should be obtained from within the future reservoir basin (Anderson and McNicol, 1989). Accordingly, the investigation of the dam site and the surrounding area should determine the availability of impervious and pervious materials for the embankment, sand and gravels for drains and filter blankets, and stone for riprap.

In some cases only one type of soil is easily obtainable for an earth dam. If this is impervious, the design will consist of a homogeneous embankment, which incorporates a small amount of permeable material in order to control internal seepage. On the other hand, where sand and gravel are in plentiful supply a very thin earth core may be built into the dam if enough impervious soil is available, otherwise an impervious membrane may be constructed of concrete or interlocking steel-sheet piles. However, since concrete can withstand very little settlement such core walls must be located on sound foundations.

Sites which provide a variety of soils lend themselves to the construction of zoned dams. The finer, more impervious materials are used to construct the core whilst the coarser materials provide strength

and drainage in the upstream and downstream zones.

Embankment soils need to develop high shear strength, low permeability and low water absorption, and undergo minimal settlement. This is achieved by compaction. The degree of compaction achieved is reflected by the dry density of the soil. The relationship between dry density and moisture content for a particular compactive effort is assessed by a compaction test.

9.6.3 River diversion

Wherever dams are built there are problems concerned with keeping the associated river under control. These have a greater influence on the design of an embankment than a concrete dam. In narrow, steep-sided valleys the river is diverted through a tunnel or conduit before the foundation treatment is completed over the floor of the river. However, the abutment sections of an embankment can be constructed in wider valleys prior to river diversion. In such instances suitable borrow materials must be set aside for the closure section as this often has to be constructed rapidly so that overtopping is avoided. But rapid placement of the closure section can give rise to differential settlement and associated cracking. Hence, extra filter drains may be required to control leakage through such cracks. Compaction of the closure section at a higher average water content means that it can adjust more easily to differential settlement without cracking.

9.6.4 Ground improvement

Grouting has proved effective in reducing percolation of water through foundations and its introduction into dam construction has allowed considerable cost saving by avoiding the use of deep cutoff and wing trenches. Consequently, sites which previously were considered unsuitable because of adverse geological conditions can now be utilized.

Initial estimates of the groutability of ground frequently have been based upon the results of pumping-in tests, in which water is pumped into the ground via a drillhole. Lugeon (1933) suggested that grouting beneath concrete gravity dams was necessary when the permeability exceeded 1 lugeon unit (i.e. a flow of 1 l/m/min at a pressure of 1 MPa). However, this standard has been relaxed in modern practice, particularly for earth dams and for foundations where seepage is acceptable in terms of lost storage and non-erodability of foundation or core materials (Houlsby, 1990).

The effect of a grout curtain is to form a wall of low permeability within the ground below the perimeter of a dam. Holes are drilled and grouted, from the base of the cutoff or heel trench downwards. Where joints are vertical it is advisable to drill groutholes at a rake of from 10 to 15° since these cut across the joints at different levels, whereas vertical holes may miss them.

The rate at which grout can be injected into the ground generally increases with an increase in the grouting pressure, but this is limited since excessive pressures cause the ground to fracture and lift. The safe maximum pressure depends on the weight of overburden, the strength of the ground, the *in situ* stresses and the pore water pressures. However, there is no simple relationship between these factors and safe maximum grouting pressure. Hydraulic fracture tests may be used, especially in fissile rocks, to determine the most suitable pressures or the pressures may be related to the weight of overburden.

Once the standard of permeability has been decided for a section of or an entire grout curtain, it is achieved by split spacing or closure methods in which primary, secondary, tertiary, etc., sequences of grouting are carried out until water tests in the groutholes approach the required standard. In multiple-row curtains the outer rows should be completed first thereby allowing the innermost row to effect closure on the outer rows. A spacing of 1.5 m between rows is usually satisfactory. The upstream row should be the tightest row, tightness decreasing downstream. Single-row curtains are usually constructed by drilling alternate holes first and then completing the treatment by intermediate holes. Ideally a grout curtain is taken to a depth where the requisite degree of tightness is available naturally. This is determined either by investigation holes sunk prior to its design or by primary holes sunk during grouting. The search usually does not go beyond a depth equal to the height of the storage head above ground surface.

Consolidation grouting is usually shallow, the holes seldom extending more than 10 m. It is intended to improve the jointed rock mass by increasing its strength and reducing its permeability. It also improves the contact between concrete and rock, and makes good any slight loosening of the rock surface due to blasting operations. In addition, it affords a degree of homogeneity to the foundation which is desirable if differential settlement and unbalanced stresses are to be avoided. In other words the grout increases rock stiffness and attempts to bring Young's modulus to the required high uniform values. Holes are usually drilled normal to the foundation surface but in certain instances they may be orientated to intersect specific features. They are set out on a grid pattern at 3–14 m centres depending on the nature of the rock. Consolidation grouting must be completed before the construction of a dam begins.

Casagrande (1961) cast doubts on the need for grout curtains, maintaining that a single-row grout curtain constructed prior to reservoir filling is frequently inadequate. He also maintained that expensive grouting was useless as far as reducing water pressures was concerned and that drainage systems were the only efficient method of controlling the piezometric level and therefore uplift forces along the dam foundation. He further maintained that drainage is the only efficient treatment available for rock of low hydraulic conductivity, that is, rock with fine fissures. Drainage can control the hydraulic potential on the downstream side of a dam thus achieving what is required of a grout curtain, except, of course, that drainage does not reduce the amount of leakage. However, leakage is not of consequence in most rock masses where the hydraulic conductivity is low. Casagrande contended that for fissured rocks of low permeability (less than 5 lugeon units) drainage is generally essential, whereas grouting constitutes a wasted effort. Conversely, if the permeability is high (more than 50 lugeon units) grouting is necessary to control groundwater leakage beneath a dam.

9.7 FOUNDATIONS FOR BUILDINGS

9.7.1 Types of foundation structure

The design of foundations embodies three essential operations: (i) calculating the loads to be transmitted by the foundation structure to the soils or rocks supporting it; (ii) determining the engineering performance of these soils and rocks; and (iii) designing a suitable foundation.

Footings distribute the load to the ground over an area sufficient to suit the pressures to the properties of the soil or rock. Their size is therefore governed by the strength of the foundation materials. If the footing supports a single column it is known as a spread or pad footing whereas a footing beneath a wall is referred to as a strip or continuous footing.

The amount and rate of settlement of a footing due to a given load per unit area of its base, is a function of the dimensions of the base, and of the compressibility and permeability of the foundation materials located between the base and a depth which is at least one and a half times the width of the base. In addition, if footings are to be constructed on cohesive soil, it is necessary to determine whether or not the soil is likely to swell or shrink according to any seasonal variations. Fortunately significant variations below a depth of about 2 m are rather rare.

Footings usually provide the most economical type of foundation structure but the allowable bearing capacity must be chosen to provide an adequate factor of safety against shear failure in the soil and to ensure that settlements are not excessive. Settlement for any given pressure increases with the width of footing in almost direct proportion on clays and to a lesser degree on sands.

A raft permits the construction of a satisfactory foundation in materials whose strength is too low for the use of footings. The chief function of a raft is to spread the building load over as great an area of ground as possible and thus reduce the bearing pressure to a minimum. Also, a raft provides a degree of rigidity which reduces differential movements in the superstructure. The settlement of a raft foundation does not depend on the weight of the building which is supported by the raft. It depends on the difference between this weight and the weight of the soil which is removed prior to the construction of the raft, provided the heave produced by the excavation is inconsequential. A raft can be built at a sufficient depth so that the weight of soil removed equals the weight of the building. Hence, such rafts are sometimes called floating foundations. The success of this type of foundation structure in overcoming difficult soil conditions has led to the use of deep-raft and rigid frame basements for a number of high buildings on clay.

When the soil immediately beneath a proposed structure is too weak or too compressible to provide adequate support, the loads can be transferred to more suitable material at greater depth by means of piles. Such bearing piles must be capable of sustaining the load with an adequate factor of safety, without allowing settlement detrimental to the structure to occur. Although these piles derive their carrying capacity from end-bearing at their bases, friction along their sides also contributes towards this. Indeed, friction is likely to be the predominant

factor for piles in clays and silts whilst end-bearing provides the carrying capacity for piles terminating in or on gravel or rock.

Piles may be divided into three main types, according to the effects of their installation: (i) displacement piles; (ii) small displacement piles; and (iii) non-displacement piles (Anon, 1986). Displacement piles are installed by driving and so their volume has to be accommodated below ground by vertical and lateral displacements of soil which may give rise to heave or compaction, this could have detrimental effects upon neighbouring structures. Driving may also cause piles which are already installed to lift. Driving piles into clay may affect its consistency, that is, the penetration of piles combined with the vibrations set up by the falling hammer, destroys the structure of the clay and initiates a new process of consolidation which drags the piles in a downward direction. Sensitive clays are affected in this way whilst insensitive clays are not. Small displacement piles include some piles which may be used in soft alluvial ground of considerable depth. They also may be used to withstand uplift forces. They are not suitable in stiff clays or gravels. Non-displacement piles are formed by boring and the hole may be lined with casing which may or may not be left in place. When working near existing structures which are founded on loose sands or silts, particularly if these are saturated, it is essential to avoid using methods which cause dangerous vibrations and may give rise to a quick condition.

For practical purposes the ultimate bearing capacity may be taken as that load which causes the head of the pile to settle 10% of the pile diameter. The ratio between the settlement of a pile foundation and that of a single pile acted upon by the design load can have almost any value. This is due to the fact that the settlement of an individual pile depends only on the nature of the soil in direct contact with the pile, whereas the settlement of a pile foundation also depends on the number of piles and on the compressibility of the strata located between the level of the ends of the piles and the surface of the bedrock.

9.7.2 Bearing capacity

Foundation design is primarily concerned with ensuring that movements of a foundation are kept within limits which can be tolerated by the proposed structure without adversely affecting its functional requirements. Hence, the design of a foundation structure requires an understanding of the local geological and groundwater conditions, and more particularly an appreciation of the various types of ground movement that can occur.

In order to avoid shear failure or substantial shear deformation of the ground, the foundation pressures used in design should have an adequate factor of safety when compared with the ultimate bearing capacity of the foundation. The ultimate bearing capacity is the value of the loading intensity which causes the ground to fail in shear. If this is to be avoided, then a factor of safety must be applied to the ultimate bearing capacity, the value obtained being the safe bearing capacity. But even this value may still mean that there is a risk of excessive or differential settlement. Thus the allowable bearing capacity is the value which is used in design, this taking into account all possibilities of ground movement, and so its value is normally less than that of the safe bearing capacity. The value of ultimate bearing capacity depends on the type of foundation structure as well as the soil properties, therefore the

dimensions, shape, and depth at which a footing is placed all influence the bearing capacity. More specifically the width of the foundation is important in cohesionless sands — the greater the width the larger the bearing capacity, whilst in saturated clays it is of little effect. With uniform soil conditions the ultimate bearing capacity increases with depth of installation of the foundation structure. This increase is associated with the confining effects of the soil, the decreased overburden pressure at foundation level, and with the shear forces that can be mobilized between the sides of the foundation structure and the ground. The presumed bearing values for various types of soil and rock are given in Table 9.8.

The development of foundation failure involves first, the soil beneath the foundation being forced downwards in a wedge-shaped zone (Fig. 9.22). Consequently, the soil beneath the wedge is forced downwards and outwards, elastic bulging and distortion taking place within the soil mass. Second, the soil around the foundation perimeter pulls away from the foundation and the shear forces propagate

Table 9.8 Presumed allowable bearing values under static loading. (After Anon, 1986.)

Category	Types of rocks and soils	Presumed allowable bearing value (kPa)	Remarks
Rocks	Strong igneous and gneissic rocks in sound condition	10 000	These values are based on the assumption that the foundations are taken down to unweathered rock
	Strong limestones and strong sandstones	4 000	
	Schists and slates	3 000	
	Strong shales, strong mudstones and strong siltstones	2 000	
Non-cohesive soils	Dense gravel, or dense sand and gravel	>600	Width of foundation not less than 1 m. Groundwater level assumed to be at a depth not less than below the base of the foundation
	Medium dense gravel, or medium dense sand and gravel	<200−600	
	Loose gravel, or loose sand and gravel	<200	
	Compact sand	>300	
	Medium dense sand	100−300	
	Loose sand	<100 Value depending on degree of looseness	
Cohesive soils	Very stiff boulder clays and hard clays	300−600	Susceptible to long-term consolidation settlement
	Stiff clays	150−300	
	Firm clays	75−150	
	Soft clays and silts	<75	

Note. These values are for preliminary design purposes only, and may need alteration upwards or downwards. No addition has been made for the depth of embedment of the foundation.

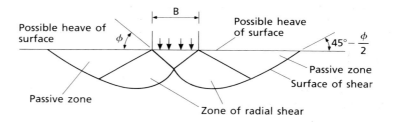

Fig. 9.22 Foundation failure.

outwards from the apex of the wedge. This forms the zone of radial shear in which plastic failure by shear occurs. If the soil is very compressible or can endure large strains without plastic flow the failure is confined to fan-shaped zones of local shear. The foundation displaces downwards with little load increase. On the other hand, if the soil is more rigid, the shear zone propagates outwards until a continuous surface of failure extends to ground surface and the surface heaves.

The weight of the material in the passive zone resists the lifting force and provides the reaction through the other two zones which counteracts downwards motion of the foundation structure. Thus, the bearing capacity is a function of the resistance to uplift of the passive zone. This, in turn, varies with the size of the zone (which is a function of the internal angle of friction), with the unit weight of the soil and with the sliding resistance along the lower surface of the zone (which is a function of the cohesion, internal angle of friction, and unit weight of the soil). A surcharge placed on the passive zone or increasing the depth of the foundation therefore increases the bearing capacity.

The stress distribution due to a structure declines rapidly with depth within the soil. It should be determined in order to calculate bearing capacity and settlement at given depths.

The pressure acting between the bottom of a foundation structure and the soil is the contact pressure. The assumption that a uniformly loaded foundation structure transmits the load uniformly so that the ground is uniformly stressed is by no means valid, for example, the intensity of the stresses at the edges of a rigid foundation structure on hard clay is theoretically infinite. In fact, of course, the clay yields slightly and so reduces the stress at the edges. As the load is increased more and more local yielding of the ground material takes place until, when the loading is close to that which would cause failure, the distribution is probably very nearly uniform. Therefore at working loads a uniformly loaded rigid

foundation structure on clay imposes a widely varying contact pressure. On the other hand, a rigid footing on the surface of dry sand imposes a parabolic distribution of pressure. Since there is no cohesion in such material, no stress can develop at the edges of a footing. If the footing is below the surface of the sand, the pressure at the edge is no longer zero but increases with depth. The pressure distribution tends therefore to become more nearly uniform as the depth increases. If a footing is perfectly flexible, then it will distribute a uniform load over any type of foundation material.

The ultimate bearing capacity of foundations on cohesionless deposits depends on the width and depth of placement of the foundation structure as well as the angle of shearing resistance. The position of the water table in relation to the foundation structure has an important influence on the ultimate bearing capacity. High groundwater levels lower the effective stresses in the ground so that the ultimate bearing capacity is reduced by anything up to 50%. Generally speaking, gravels and dense sands afford good foundations. It is possible to estimate the bearing capacity of such soils from penetration tests, either static or dynamic, or plate-load tests.

The ultimate bearing capacity of foundations on clay soils depends on the shear strength of the soil and the shape and depth at which the foundation structure is placed. The shear strength of clay is, in turn, influenced by its consistency. Although there is a small decrease in the moisture content of a clay beneath a foundation structure which gives rise to a small increase in soil strength, this is of no importance as far as estimation of the factor of safety against shear is concerned. In relation to applied stress, saturated clays behave as purely cohesive materials provided that no change of moisture content occurs. Thus, when a load is applied to saturated clay it produces excess pore water pressures which are not quickly dissipated. In other words the angle of shearing resistance is equal to zero. The assumption that $\phi = 0$ forms the basis of all normal calculations of ultimate bearing capacity of clays. The strength may then be taken as the undrained shear strength or one half the unconfined compressive strength. To the extent that consolidation does occur, the results of analyses based on the premise that $\phi = 0$ are on the safe side. Only in special cases, with prolonged loading periods or with very silty clays, is the assumption sufficiently far from the truth to justify a more elaborate analysis.

For all types of foundation structures on clays the factors of safety must be adequate against bearing capacity failure. Experience has indicated that it is desirable to use a factor of safety of 3, yet although this means that complete failure almost invariably is ruled out, settlement may still be excessive. It is therefore necessary to give consideration to the settlement problem if bearing capacity is to be viewed correctly. More particularly it is important to make a reliable estimate of the amount of differential settlement that may be experienced by the structure. If the estimated differential settlement is excessive it may be necessary to change the layout or type of foundation structure.

If a rock mass contains few defects, the allowable contact pressure at the surface may be taken conservatively as the unconfined compressive strength of the intact rock. Most rock masses, however, are affected by joints or weathering which may significantly alter their strength and engineering behaviour. Table 9.9 gives values of allowable contact pressure for jointed rocks based on their RQD. If design is based on these values the settlement of foundations should not exceed 12.7 mm even for large loaded areas. The RQD should be the average within a depth below foundation level equal to the width of the foundation, provided the RQD is fairly uniform

within that depth. If the upper part of the rock mass, within a depth equal to about quarter the width of the foundation, is of low quality the value for this part should be used or the inferior rock should be removed.

The great variation in the physical properties of weathered rock and the non-uniformity of the extent of weathering even at a single site permit few generalizations concerning the design and construction of foundation structures. The depth to bedrock and the degree of weathering must be determined. If the weathered residuum plays the major role in the regolith, rock fragments being of minor consequence, then design of rafts or footings should be according to the matrix material. Piles can provide support at depth.

9.7.3 Settlement

The average values of settlement beneath a structure together with the individual settlements experienced by its various parts influence the degree to which the structure serves its purpose. The damage attributable to settlement can range from complete failure of the structure to slight disfigurement (Fig. 9.23).

If cohesionless sediments are densely packed they will be almost incompressible. Loosely packed sand located above the water table will undergo some settlement but is otherwise stable. Where foundation level is below the water table, greater settlement is likely to be experienced. Additional settlement may occur if the water table fluctuates or the ground is subject to vibrations. Settlement commonly is relatively rapid but there can be a significant time lag when stresses are large enough to produce appreciable grain fracturing. Nonetheless settlement in sands and gravels is frequently substantially complete by the end of the construction period.

Settlement can present a problem in clayey soils so that the amount which is likely to take place when they are loaded needs to be determined. Settlement invariably continues after the construction period, often for several years. Immediate or elastic settlement is that which occurs under constant volume (undrained) conditions when clay deforms to accommodate the imposed shear stresses. Primary consolidation in a clay takes place due to the void space being gradually reduced as the pore water and/or air are expelled therefrom on loading. The rate at which this occurs depends on the rate at which the excess pore water pressure induced by a structural load is dissipated, thereby allowing the

Table 9.9 Allowable contact pressure for jointed rock*

RQD	Allowable contact pressure (MPa)
100	32.2
90	21.5
75	12.9
50	7.0
25	3.2
0	1.1

* If the value of the allowable contact pressure exceeds the unconfined compressive strength of intact samples then it should be taken as the unconfined compressive strength.

Fig. 9.23 Settlement of buildings on soft alluvial soils, Amsterdam, the Netherlands.

structure to be supported entirely by the soil skeleton. Consequently, the permeability of the clay is all important. After a sufficient time has elapsed excess pore water pressures approach zero but a deposit of clay may continue to decrease in volume. This is referred to as secondary consolidation and involves compression of the soil fabric.

Settlement is rarely a limiting condition in foundations on most fresh rocks and does not entail special study except in the case of special structures where settlements must be small. The problem then generally resolves itself into one of reducing the unit-bearing load by widening the base of structures. In certain cases differential settlements are provided for by designing articulated structures capable of taking differential movements of individual sections without damaging the structure. On the other hand, appreciable settlements may take place in low grade compaction shales.

Generally, uniform settlements can be tolerated without much difficulty, but large settlements are inconvenient and may cause serious disturbance to

services even where there is no evident damage to the structure. However, differential settlement is of greater significance than maximum settlement since the former is likely to distort or even shear a structure. Buildings which suffer large maximum settlement are also likely to experience large differential settlement. Both should therefore be avoided.

Burland and Wroth (1975) accepted a safe limit for angular distortion (difference in settlement between two points) of 1:500, as satisfactory for framed buildings but stated that it was unsatisfactory for buildings with load-bearing walls. Damage in the latter has occurred with very much smaller angular distortions. The rate at which settlement occurs also influences the amount of damage suffered.

For most buildings it is the relative deflections which occur after completion that cause damage. Therefore the ratio between the immediate and total settlement is important. In overconsolidated clays this averages about 0.6 whilst for normally consolidated clay it is usually less than 0.2. This low value coupled with larger total settlement makes the problems of design for normally consolidated clays much more severe than for overconsolidated clays.

Settlements may be reduced: (i) by the correct design of the foundation structure, which may include larger or deeper foundations; (ii) if the site is preloaded or surcharged prior to construction; or (iii) if the soil is subjected to dynamic compaction or vibrocompaction. It is advantageous if the maximum settlement of large structures is reached earlier than later. The installation of vertical drains, which provide shorter drainage paths for the escape of water to strata of higher permeability, is one means by which this can be achieved. Vertical drains may effect up to 80% of the total settlement in cohesive soils during the construction stage. Differential settlement also can be accommodated by methods similar to those used to accommodate subsidence (Anon, 1975b).

9.7.4 Subsidence

Subsidence can be regarded as the vertical component of ground movement caused by mining operations although there is a horizontal component. Subsidence can and does have serious effects on buildings, services, and communications. It can be responsible for flooding, lead to the sterilization of land, or call for extensive remedial measures or special constructional design in site development. An account of subsidence is provided in Chapter 8.

9.7.5 Methods of grout treatment

In recent years there has been an increase in the extent to which the various methods of grout treatment have been used to improve subsurface conditions. Some of these techniques are not new but in the past they were used more as desperate remedies for dealing with unforeseen problems connected with poor ground conditions whilst today they are recognized as part of a normally planned construction process.

Grouting refers to the process of injecting setting fluids into fissures, pores, and cavities in the ground. It may either be preplanned or an emergency expedient. The process is widely used in foundation engineering in order to reduce seepage of water or to increase the mechanical performance of the soils or rocks concerned.

If the sealing and strengthening actions are to be successful, then grouting must extend a considerable distance into the formation. This is achieved by injecting the grout into a special array of boreholes and is referred to as permeation grouting. Permeation grouting is the most commonly used method of grouting in which the groutability and therefore the choice of grout is influenced by the pore size of the ground to be treated, which is approximately related to particle grading. Normally cement or clay–cement grouts are used in coarser soils and clay–chemical or chemical grouts are used in finer-grained soils. The limits for particulate grouts are generally regarded as a 10:1 size factor between the D_{15} of the grout and the D_{15} size of the granular system to be injected. Generally, cement grouts are limited to soils with pore dimensions greater than 0.2 mm (Table 9.10), i.e. they will not permeate fine sands. Because chemical grouts are non-particulate, their penetrability depends primarily on their viscosity. Hence, where it is necessary to use chemical grouts, especially for larger jobs, cheaper high-viscosity grouts may be used to fill the larger voids and more costly low-viscosity grouts to fill voids too small to be penetrated by the initial grouting.

Cement grout cannot enter a fissure smaller than about 0.1 mm. In fissured rocks the D_{85} of the grout must be smaller than one-third the fissure width.

Table 9.10 Types of grout. (After Anon, 1986.)

Ground	Typical grouts used	Examples
Alluvials Open gravels Gravels Coarse sands Medium sands	Suspension	Cement suspensions with particle size of about 0.55 mm Cement clay, clay treated with reagents Separated clay and reagents, bentonitic clays with sodium silicate and deflocculants (clay gels) Two-shot sodium silicate based systems for conferring strength Bituminous emulsions with fillers and emulsion breaker
Coarse sands Medium sands	Colloidal solutions	Single-shot silicate based systems for strength (silicate–organic ester) Single-shot lignin based grouts for moderate strength and impermeability Silicate–metal salt single-shot system, e.g. sodium silicate–sodium aluminate; sodium silicate–sodium bicarbonate Water soluble precondensates, e.g. urea-formaldehyde Oil based elastomers (high viscosity)
Fine sands Silts	Solutions	Water soluble polysaccharides with metal salt to give insoluble precipitate Water soluble acrylamide, water soluble phenoplasts
Fissured rocks Open jointed Medium jointed	Suspensions	Cement–sand, cement, cement clay
Medium jointed Fine jointed	Solutions	Oil based elastomers, non-water soluble polyesters, epoxides, and range of water soluble polymer systems given above Hair cracks in concrete would be treated with a high strength low viscosity polyester or epoxy resin

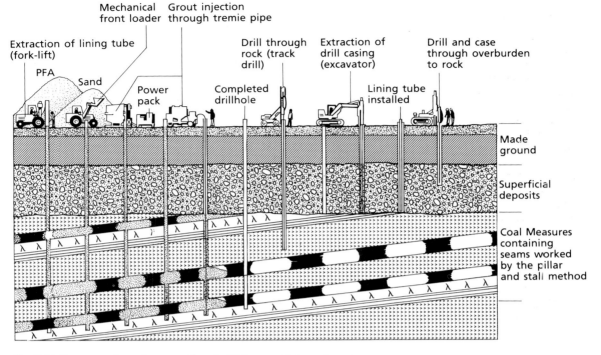

Fig. 9.24 Sequence of grouting operations to fill large voids in coal seams.

Experience shows that there is an upper limit to this ratio, as large quantities of grout have been lost from sites via open fissures. The shape of an opening also affects groutability. In order that the grout can achieve effective adhesion the sides of the fissure or voids must be clean. If they are coated with clay, then they need to be washed prior to grouting. Cavities in rocks may have to be filled with bulk grouts (usually mixtures of cement, pulverized fly-ash, and sand; gravel may be added when large openings need filling (Fig. 9.24)) or foam grouts (cement grout to which a foaming agent is added).

Vibroflotation is used to improve poor ground below foundation structures. The process may reduce settlement by more than 50% and the shearing strength of treated soils is increased substantially. Vibrations of appropriate form can eliminate intergranular friction of cohesionless soils so those initially loosely packed can be converted into a dense state. A vibroflot is used to penetrate the soil and can also operate efficiently below the water table. The best results have been obtained in fairly coarse sands which contain little or no silt or clay, since both reduce the effectiveness of the vibroflot.

However, today it is more usual to form columns of coarse backfill, formed at individual compaction centres, to stiffen soils. The vibroflot is used for compacting these columns which, in turn, effect a reduction in settlement (Fig. 9.25). Since the granular backfill replaces the soil this process is sometimes known as vibroreplacement. Vibroreplacement is commonly used in soft normally consolidated compressible clays, saturated silts, and alluvial and estuarine soils. Stone columns have been formed successfully in soils with undrained cohesive strengths as low as 7 kPa. Vibrodisplacement involves the vibroflot penetrating the ground by shearing and displacing the ground around it. It is accordingly restricted to strengthening insensitive clay soils which have sufficient cohesion to maintain a stable borehole, that is, to those over 20 kPa undrained strength. These soils require treatment primarily to boost their bearing capacity, the displacement method inducing some measurable increase in the strength of the clay between the columns.

Dynamic compaction brings about an improvement in the mechanical properties of a soil by the

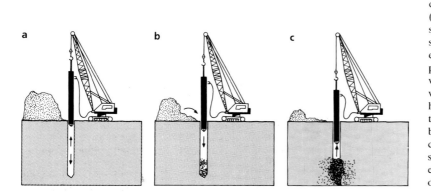

Fig. 9.25 Formation of a stone column by vibrocompaction: (a) sinking the vibrator into the subsoil up to the depth where sufficient load bearing capacity is encountered; (b) aggregates are placed into the hole made by the vibrator and after each filling the vibrator is sunk again into the hole; (c) it is necessary to repeat this process as many times as may be required to achieve a degree of compaction of the surrounding subsoil and the aggregate as to ensure that no further penetration of the vibrator can be effected.

Fig. 9.26 Dynamic compaction.

repeated application of very high intensity impacts to the surface. This is achieved by dropping a large weight, typically 10 to 20 tonnes, from crawler cranes, from heights of 15 to 40 m, at regular intervals across the surface (Fig. 9.26). Repeated passes are made over a site, although several tampings may be made at each imprint during a pass. Each imprint is back-filled after tamping. The first pass at widely spaced centres improves the bottom layer of the treatment zone and subsequent passes then consolidate the upper layers. In finer materials the increased pore water pressures must be allowed to dissipate between passes, which may take several weeks.

Care must be taken in establishing the treatment pattern, tamping energies, and the number of passes for a particular site and this should be accompanied by *in situ* testing as the work proceeds. Coarse granular fill requires more energy to overcome the possibility of bridging action, for similar depths, than finer material.

Suggestions for further reading

General engineering geology

Anon. (1976). *Manual of Applied Geology for Engineering*. Institution of Civil Engineers, Telford Press, London.

Attewell, P.B. and Farmer, I.W. (1976). *Principles of Engineering Geology*. Chapman & Hall, London.

Bell, F.G. (1980). *Engineering Geology and Geotechnics*. Butterworths, London.

Bell, F.G. (1983). *Fundamentals of Engineering Geology*. Butterworths, London.

Krynine, P.D. and Judd, W.R. (1956). *Principles of Engineering Geology and Geotechnics*. McGraw-Hill, New York.

Legget, R.F. and Hatheway, A.W. (1988). *Geology and Engineering* (3rd edn). McGraw-Hill, New York.

Legget, R.F. and Karrow, P.F. (1983). *Handbook of Geology in Civil Engineering*. McGraw-Hill, New York.

Paige, S. (ed.) (1950). *Applications of Geology to Engineering Practice*, Berkey Volume. Geological Society of America, New York.

Rahn, P.R. (1986). *Engineering Geology: An Environmental Approach*. Elsevier, New York.

Trask, P.D. (ed.) (1950). *Applied Sedimentation*. Wiley, New York.

Zaruba, Q. and Mencl, V. (1976). *Engineering Geology*. Academia, Prague.

General geology

Billings, M.P. (1962). *Structural Geology*. Prentice-Hall, Englewood Cliffs, New Jersey.

Dunbar, C.O. and Rodgers, J. (1957). *Principles of Stratigraphy*. Wiley, New York.

Fry, N. (1984). *The Field Description of Metamorphic Rocks*. The Geological Society, London.

Griffiths, J.C. (1967). *Scientific Method in Analysis of Sediments*. McGraw-Hill, New York.

Grim, R.E. (1962). *Applied Clay Mineralogy*. McGraw-Hill, New York.

Hobbs, B., Means, W.D., and Williams, P.F. (1976). *An Outline of Structural Geology*. Wiley, New York.

Duff, P.M.L.D. (ed.) (1993). *Holmes' Principles of Physical Geology* (4th edn). Chapman and Hall, London.

Krumbein, W.C. and Sloss, L.L. (1963). *Stratigraphy and Sedimentation*. Freeman, San Francisco.

Lahee, F.H. (1950). *Field Geology*. McGraw-Hill, New York.

Milner, H.B. (1962). *Sedimentary Petrography* (2 vols). Allen & Unwin, London.

Ragan, D.M. (1973). *Structural Geology: An Introduction to Geometrical Techniques*. Wiley, New York.

Ramsey, J.G. (1967). *Folding and Fracturing of Rocks*. McGraw-Hill, London.

Reid, H.H. and Watson, J. (1962). *Introduction to Geology*, Vol. 1. Macmillan, London.

Shrock, R.R. (1948). *Sequence in Layered Rocks*. McGraw-Hill, New York.

Thorpe, R.S. and Brown, G.C. (1985). *The Field Description of Igneous Rocks*. The Geological Society, London.

Tucker, M.E. (1982). *The Field Description of Sedimentary Rocks*. The Geological Society, London.

Surface processes

Bagnold, R.A. (1941). *Physics of Wind Blown Sand*. Methuen, London.

Coates, D.R. (ed.) (1976). *Geomorphology and Engineering*. Allen & Unwin, London.

Cooke, R.M. and Warren, A. (1973). *Geomorphology in Deserts*. Batsford, London.

Embleton, C. and King, C.A.M. (1968). *Glacial and Periglacial Geomorphology*. Arnold, London.

Flint, R.F. (1967). *Glacial and Pleistocene Geology*. Wiley, New York.

Fookes, P.G. and Vaughan, P.R. (eds) (1986). *A Handbook of Engineering Geomorphology*. Surrey University Press, London.

Ford, D.C. and Williams, P.W. (1989). *Karst Geomorphology and Hydrology*. Unwin Hyman, London.

Glennie, K.W. (1970). *Desert Sedimentary Environments*. Elsevier, Amsterdam.

Hails, J.R. (ed.) (1977). *Applied Geomorphology*. Elsevier, Amsterdam.

King, C.A.M. (1976). *Beaches and Coasts*. Arnold, London.

Komar, P.O. (1976). *Beach Processes and Sedimentation*. Prentice-Hall, Englewood Cliffs, New Jersey.

Leopold, L.B., Wolman, M.G., and Miller, L.P. (1964). *Fluvial Processes in Geomorphology*. Freeman, San Francisco.

Muller, S.W. (1947). *Permafrost or Permanently Frozen Ground and Related Engineering Problems*. Edward Bros, Ann Arbor, Michigan.

Ollier, C.D. (1984). *Weathering*. Longman, Harlow.

Schuster, R.L. and Krizek, R.J. (eds) (1978). *Landslides: Analysis and Control*. Transportation Research Board, Special Report 176, National Academy of Science, Washington, DC.

Silvester, R. (1974). *Coastal Engineering*. Elsevier,

Amsterdam.

Veder, C. (1981). *Landslides and their Stabilization.* Springer-Verlag, New York.

Verstappen, J.T. (1983). *Applied Geomorphology: Geomorphological Surveys for Environmental Development.* Elsevier, Amsterdam.

Ward, R. (1978). *Floods.* Macmillan, London.

Washburn, A.L. (1973). *Periglacial Processes and Environments.* Arnold, London.

Zaruba, Q. and Mencl, V. (1985). *Landslides and their Control.* Elsevier, Amsterdam.

Groundwater

Brassington, R. (1988). *Field Hydrogeology.* The Geological Society, London.

Cripps, J.C., Bell, F.G., and Culshaw, M.G. (eds) (1986). *Groundwater in Engineering Geology.* Engineering Geology Special Publication No. 3. The Geological Society, London.

Davis, S.N. and De Weist, R.J.H. (1966). *Hydrogeology.* Wiley, New York.

De Weist, R.J.H. (1967). *Geohydrology.* Wiley, New York.

Fetter, C.W. (1980). *Applied Hydrogeology.* Merrill, Columbus, Ohio.

Freeze, R.A. and Cherry, J.A. (1979). *Groundwater.* Prentice-Hall, Englewood Cliffs, New Jersey.

Hamill, L. and Bell, F.G. (1986). *Groundwater Resource Development.* Butterworths, London.

Price, M. (1985). *Introducing Groundwater.* Allen & Unwin, London.

Todd, D.K. (1980). *Ground Water Hydrology* (2nd edn). Wiley, New York.

Verruijt, A. (1970). *The Theory of Groundwater Flow.* Macmillan, London.

Walton, W.C. (1970). *Groundwater Resource Evaluation.* McGraw-Hill, New York.

Construction materials

Collis, L. and Fox, R.A. (eds) (1985). *Aggregates: Sand, Gravel and Crushed Rock Aggregates.* Engineering Geology Special Publication No. 1. The Geological Society, London.

Fookes, P.G. and Hawkins, A.B. (eds) (1993). *Handbook of Engineering Geomaterials.* Blackie, Glasgow.

Keeling, P.S. (1969). *The Geology and Mineralogy of Brick Clays.* Brick Development Association, Stoke-on-Trent.

Prentice, J.E. (1990). *Geology of Construction Materials,* Chapman & Hall, London.

Simpson, J.L. and Horrobin, P.J. (1970). *The Weathering and Performance of Building Materials.* Medical and Technical Publishing Co., Aylesbury.

Winkler, E.M. (1973). *Stone: Properties, Durability in Man's Environment.* Springer-Verlag, New York.

Description and behaviour of soils and rocks

Anon. (1986). *Guide to Rock and Soil Descriptions.* Geoguide 3. Geotechnical Control Office, Government Publication Centre, Hong Kong.

Bell, F.G. (1992). *Engineering Properties of Soils and Rocks* (3rd edn). Butterworths-Heinemann, Oxford.

Brand, E.W. and Brenner, R.P. (eds) (1981). *Soft Clay Engineering.* Elsevier, Amsterdam.

Cripps, J.C., Culshaw, M.G., Coulthard, J.M., and Henscher, S. (1992). *Engineering Geology of Weak Rock,* Engineering Geology Special Publication No. 8. Balkema, Rotterdam.

Deere, D.U. and Miller, R.P. (1966). *Engineering Classification and Index Properties for Intact Rock.* Technical Report AFWL-TR-116, Air Force Weapons Laboratory, Kirtland Air Force Base, New Mexico.

Farmer, I.W. (1983). *Engineering Behaviour of Rocks.* Chapman & Hall, London.

Forster, A., Culshaw, M.G., Cripps, J.C., Moon, C.F., and Little, J.A. (eds) (1991). *Quaternary Engineering Geology.* Engineering Geology Special Publication No. 7. The Geological Society, London.

Gillot, J.E. (1968). *Clay in Engineering Geology.* Elsevier, Amsterdam.

Goodman, R.E. (1988). *Introduction to Rock Mechanics* (2nd edn). Wiley, New York.

Mitchell, J.K. (1988). *Fundamentals of Soil Behaviour* (2nd edn). Wiley, New York.

West, G. (1991). *The Field Description of Engineering Soils and Rocks.* The Geological Society, London.

Site investigation and engineering geological maps

Allum, J.A.E. (1966). *Photogeology and Regional Mapping.* Pergamon, London.

Anon. (1981). *Code of Practice for Site Investigations. BS 5930.* British Standards Institution, London.

Bell, F.G. (1975). *Site Investigations in Areas of Mining Subsidence.* Newnes-Butterworths, London.

Bell, F.G., Culshaw, M.G., Cripps, J.C., and Coffey, R. (eds) (1990). *Field Instrumentation and Engineering Geology.* Engineering Geology Special Publication No. 6. The Geological Society, London.

Butler, B.C.M. and Bell, J.D. (1988). *Interpretation of Geological Maps.* Longman Scientific and Technical, Harlow.

Carter, M. and Symons, M.W. (1989). *Site Investigations and Foundations Explained.* Pentech Press, London.

Clayton, C.R.I., Simons, N.W., and Matthews, M.C. (1982). *Site Investigations: A Handbook for Engineers.* Granada, London.

Dearman, W.R. (1991). *Engineering Geological Maps.* Butterworths-Heinemann, Oxford.

Demek, J. (1972). *Manual of Detailed Geomorphological Mapping.* Academia, Prague.

Griffiths, D. and King, R.E. (1984). *Applied Geophysics*

for Engineers and Geologists. Pergamon Press, Oxford.

Hanna, T.H. (1985). *Field Instrumentation in Geotechnical Engineering*. Trans Tech Publications, Clausthal-Zellerfeld.

Hawkins, A.B. (ed.) (1986). *Site Investigation Practice: Assessing BS 5930*. Engineering Geology Special Publication No. 2, The Geological Society, London.

Hvorslev, M.J. (1949). *Subsurface Exploration and Sampling for Civil Engineering Purposes*. American Society of Civil Engineers Report. Waterways Experimental Station, Vicksburg.

Kennie, T.J.M. and Matthews, M.C. (eds) (1985). *Remote Sensing in Civil Engineering*. Surrey University Press, London.

Maltman, A. (1990). *Geological Maps: An Introduction*. Open University Press, Milton Keynes.

Milson, J. (1989). *Field Geophysics*. The Geological Society, London.

Mitchell, C.W. (1973). *Terrain Evaluation*. Longmans, London.

Sabins, F.F. (1978). *Remote Sensing — Principles and Interpretation*. Freeman, San Francisco.

Weltman, A.J. and Head, J.M. (1983). *Site Investigation Manual*. Special Publication No. 25. Construction Industry Research and Information Association, London.

Geology and planning

Anon. (1978). *The Assessment and Mitigation of Earthquake Risk*. UNESCO, Paris.

Anon. (1985). *Geology for Urban Planning*. Economic Commission for Asia and the Pacific, United Nations, New York.

Bolt, B.A., Horn, W.L., McDonald, G.A., and Scott, R.F. (1975). *Geological Hazards*. Springer-Verlag, New York.

Brunsden, D. and Prior, D.B. (1984). *Slope Instability*. Wiley-Interscience, Chichester.

Canter, L. (1977). *Environmental Impact Assessment*. McGraw-Hill, New York.

Coates, D.R. (1981). *Environmental Geology*. Wiley, New York.

Cooke, R.U. and Doornkamp, J.C. (1974). *Geomorphology in Environmental Management*. Clarendon Press, Oxford.

Crawford, J.F. and Smith, P.G. (1984). *Landfill Technology*. Butterworths, London.

Culshaw, M.G., Bell, F.G., Cripps, J.C., and O'Hara, M. (eds) (1987). *Planning and Engineering Geology*. Engineering Geology Special Publication No. 4. The Geological Society, London.

Flawn, P.T. (1975). *Environmental Geology*. Harper & Row, New York.

Griggs, G.B. and Gilchrist, J.A. (1983). *Geologic Hazards, Resources and Environmental Planning*. Wadsworth, Belmont, California.

Holzer, T.L. (ed.) (1984). *Man-induced Land Subsidence*.

Reviews in Engineering Geology. American Geological Society, New York.

Howard, A.D. and Remson, I. (1978). *Geology in Environmental Planning*. McGraw-Hill, New York.

Keller, E.A. (1981). *Environmental Geology*. Merrill, Columbus, Ohio.

Krauskopf, K.B. (1988). *Radioactive Waste Disposal and Geology*. Chapman & Hall, London.

Legget, R.F. (1973). *Cities and Geology*. McGraw-Hill, New York.

Levenson, D. (1980). *Geology and the Urban Environment*. Oxford University Press, New York.

McHarg, I.L. (1969). *Design with Nature*. National History Press, Garden City, New York.

Porteous, A. (ed.) (1985). *Hazardous Waste Management Handbook*. Butterworths, London.

Tan, B.K. and Ran, J.L. (eds) (1986). *Landplan II: Role of Geology in Planning and Development of Urban Centres in Southeast Asia*. Association of Geoscientists for International Development, Report Series No. 12, Bangkok.

Utgard, R.O., McKenzie, G.D., and Foley, D. (eds) (1978). *Geology in the Urban Environment*. Burgess, Minneapolis.

Whittacker, B.N. and Reddish, D.J. (1989). *Subsidence, Occurrence, Prediction and Control*. Elsevier, Amsterdam.

Geology and construction

Attewell, P.B. and Taylor, R.K. (eds) (1984). *Ground Movements and their Effects on Structures*. Surrey University Press, London.

Bell, F.G. (1978). *Foundation Engineering in Difficult Ground*. Butterworths, London.

Bell, F.G. (ed.) (1987). *Ground Engineer's Reference Book*. Butterworths, London.

Bell, F.G. (ed.) (1992). *Engineering in Rock Masses*. Butterworths-Heinemann, Oxford.

Bell, F.G. (1993). *Engineering Treatment of Soils*. Spon, London.

Bell, F.G., Cripps, J.C., Culshaw, M.G., and Lovell, M.A. (eds) (1988). *Engineering Geology of Ground Movements*. Engineering Geology Special Publication No. 5. The Geological Society, London.

Brady, B.H.G. and Brown, E.T. (1985). *Rock Mechanics for Underground Mining*. Allen & Unwin, London.

Bromhead, E.N. (1986). *The Stability of Slopes*. Blackie, Glasgow.

Church, H.K. (1981). *Excavation Handbook*. McGraw-Hill, New York.

Hammer, M.J. and MacKichan, K.A. (1981). *Hydrology and Quality of Water Resources*. Wiley, New York.

Langefords, U. and Kilstrom, K. (1973). *The Modern Technique of Rock Blasting*. Wiley, New York.

Linsley, R.K. and Franzini, J.B. (1972). *Water Resources Engineering*. McGraw-Hill, New York.

McGregor, K. (1967). *The Drilling of Rock*. C R Books Ltd, A Maclaren Co., London.

Megaw, T.M. and Bartlett, J.V. (1982). *Tunnels: Planning and Design*, Vol. 2. Methuen, London.

Sherard, J.L., Woodward, R.L., Gizienski, S.F., and Clevenger, W.A. (1967). *Earth and Earth Rock Dams*. Wiley, New York.

Szechy, K. (1966). *The Art of Tunnelling*. Akademiai Kiado, Budapest.

Tomlinson, M.J. (1986). *Foundation Design and Construction*, (5th edn). Longmans, London.

Walters, R.C.S. (1962). *Dam Geology*. Butterworths, London.

References

Algermissen, S.T. and Perkins, D.M. (1976). United States Geological Survey Open File Report 76−716.

Al-Jassar, S.H. and Hawkins, A.B. (1979). Geotechnical properties of the Carboniferous Limestone of the Bristol area, *Proceedings 4th International Congress Rock Mechanics (ISRM), Montreaux.* A.A. Balkema, Rotterdam, pp. 3−14.

Ambraseys, N.N. (1974). Notes on engineering seismology. In *Engineering Seismology and Earthquake Engineering* (ed. J. Solnes). NATO Advanced Studies Institute Series, Applied Sciences, **3**, 33−54.

Andersland, O.B. and Anderson, D.M. (eds) (1978). *Geotechnical Engineering for Cold Regions.* McGraw-Hill, New York.

Anderson, J.G.C. and McNicol, R. (1989). The engineering geology of the Kielder Dam. *Quarterly Journal of Engineering Geology*, **22**, 111−30.

Anon. (1970). Logging of cores for engineering purposes, Engineering Group Working Party Report. *Quarterly Journal of Engineering Geology*, **3**, 1−24.

Anon. (1972). The preparation of maps and plans in terms of engineering geology, Engineering Group Working Party Report. *Quarterly Journal of Engineering Geology*, **5**, 293−381.

Anon. (1975a). *Methods for Sampling and Testing Mineral Aggregates, Sands and Fillers, BS 812.* British Standards Institution, London.

Anon. (1975b). *Subsidence Engineer's Handbook.* National Coal Board, London.

Anon. (1976). *Engineering Geology Maps. A Guide to their Preparation.* UNESCO, Paris.

Anon. (1977a). The description of rock masses for engineering purposes. Engineering Group Working Party Report, *Quarterly Journal of Engineering Geology*, **10**, 43−52.

Anon. (1977b). *Ground Subsidence.* Institution of Civil Engineers, Telford Press, London.

Anon. (1979). Classification of rocks and soils for engineering geological mapping, Part 1 − Rock and soil materials. Report of the Commission of Engineering Geological Mapping, *Bulletin of the International Association of Engineering Geology*, No. 19, 364−71.

Anon. (1981a). *Specification for Aggregates from Natural Sources for Concrete, BS 882.* British Standards Institution, London.

Anon. (1981b). Basic geotechnical description of rock masses. International Society of Rock Mechanics Commission on the Classification of Rocks and Rock Masses. *International Journal of Rock Mechanics and Mining Sciences and Geomechanical Abstracts*, **18**, 85−110.

Anon. (1981c). *Code of Practice on Site Investigations, BS. 5930.* British Standards Institution, London.

Anon. (1982). *The Treatment of Disused Mine Shafts and Adits.* National Coal Board, London.

Anon. (1983a). *The Selection of Natural Building Stone.* Digest 260, Building Research Establishment. HMSO, London.

Anon. (1983b). *Code of Practice for Test Pumping Water Wells, BS 6316.* British Standards Institution, London.

Anon. (1985). *Specification for Clay Bricks, BS 3921.* British Standards Institution, London.

Anon. (1986). *Code of Practice for Foundations, BS 8004.* British Standards Institution, London.

Anon. (1988). Engineering geophysics. Engineering Group Working Party Report. *Quarterly Journal of Engineering Geology*, **21**, 207−73.

Anon. (1990). Tropical residual soils. Engineering Group Working Party Report. *Quarterly Journal of Engineering Geology*, **23**, 1−101.

Atkinson, T. (1970). Ground preparation by ripping in open pit mining. *Mining Magazine*, **122**, 458−69.

Atkinson, T. (1971). Selection of open pit excavating and loading equipment. *Transactions of the Institution of Mining and Metallurgy, Section A, Mining Industry*, **80**, A101−29.

Attewell, P.B. (1971). Geotechnical properties of the Great Limestone in northern England. *Engineering Geology*, **5**, 89−112.

Audric, T. and Bouquier, L. (1976). Collapsing behaviour of some loess soils from Normandy. *Quarterly Journal of Engineering Geology*, **9**, 265−78.

Bagnold, R.A. (1941). *Physics of Wind Blown Sand.* Methuen, London.

Baillieul, T.A. (1987). Disposal of high-level nuclear waste in America. *Bulletin of the Association of Engineering Geologists*, **24**, 207−16.

Baker, P.E. (1979). Geological aspects of volcano prediction. *Journal of the Geological Society*, **136**, 341−6.

Baker, R.F. and Marshall, H.E. (1958). Control and correction. In *Landslides in Engineering Practice* (ed. E.B. Eckel). Committee on Landslide Investigations, Highway Research Board, Special Report 29, Washington, DC, pp. 150−88.

Balk, R. (1938). *Structural Behaviour of Igneous Rocks.* Geological Society of America, Memoir 6, New York.

Barber, C. (1982). *Domestic Waste and Leachate.* Notes on Water Research No. 31. Water Research Centre, Medmenham.

Barrow, G. (1912). On the geology of lower Deeside and the southern Highland border. *Proceedings of the Geologists Association*, **23**, 268–84.

Barton, N. (1976). The shear strength of rock and rock joints. *International Journal of Rock Mechanics and Mining Sciences and Geomechanical Abstracts*, **13**, 255–79.

Barton, N. (1978). Suggested methods for the quantitative description of discontinuities in rock masses. International Society of Rock Mechanics Commission on Standardization of Laboratory and Field Tests. *International Journal of Rock Mechanics and Mining Sciences and Geomechanical Abstracts*, **15**, 319–68.

Barton, N. (1988). Rock mass classification and tunnel reinforcement selection using the Q-system. *Proceedings of the Symposium on Rock Classification for Engineering Purposes*. American Society for Testing Materials, Special Technical Publication 984, Philadelphia, pp. 59–88.

Barton, N., Lien, R., and Lunde, J. (1975). *Engineering Classification of Rock Masses for the Design of Tunnel Support*. Norwegian Geotechnical Institute, Oslo, Publication 106.

Beck, B.F. (ed.) (1984). *Sinkholes: Their Geology, Engineering and Environmental Impact*. Balkema, Rotterdam.

Bell, D.H. and Wilson, D.D. (1988). Hazardous waste disposal in New Zealand. *Bulletin of the International Association of Engineering Geology*, **37**, 15–26.

Bell, F.G. (1978a). Some petrographic factors relating to porosity and permeability in the Fell Sandstones of Northumberland. *Quarterly Journal of Engineering Geology*, **11**, 113–26.

Bell, F.G. (1978b). The physical and mechanical properties of the Fell Sandstones. *Engineering Geology*, **12**, 1–29.

Bell, F.G. (1981a). A survey of the physical properties of some carbonate rocks. *Bulletin of the International Association of Engineering Geology*, No. 24, 105–10.

Bell, F.G. (1981b). Geotechnical properties of evaporites. *Bulletin of the International Association of Engineering Geology*, No. 24, 137–44.

Bell, F.G. (1988a). The history and techniques of coal mining and the associated effects and influence on construction. *Bulletin Association of Engineering Geologists*, **24**, 471–504.

Bell, F.G. (1988b). Ground movements associated with the withdrawal of fluids. In *Engineering Geology of Underground Movements*, Engineering Geology Special Publication No. 5 (eds F.G. Bell, J.C. Cripps, M.G. Culshaw, and M.A. Lovell). The Geological Society, London, pp. 367–76.

Bell, F.G. (1991). A survey of the geotechnical properties of some till deposits on the north coast of Norfolk. In *Quaternary Engineering Geology*, Engineering Geology Special Publication No. 7 (eds A. Forster, M.G. Culshaw, J.C. Cripps, J.A. Little, and C.F. Moon). The Geological Society, London, pp. 103–10.

Bell, F.G. (1992a). An investigation of a site in Coal Measures for brickmaking materials: an illustration of procedures. *Engineering Geology*, **32**, 39–52.

Bell, F.G. (1992b). The durability of sandstone as building stone, especially in urban environments. *Bulletin Association of Engineering Geologists*, **24**, 49–60.

Bell, F.G. (1992c). A history of salt mining in mid-Cheshire, England, and its influence on planning. *Bulletin of the Association of Engineering Geologists*, **29**, 371–86.

Bell, F.G. (1993a). Construction stone. In *Handbook of Engineering Geomaterials* (eds P.G. Fookes and A.B. Hawkins). Blackie, Glasgow, Chapter 14.

Bell, F.G. (1993b). The durability of carbonate rock as building stone with comments on its preservation. *Journal of Environmental Geology and Water Sciences*, **19** (in press).

Bell, F.G. and Cashman, P.M. (1986). Control of groundwater by groundwater lowering. In *Groundwater in Engineering Geology*, Engineering Geology Special Publication No. 3 (eds J.C. Cripps, F.G. Bell, and M.G. Culshaw). The Geological Society, London, pp. 471–86.

Bell, F.G. and Coulthard, J.M. (1990). Stone preservation with illustrative examples from the United Kingdom. *Journal of Environmental Geology and Water Science*, **16**, 75–81.

Bell, F.G. and Coulthard, J.M. (1991). The Laminated Clays of Teesside. In *Quaternary Engineering Geology*, Engineering Geology Special Publication No. 7 (eds A. Forster, M.G. Culshaw, J.C. Cripps, C.F. Moon, and J.A. Little). The Geological Society, London, pp. 339–48.

Bell, F.G. and Culshaw, M.G. (1992). Some physical properties of weak sandstones of Triassic age from the English Midlands. In *Engineering Geology of Weak Rocks*, Engineering Geology Special Publication No. 8 (eds J.C. Cripps, M.G. Culshaw, J.M. Coulthard, and S.R. Hencher). The Geological Society, Balkema, Rotterdam, pp. 139–48.

Bell, F.G. and Forster, A. (1991). The geotechnical characteristics of the till deposits of Holderness. In *Quaternary Engineering Geology*, Engineering Geology Special Publications No. 7 (eds A. Forster, M.G. Culshaw, J.C. Cripps, J.A. Little, and C.F. Moon). The Geological Society, London, pp. 111–18.

Bell, F.G. and Fox, R.M. (1991). The effects of mining subsidence on discontinuous rock masses and the influence on foundations: the British experience. *The Civil Engineer in South Africa*, **33**, 201–10.

Bell, F.G. and Kerr, A. (1993). Coal mining and water quality with illustrations from Britain. *Proceedings International Conference on Environmental Management, Geowater and Engineering Aspects, Wollongong*, (eds R.N. Choudhury and S. Sivakumar). Balkema, Rotterdam, pp. 607–14.

Bell, F.G. and Mitchell, J.K. (1986). Control of groundwater by exclusion. In *Groundwater in Engineering Geology*, Engineering Geology Special Publication No. 3 (eds J.C. Cripps, F.G. Bell, and M.G. Culshaw). The Geological Society, London, pp. 429–43.

Bell, F.G., Cripps, J.C., Edmunds, C.N., and Culshaw, M.G. (1990). The influence of the fabric of chalk on its engineering behaviour. *Proceedings of the International Chalk Symposium*, Brighton. Institution of Civil Engineers/Institution of Geologists, pp. 187–94.

Bell, F.G., Jermy, C.A., and Mortimer, B. (1991). Dispersive soils: a brief review and some South African experiences. *Proceedings of the 9th Asian Conference on Soil Mechanics and Foundation Engineering, Bangkok*, pp. 129–32.

Bernknopf, R.L., Campbell, R.H., Brookshire, D.S., and Shapiro, C.D. (1988). A probabilistic approach to landslide hazard mapping in Cincinnati, Ohio, with applications for economic evaluation. *Bulletin of the Association of Engineering Geologists*, **26**, 39–56.

Berry, P.L. and Poskitt, T.J. (1972). The consolidation of peat. *Geotechnique*, **22**, 27–52.

Berry, P.L., Illsley, D., and McKay, I.R. (1985). Settlement of two housing estates at St Annes due to consolidation of a near surface peat stratum. *Proceedings of the Institution of Civil Engineers*, Part 1, **77**, 111–36.

Bezuidenhout, C.A. and Enslin, J.F. (1970). Surface subsidence and sinkholes in dolomitic areas of the Far West Rand, Transvaal, South Africa. *Proceedings of the 1st International Symposium on Land Subsidence, Tokyo*, International Association of Hydrological Sciences. UNESCO Publication No. 88, **2**, 482–95.

Bieniawski, Z.T. (1974). Geomechanics classification of rock masses and its application to tunnelling. *Proceedings of the 3rd Congress of the International Society of Rock Mechanics, Denver*, **1**, 27–32.

Bieniawski, Z.T. (1989). *Engineering Rock Mass Classifications*. Wiley-Interscience, New York.

Bishop, A.W. (1973). The stability of tips and spoil heaps. *Quarterly Journal of Engineering Geology*, **6**, 335–76.

Blight, G.E. (1990). Construction in tropical soils. *Proceedings of the 2nd International Conference on Geomechanics in Tropical Soils, Singapore*. Balkema, Rotterdam, **2**, 449–68.

Blikra, L.H. (1990). Geological mapping of rapid mass movement deposits as an aid to land-use planning. *Engineering Geology*, **29**, 365–76.

Bliss, J.C. and Rushton, K.R. (1984). The reliability of packer tests for estimating the hydraulic conductivity of aquifers. *Quarterly Journal of Engineering Geology*, **17**, 81–91.

Booth, B. (1979). Assessing volcanic risk. *Journal of the Geological Society*, **136**, 331–40.

Brand, E.W. (1985). Geotechnical engineering in tropical soils. *Proceedings of the 1st International Conference on Tropical Laterite and Saprolitic Soils, Brasilia*, **3**, 23–100.

Brassington, F.C. and Walthall, S. (1985). Field techniques using borehole packers in hydrogeological investigations. *Quarterly Journal of Engineering Geology*, **18**, 181–94.

Briggs, R.P. (1987). Engineering geology, and seismic and volcanic hazards in the Hijaz railway region — Syria, Jordan and Saudi Arabia. *Bulletin of the Association of Engineering Geologists*, **24**, 403–23.

Brink, A.B.A., Mabbutt, J.A., Webster, R., and Beckett, P.H.T. (1966). *Report of the Working Group on Land Classification and Data Storage*. Military Engineering Experimental Establishment, Report No. 940, Christchurch.

Brook, C.A. and Alison, T.L. (1986). Fracture mapping and ground subsidence susceptibility modelling in covered karst terrain: the example of Dougherty, County Georgia. *Proceedings of the 3rd International Symposium on Land Subsidence, Venice*, International Association of Hydrological Sciences, Publication No. 151, pp. 595–606.

Brown, R.H., Konoplyanstev, A.A., Ineson, J., and Kovalevsky, V.S. (eds) (1972). *Groundwater Studies: An International Guide for Research and Practice*. Studies and Reports in Hydrology, No. 7. UNESCO, Paris.

Brunsden, D., Doornkamp, J.C., Fookes, P.G., and Jones, D.K.C. (1975). Large scale geomorphological mapping and highway engineering design. *Quarterly Journal of Engineering Geology*, **8**, 227–53.

Burland, J.B. (1990). On the compressibility and shear strength of natural clays. *Geotechnique*, **40**, 329–78.

Burland, J.B. and Lord, J.A. (1970). The load deformation behaviour of Middle Chalk at Mundford, Norfolk: a comparison between full-scale performance and *in situ* laboratory measurements. In *Site Investigations in Soils and Rocks*, British Geotechnical Society. Pentech Press, London, pp. 3–16.

Burland, J.B. and Wroth, C.P. (1975). Settlement of buildings and associated damage. In *Settlement of Structures*, British Geotechnical Society. Pentech Press, London, pp. 611–54.

Burland, J.B., Longworth, T.I., and Moore, J.T. (1977). A study of ground movement and progressive failure caused by deep excavation in Oxford Clay. *Geotechnique*, **27**, 557–91.

Burton, L., Kates, R.W., and White, G.F. (1978). *The Environment as Hazard*. Oxford University Press, New York.

Casagrande, A. (1932). Discussion on frost heaving. *Proceedings Highway Research Board*, Bulletin No. 12, Washington DC, p. 169.

Casagrande, A. (1936). Characteristics of cohesionless soils affecting the stability of slopes and earth fills. *Journal of the Boston Society of Civil Engineers*, **23**, 3–32.

Casagrande, A. (1948). Classification and identification of soils. *Transactions of the American Society of Civil Engineers*, **113**, 901–92.

Casagrande, A. (1961). Control of seepage through foundations and abutment dams. *Geotechnique*, **11**, 161–81.

Chandler, R.J. (1969). The effect of weathering on the shear strength properties of Keuper Marl. *Geotechnique*, **19**, 321–34.

Chowdhury, R.N. (1978). *Slope Analysis*. Elsevier, Amsterdam.

Clark, L. (1977). The analysis and planning of step drawdown tests. *Quarterly Journal of Engineering Geology*, **10**, 125–43.

Clayton, K.M. (1990). Sea-level rise and coastal defences in the U.K. *Quarterly Journal of Engineering Geology*, **23**, 283–88.

Cleary, W.J. and Hosier, P.E. (1987). North Carolina coastal geologic hazards: an overview. *Bulletin of the Association of Engineering Geologists*, **24**, 469–88.

Clevenger, M.A. (1958). Experience with loess as a foundation material. *Transactions of the American Society of Civil Engineers*, **123**, 151–80.

Clough, G.W. (1981). Innovations in tunnel construction and support techniques. *Bulletin of the Association of Engineering Geologists*, **18**, 151–67.

Cooke, R.U. and Doornkamp, J.C. (1974). *Geomorphology in Environmental Management*. Clarendon Press, Oxford.

Cooke, R.U., Goudie, A.S., and Doornkamp, J.C. (1978). Middle East — review and bibliography of geomorphological contributions. *Quarterly Journal of Engineering Geology*, **11**, 9–18.

Cottechia, V. (1978). Systematic reconnaissance mapping and registration of slope movements. *Bulletin of the International Association of Engineering Geology*, **14**, 325–46.

Coulthard, J.M. and Bell, F.G. (1992). The influence of weathering on the engineering properties of the Lower Lias Clay. *Engineering Geology of Weak Rocks*, Engineering Geology Special Publication No. 8 (eds J.C. Cripps, M.G. Culshaw, J.M. Coulthard, and S.R. Hencher). The Geological Society, Balkema, Rotterdam, pp. 183–92.

Cratchley, C.R., McCann, D.M., and Ates, M. (1976). Application of geophysical techniques to the location of weak tunnelling ground with an example from the Foyers hydroelectric scheme, Loch Ness. *Transactions of the Institution of Mining and Metallurgy*, Section A, **85**, A127–A135.

Cripps, J.C. and Taylor, R.K. (1981). The engineering properties of mudrocks. *Quarterly Journal of Engineering Geology*, **14**, 325–46.

Cripps, J.C., Deaves, A., Culshaw, M.G., and Bell, F.G. (1988). Geological controls on the flow of groundwater into underground excavations. *Proceedings of the 3rd International Mine Water Congress, Melbourne*. Australian Institution of Mining and Metallurgy, pp. 76–86.

Cripps, J.C., McCann, D.M., Culshaw, M.G. and Bell, F.G. (1988). An assessment of geophysical methods for the exploration of mine shafts. *Proceedings International Conference on Mineworkings-'88, Edinburgh*, Engineering Technics Press, Edinburgh, pp. 53–60.

Culshaw, M.G. and Bell, F.G. (1992). The rockfalls of James Valley, St Helena. *Proceedings of the 6th International Symposium in Landslides, Christchurch*, (ed. D.H. Bell). Balkema, Rotterdam, **2**, 925–35.

Culshaw, M.G., Foster, A., Cripps, J.C., and Bell, F.G. (1990). Applied geology maps for land-use planning in Great Britain. *Proceedings of the 6th Congress of the International Association of Engineering Geology, Amsterdam* (ed. D. Price). Balkema, Rotterdam, **1**, 85–93.

Darcy, H. (1856). *Les Fontaines Publiques de la Ville de Dijon*. Dalmont, Paris.

Darracott, B.W. and Lake, M.I. (1981). An initial appraisal of ground probing radar for site investigation in Britain. *Ground Engineering*, **14**(3), 14–18.

Davis, W.M. (1909). *Geographical Essays*. Dover, New York.

Dearman, W.R. (1974). Weathering classification in the characterization of rock for engineering purposes in British practice. *Bulletin of the International Association of Engineering Geology*, No. 9, 33–42.

Dearman, W.R. (1991). *Engineering Geological Mapping*. Butterworths-Heinemann, Oxford.

De Beer, E., Fagnoul, A., Lonsberg, E., Nuyens, J., and Maetens, J. (1980). A review of the engineering geological mapping in Belgium. *Bulletin of the International Association of Engineering Geology*, No. 21, 91–8.

Deere, D.U. (1964). Technical description of cores for engineering purposes. *Rock Mechanics and Engineering Geology*, **1**, 18–22.

Deere, D.U. and Miller, R.P. (1966). *Engineering Classification and Index Properties for Intact Rock*. Technical Report No AFWL-TR-65-116, Air Force Weapons Laboratory, Kirtland Air Base, New Mexico.

De Graft-Johnson, J.W.S., Bhatia, H.S. and Yebod, S.L. (1973). Geotechnical properties of Accra Shales. *Proceedings 8th International Conference Soil Mechanics and Foundation Engineering, Moscow*, **2**, 97–104.

Dobereiner, L. and De Freitas, M.H. (1986). Geotechnical properties of weak sandstones. *Geotechnique*, **36**, 79–94.

Eriksson, L.G. (1989). Underground disposal of high-level radioactive waste in the United States of America. *Bulletin of the International Association of Engineering Geology*, No. 39, 35–52.

Evans, D.M. (1966). Man-made earthquakes in Denver. *Geotimes*, **10**, 11–18.

Eyles, N. and Sladen, J.A. (1981). Stratigraphy and geotechnical properties of weathered lodgement till in Northumberland, England. *Quarterly Journal of Engineering Geology*, **14**, 129–42.

Fasiska, E., Wagenblast, H., and Dougherty, M.T. (1974). The oxidation mechanism of sulphide minerals. *Bulletin of Association Engineering Geologists*, **11**, 75–82.

Feda, J. (1988). Collapse of loess on wetting. *Engineering Geology*, **25**, 263–9.

Fisk, N.H. (1944). *Geological Investigation of the Alluvial*

Valley of the Lower Mississippi. United States Corps of Engineers, Mississippi River Commission, Vicksburg.

Fleuty, M.J. (1964). The description of folds. *Proceedings Geologists Association*, **75**, 461–9.

Fobe, B. and Goossens, M. (1990). The groundwater vulnerability map for the Flemish region: its principles and uses. *Engineering Geology*, **29**, 355–63.

Folk, R.L. (1973). Carbonate petrology in the post-Sorbian age. In *Evolving Concepts in Sedimentology* (ed. R.N. Ginsberg). Johns Hopkins University Press, Baltimore.

Fookes, P.G. (1991). Geomaterials. *Quarterly Journal of Engineering Geology*, **24**, 3–16.

Fookes, P.G. and Hawkins, A.B. (1988). Limestone weathering: its engineering significance and a proposed classification scheme. *Quarterly Journal of Engineering Geology*, **21**, 7–32.

Fookes, P.G. and Higginbottom, I.E. (1975). The classification and description of near-shore carbonate sediments for engineering purposes. *Geotechnique*, **25**, 406–11.

Fookes, P.G. and Sweeney, M. (1976). Stabilization and control of local rock falls and degrading rock slopes. *Quarterly Journal of Engineering Geology*, **9**, 37–56.

Fookes, P.G., Dearman, W.R. and Franklin, J.A. (1972). Some engineering aspects of weathering with field examples from Dartmoor and elsewhere. *Quarterly Journal of Engineering Geology*, **3**, 1–24.

Fookes, P.G., Gourley, C.S., and Ohikere, E. (1988). Rock weathering in engineering time. *Quarterly Journal of Engineering Geology*, **21**, 33–57.

Forster, A. and Culshaw, M.G. (1990). The use of site investigation data for the preparation of engineering geological maps and reports for use by planners and civil engineers. *Engineering Geology*, **29**, 347–54.

Foster, S.S.D. and Crease, R.I. (1974). Nitrate pollution of chalk groundwater in East Yorkshire — a hydrogeological appraisal. *Journal Institution Water Engineers and Scientists*, **28**, 178–94.

Franklin, J.L. and Broch, E. (1972). The point load test. *International Journal of Rock Mechanics and Mining Science*, **9**, 669–97.

Franklin, J.L., Broch, E., and Walton, G. (1971). Logging the mechanical character of rock. *Transactions of the Institution of Mining and Metallurgy*, **81**, Mining Section, A1–9.

Franklin, J.L., Manailoglou, J., and Sherwood, D. (1974). Field determination of direct shear strength. *Proceedings of the 3rd Congress of the International Society of Rock Mechanics, Denver*, **2**, 233–40.

French, W.J. (1991). Concrete petrography: a review. *Quarterly Journal of Engineering Geology*, **24**, 17–48.

Gabrysch, R.K. (1976). Land surface subsidence in the Houston–Galveston region, Texas. *Proceedings of the 2nd International Association of Hydrological Sciences*, UNESCO Publication No. 121, pp. 16–24.

George, H. (1986). Characteristics of varved clays of the Elk Valley, British Columbia, Canada. *Engineering Geology*, **23**, 59–74.

Gerber, A. and Harmse, H.J. von M. (1987). Proposed procedure for identification of dispersive soils by chemical testing. *The Civil Engineer in South Africa*, **29**, 397–9.

Gidigasu, M.D. (1974). Degree of weathering and identification of laterite materials for engineering purposes. *Engineering Geology*, **8**, 213–66.

Gillott, J.E. (1979). Fabric, composition and properties of sensitive soils from Canada, Alaska and Norway. *Engineering Geology*, **14**, 149–72.

Gillott, J.E. and Swenson, E.G. (1969). Mechanism of alkali carbonate reaction. *Quarterly Journal of Engineering Geology*, **2**, 7–24.

Gogte, B.S. (1973). An evaluation of some common Indian rocks with special reference to alkali aggregate reactions. *Engineering Geology*, **7**, 135–54.

Grabowska-Olszewska, B. (1988). Engineering geological problems of loess in Poland. *Engineering Geology*, **25**, 177–99.

Grainger, P. and Harris, J. (1986). Weathering and slope stability of Upper Carboniferous mudrocks in south west England. *Quarterly Journal of Engineering Geology*, **19**, 155–73.

Grainger, P. and McCann, D.M. (1978). Interborehole acoustic measurements in site investigation. *Quarterly Journal of Engineering Geology*, **10**, 241–56.

Gray, D.A., Mather, J.D., and Harrison, I.B. (1974). Review of groundwater pollution for waste sites in England and Wales with provisional guidelines for future site selection. *Quarterly Journal of Engineering Geology*, **7**, 181–96.

Grim, R.E. (1952). *Applied Clay Mineralogy*. McGraw Hill, New York.

Gupta, R.P. and Joshi, B.C. (1990). Landslide hazard zoning using the GIS approach — a case study from the Ramganga catchment, Himalayas. *Engineering Geology*, **28**, 119–31.

Hallbauer, D.K., Nieble, C., Berard, J., Rummel, F., Houghton, A.M., Broch, E., and Szlavin, J. (1978). Suggested methods for petrographic description. International Society of Rock Mechanics, Commission on Standardization of Laboratory and Field Tests. *International Journal of Rock Mechanics and Mining Science and Geomechanical Abstracts*, **15**, 41–5.

Hanrahan, E.T. (1954). An investigation of some physical properties of peat. *Geotechnique*, **4**, 108–23.

Hansmire, W.H. (1981). Tunneling and excavation in soft rock and soil. *Bulletin of the Association of Engineering Geologists*, **18**, 77–89.

Harker, A. (1939). *Metamorphism*. Methuen, London.

Hawkins, A.B. (1986). Rock descriptions. In *Site Investigation Practice: Assessing BS 5930*, Engineering Geology Special Publication No. 2 (ed. A.B. Hawkins). The Geological Society, London, pp. 59–66.

Hawkins, A.B. and McConnell, B.J. (1991). Influence of geology on geomechanical properties of sandstones. *Proceedings of the 7th International Congress Rock*

Mechanics (ISRM), Aachen. Balkema, Rotterdam, **1**, 257–60.

Hawkins, T.R.W. and Chadha, D.S. (1990). Locating the Sherwood Sandstone aquifer with the aid of resistivity surveying in the Vale of York. *Quarterly Journal of Engineering Geology*, **23**, 229–42.

Hays, W.W. and Shearer, C.F. (1981). Suggestions for improving decision making to face geologic and hydrologic hazards — earth science consideration. *United States Geological Survey, Professional Paper 1240–B*, B103–8.

Hjulstrum, F. (1935). Studies of the morphological activity of rivers as illustrated by the river Fynis. *Uppsala University Geological Institute, Bulletin*, **25**, 221–557.

Hobbs, N.B. (1975). Factors affecting the prediction of settlement of structures on rocks with particular reference to the Chalk and Trias. In *Settlement of Structures*. British Geotechnical Society. Pentech Press, London, pp. 579–610.

Hobbs, N.B. (1986). Mire morphology and the properties and behaviour of some British and foreign peats. *Quarterly Journal of Engineering Geology*, **19**, 7–80.

Hoek, E. (1983). Strength of rock masses. *Geotechnique*, **33**, 187–223.

Hoek, E. and Bray, J.W. (1981). *Rock Slope Engineering* (3rd edn). Institution of Mining and Metallurgy, London.

Hoek, E. and Brown, E.T. (1980). *Underground Excavations in Rock*. Institution of Mining and Metallurgy, London.

Holzer, T.L., Davis, S.N., and Lofgren, B.E. (1979). Faulting caused by groundwater extraction in south central Arizona. *Journal of Geophysical Research*, **84**, 603–12.

Horn, A. (1982). Swell and creep properties of an African black clay. *Proceedings of the American Society of Civil Engineers, Geotechnical Engineering Division*, Speciality Conference on Engineering and Construction in Tropical and Residual Soils, Honolulu, pp. 199–215.

Horton, R.E. (1945). Erosion development of streams and their drainage basins: hydrophysical approach to quantitative morphology. *Bulletin of the Geological Society of America*, **56**, 275–370.

Houlsby, A.C. (1990). *Construction and Design of Cement Grouting*. Wiley-Interscience, New York.

Hvorslev, M.J. (1949). *Subsurface Exploration and Sampling of Soils for Civil Engineering Purposes*. American Society of Civil Engineers Report, Waterways Experimental Station, Vicksburg.

Iliev, I.G. (1967). An attempt to estimate the degree of weathering of intrusive rocks from their physicomechanical properties. *Proceedings of the 1st Congress of the International Society of Rock Mechanics, Lisbon*, **1**, 109–14.

Ineson, J. and Gray, D.A. (1963). Electrical investigations of borehole fluids. *Journal of Hydrology*, **1**, 204–18.

Ingle, J.G. (1966). *Movement of Beach Sand*. Elsevier, Amsterdam.

Irfan, T.Y. and Dearman, W.R. (1978a). Engineering classification and index properties of weathered granite. *Bulletin of the International Association of Engineering Geology*, No. 17, 79–90.

Irfan, T.Y. and Dearman, W.R. (1978b). The engineering petrography of a weathered granite in Cornwall, England. *Quarterly Journal of Engineering Geology*, **11**, 233–44.

Jaeger, C. (1965). The Vajont rock slide. *Water Power*, **17**, 110–11, 142–4.

James, A.N. and Kirkpatrick, I.M. (1980). Design of foundations of dams containing soluble rocks and soils. *Quarterly Journal Engineering Geology*, **13**, 189–98.

Jammal, S.E. (1986). The Winter Park sinkhole and Central Florida sinkhole type subsidence. *Proceedings of the 3rd International Symposium on Land Subsidence, Venice*, International Association of Hydrological Sciences, Publication No. 151, pp. 585–94.

Jermy, C.A. and Bell, F.G. (1991). Coal bearing strata and the stability of coal mines in South Africa. *Proceedings of the 7th International Congress on Rock Mechanics (ISRM), Aachen*. Balkema, Rotterdam, **2**, 1125–31.

Johnson, D.W. (1919). *Shoreline Processes and Shoreline Development*. Wiley, New York.

Jones, D.K.C., Cooke, R.U., and Warren, A. (1986). Geomorphological investigation, for engineering purposes, of blowing sand and dust hazard. *Quarterly Journal of Engineering Geology*, **19**, 251–70.

Jumikis, A. (1968). *Soil Mechanics*. Van Nostrand, Princeton.

Kenny, R. (1990). Hydrogeomorphic flood hazard evaluation for semi-arid environments. *Quarterly Journal of Engineering Geology*, **23**, 333–6.

Kerr, R.C. and Nigra, J.O. (1952). Eolian sand control. *Bulletin of the American Association of Petroleum Geologists*, **36**, 1541–73.

King, L.C. (1963). *South African Scenery — a Textbook of Geomorphology*. Oliver & Boyd, Edinburgh.

Kirsten, H.A.D. (1988). Case histories of groundmass characterization for excavatability. *Proceedings of the Symposium on Rock Classification for Engineering Purposes*, American Society for Testing Materials, Special Technical Publication 984, pp. 102–20.

Knill, J.L. (1971). Assessment of reservoir feasibility. *Quarterly Journal of Engineering Geology*, **4**, 355–72.

Kockelman, W.J. (1988). Reducing landslide hazards. In *Proceedings of the Symposium on Environmental Geotechnics and Problematic Soils and Rocks, Bangkok* (eds A.S. Bala-subramaniam, S. Chandra, D.T. Bergado, and P. Nutalaya). Balkema, Rotterdam, pp. 383–405.

Krumbein, W.C. (1941). Measurement and geological significance of shape and roundness of sedimentary particles. *Journal of Sedimentary Petrology*, **11**, 64–72.

Kummerle, R.P. and Benvie, D.A. (1988). Geologic considerations in rock excavations. *Bulletin of the Association of Engineering Geology*, **25**, 105–20.

Kusznir, N.J., Ashwin, D.P., and Bradley, A.G. (1980). Seismicity in the north Staffordshire coalfield, England.

International Journal of Rock Mechanics and Mining Sciences and Geomechanical Abstracts, **17**, 45–55.

Langer, M. (1989). Waste disposal in the Federal Republic of Germany: concepts, criteria, scientific investigations. *Bulletin of the International Association of Engineering Geology*, No. 39, 53–8.

Langer, M. and Wallner, M. (1988). Solution — mined salt caverns for the disposal of hazardous chemical wastes. *Bulletin of the International Association of Engineering Geology*, No. 37, 61–70.

Lauffer, M. (1958). Gebirgsklassifizierung fur den stollen-hua. *Geologie und Bauwesen*, **24**, 46–51.

Leatherman, S.P. (1979). Beach and dune interactions during storm conditions. *Quarterly Journal of Engineering Geology*, **12**, 281–90.

Lee, S.G. and De Freitas, M.H. (1989). A revision of the description and classification of weathered granite and its application to granites in Korea. *Quarterly Journal of Engineering Geology*, **22**, 31–48.

Lefebvre, G., Langios, P., Lupien, C., and Lavelle, J. (1984). Laboratory testing on *in situ* behaviour of peat as embankment foundation. *Canadian Geotechnical Journal*, **21**, 322–37.

Legget, R.F. (1987). The value of geology in planning. In *Planning and Engineering Geology*, Engineering Geology Special Publication No. 4 (eds M.G. Culshaw, F.G. Bell, J.C. Cripps, and M. O'Hara). The Geological Society, London, pp. 53–8.

Leggo, P.J. (1982). Geological applications of ground impulse radar. *Transactions of the Institution of Mining and Metallurgy, Section B — Applied Earth Science*, **91**, B1–5.

Leopold, L.B., Clarke, F.E., Hanshaw, B.B. and Baisley, J.R. (1971). *A Procedure for Evaluating Environmental Impact*. Department of Interior, U.S. Geological Survey Circular 745, Washington, D.C.

Lewis, W.V. (1971). The formation of Dungeness Foreland. In *Applied Coastal Geomorphology* (ed. J. Steers). Macmillan, London.

Lin, Z.G. and Wang, S.J. (1988). Collapsibility and deformation characteristics of deep-seated loess in China. *Engineering Geology*, **25**, 271–82.

Little, A.L. (1969). The engineering classification of residual tropical soils. *Proceedings of the 7th International Conference on Soil Mechanics and Foundation Engineering*, Mexico, **1**, 1–10.

Locat, J., Lefebvre, G., and Ballivy, G. (1984). Mineralogy, chemistry and physical properties interrelationships of some sensitive clays from eastern Canada. *Canadian Geotechnical Journal*, **21**, 530–40.

Lofgren, B.E. (1968). Analysis of stress causing land subsidence. *United States Geological Survey, Professional Paper No. 600-B*, B219–25.

Lovegrove, G.W. and Fookes, P.G. (1972). The planning and implementation of a site investigation for a highway in tropical conditions in Fiji. *Quarterly Journal of Engineering Geology*, **5**, 43–68.

Lovelock, P.E.R., Price, M., and Tate, T.K. (1975). Groundwater conditions in the Penrith Sandstone at Cliburn, Westmoreland. *Journal of the Institution of Water Engineers*, **29**, 157–74.

Lugeon, M. (1933). *Barrage et Geologie*. Dunod, Paris.

Lumb, P. (1983). Engineering properties of fresh and decomposed igneous rocks from Hong Kong. *Engineering Geology*, **19**, 81–94.

Lutenegger, A.J. and Hallberg, G.R. (1988). Stability of loess. *Engineering Geology*, **25**, 247–61.

Madu, R.M. (1977). An investigation into the geotechnical properties of some laterites of eastern Nigeria. *Engineering Geology*, **11**, 101–25.

Marsh, T.J. and Davies, P.A. (1984). The decline and partial recovery of groundwater levels beneath London. *Proceedings of the Institution of Civil Engineers, Part 1*, **74**, 263–76.

Marsland, A. (1986). The flood plain deposits of the lower Thames. *Quarterly Journal of Engineering Geology*, **19**, 223–47.

Martin, R.P. and Hencher, S.R. (1986). Principles for description and classification of weathered rock for engineering purposes. In *Site Investigation Practice: Assessing BS 5930*, Engineering Geology Special Publication No. 2 (ed. A.B. Hawkins). The Geological Society, London, pp. 299–308.

Martinez, M.J., Espejo, R.A., Bilbao, I.A., and Cabrera, M.D. del R. (1990). Analysis of sedimentary processes on the Las Canteras beach (Las Palmas, Spain) for its planning and management. *Engineering Geology*, **29**, 377–86.

Matheson, G.D. (1989). The collection and use of field discontinuity data in rock slope design. *Quarterly Journal of Engineering Geology*, **22**, 19–30.

McCann, D.M., Baria, R., Jackson, P.D., and Green, A.S.P. (1986). Application of cross-hole seismic measurements to site investigation. *Geophysics*, **51**, 914–25.

McCann, D.M., Culshaw, M.G., and Northmore, K.J. (1990). Rock mass assessment from seismic measurements. In *Field Testing in Engineering Geology*, Engineering Geology Special Publication, No. 6 (eds F.G. Bell, M.G. Culshaw, J.C. Cripps, and J.R. Coffey). The Geological Society, London, pp. 257–66.

McConnell, D., Mielenz, R.C., Holland, W.Y., and Greene, K.T. (1950). Petrology of concrete affected by cement aggregate reaction. In *Application of Geology to Engineering Practice* (ed. S. Paige). Berkey Volume, Memoir American Geological Society, pp. 222–50.

McGown, A. (1971). The classification for engineering purposes of tills from moraines and associated landforms. *Quarterly Journal of Engineering Geology*, **4**, 115–30.

McGown, A. and Derbyshire, E. (1977). Genetic influences on the properties of tills. *Quarterly Journal of Engineering Geology*, **10**, 389–410.

Medvedev, S.V. (1968). Measurement of ground motion and structural vibrations caused by earthquakes. *Pro-*

ceedings of the International Seminar on Earthquake Engineering, Skopje, UNESCO, pp. 35–8.

Meinzer, O. (1942). Occurrence, origin and discharge of groundwater. In Hydrology (ed. O. Meinzer). Dover, New York, pp. 385–443.

Metcalf, J.B. and Townsend, D.L. (1961). A preliminary study of the geotechnical properties of varved clays as reported in Canadian engineering case records. Proceedings of the 14th Canadian Conference on Soil Mechanics, Section 13, pp. 203–25.

Mitchell, C.W. (1973). Terrain Evaluation. Longmans, London.

Mitchell, J.K. (1986). Hazardous waste containment. In Groundwater in Engineering Geology, Engineering Geology Special Publication No. 3 (eds J.C. Cripps, F.G. Bell and M.G. Culshaw). The Geological Society, London, pp. 145–57.

Mitchell, J.K. and Sitar, N. (1982). Engineering properties of tropical residual soils. Proceedings of the Speciality Conference on Engineering and Construction in Tropical and Residual Soils, Honolulu. American Society of Civil Engineers, Geotechnical Engineering Division, pp. 30–57.

Morey, R.M. (1974). Continuous subsurface profiling by impulse radar. Proceedings of the Speciality Conference on Subsurface Exploration for Underground Excavation and Heavy Construction, Henneker. American Society of Civil Engineers, pp. 213–32.

Morfeldt, C.O. (1989). Different subsurface facilities for the geological dispersal of radioactive waste (storage cycle) in Sweden. Bulletin of the International Association of Engineering Geology, No. 39, 25–34.

Morin, E.J. (1982). Characteristics of tropical red soils. Proceedings of the Speciality Conference on Engineering and Construction in Tropical and Residual Soils, Honolulu, American Society of Civil Engineers, Geotechnical Engineering Division, pp. 172–98.

Moye, D.G. (1955). Engineering geology for the Snowy Mountain scheme. Journal of the Institution of Engineers of Australia, 27, 287–98.

Mulder, E.F.J. and Hillen, R. (1990). Preparation and application of engineering and environmental geological maps in the Netherlands. Engineering Geology, 29, 279–90.

Nickless, E.F.P. (1982). Environmental geology of the Glenrothes district, Fife Region. Description of 1:25000 sheet No. 20. Report of British Geological Survey 82.

Ola, S.A. (1978). The geology and engineering properties of black cotton soils in north eastern Nigeria. Engineering Geology, 15, 1–13.

Olivier, H.G. (1979). A new engineering–geological rock durability classification. Engineering Geology, 14, 255–79.

O'Neill, M.W. and Poormoayed, A.M. (1980). Methodology for foundations on expansive clays. Proceedings of the American Society of Civil Engineers, Journal of the Geotechnical Engineering Division, 106 (GT12),

1245–1367.

Onodera, T.F. (1963). Dynamic investigation of foundation rocks. Proceedings of the 5th Symposium on Rock Mechanics, Minnesota. Pergamon Press, New York, pp. 517–33.

Oteri, A.U.E. (1983). Delineation of saline intrusion in the Dungeness Shingle aquifer using surface geophysics. Quarterly Journal of Engineering Geology, 16, 43–52.

Oweis, I.S. and Khera, R.P. (1990). Geotechnology of Waste Management. Butterworths, London.

Peck, R.B. (1969). Deep excavation and tunnelling in soft ground. Proceedings of the 7th International Conference on Soil Mechanics and Foundation Engineering, Mexico City, 3, 225–90.

Pettijohn, F.J. (1975). Sedimentary Rocks. Harper & Row, New York.

Pettijohn, F.J., Potter, P.E., and Siever, R. (1972). Sands and Sandstones. Springer-Verlag, New York.

Poland, J.F. (1981). Subsidence in the United States due to groundwater withdrawal. Proceedings of the American Society of Civil Engineers, Journal of the Irrigation and Drainage Division, 107 (1R2), 115–35.

Popescu, M.E. (1986). A comparison between the behaviour of swelling and collapsing soils. Engineering Geology, 23, 145–63.

Price, N.L. (1966). Fault and Joint Development in Brittle and Semi-brittle Rock. Pergamon Press, Oxford.

Pyles, M.R., Mills, K., and Saunders, G. (1987). Mechanics and stability of the Lookout Creek earth flow. Bulletin of the Association of Engineering Geologists, 24, 267–80.

Quigley, R.M., Fernandez, F., and Crooks, V.E. (1988). Engineered clay liners: a short review. In Proceedings of the Symposium on Environmental Geotechnics and Problematic Soils and Rocks, Bangkok (eds A.S. Balasubramanian, S. Chandra, D.T. Bergado and P. Natalaya). Balkema, Rotterdam, pp. 63–74.

Rahn, P.H. (1984). Flood plain management program in Rapid City, South Dakota. Bulletin of the Geological Society of America, 95, 838–43.

Raybould, J.G. and Anderson, J.G. (1987). Migration of landfill gas and its control by grouting — a case history. Quarterly Journal of Engineering Geology, 20, 78–83.

Richter, C.F. (1936). An instrumental earthquake scale. Bulletin of the Seismological Society of America, 25, 1–32.

Robertson, A.M.G. (1974). Joints and gauge materials — their importance and testing. In Tunnelling in Rock (ed. Z.T. Bieniawski). South African Institute of Civil Engineers/South African National Group for Rock Mechanics/Council Scientific and Industrial Research, Pretoria, pp. 125–38.

Robinson, G.D. and Spieker, A.M. (eds) (1978). Nature to be commanded . . . earth science maps applied to land and water management. United States Geological Survey Professional Paper 966.

Sangar, F.J. and Kaplar, C.W. (1963). Plastic deformation

of frozen soils. *Proceedings International Conference on Permafrost, Lafayette*, NAS-NRC, Publication No. 1281, Washington, D.C., 305–15.

Sanglerat, G. (1972). *The Penetrometer and Soil Exploration*. Elsevier, Amsterdam.

Schuster, R.L. (1983). Engineering aspects of the 1980 Mount St. Helens eruptions. *Bulletin of the Association of Engineering Geologists*, **20**, 125–43.

Seeley, M.W. and West, D.O. (1990). Approach to geologic hazard zoning for regional planning, Inyo National Forest, California and Nevada. *Bulletin of the Association of Engineering Geologists*, **27**, 23–36.

Serafim, J.L. (1968). Influence of interstitial water on rock masses. In *Rock Mechanics in Engineering Practice* (eds K.G. Stagg and O.C. Zienkiewicz). Wiley, London, pp. 55–77.

Sharpe, C.F.S. (1938). *Landslides and Related Phenomena*. Columbia University Press, New York.

Shepard, F.P. (1963). *Submarine Geology*. Harper & Row, New York.

Sherard, J.L., Cluff, L.S., and Allen, L.R. (1974). Potentially active faults and dam foundations. *Geotechnique*, **24**, 367–429.

Sherman, L.K. (1932). Streamflow from rainfall by unit graph method. *Engineering News Record*, **108**, 501–5.

Sims, I. (1991). Quality and durability of stone for construction. *Quarterly Journal of Engineering Geology*, **24**, 67–74.

Skempton, A.W. (1953). The colloidal activity of clays. *Proceedings of the 3rd International Conference on Soil Mechanics and Foundation Engineering, Zurich*, **1**, 57–61.

Skempton, A.W. (1964). Long-term stability of clay slopes. *Geotechnique*, **14**, 77–101.

Skempton, A.W. and Northey, R.D. (1952). The sensitivity of clays. *Geotechnique*, **3**, 30–53.

Skempton, A.W., Schuster, R.L., and Petley, D.J. (1969). Joints and fissures in the London Clay at Wraysbury and Edgware. *Geotechnique*, **19**, 205–17.

Sladen, J.A. and Wrigley, W. (1983). Geotechnical properties of lodgement till — a review. In *Glacial Geology: An Introduction for Engineers and Earth Scientists* (ed. N. Eyles). Pergamon Press, Oxford, pp. 184–212.

Soderberg, A.D. (1979). Expect the unexpected: foundations for dams in karst. *Bulletin of the Association of Engineering Geologists*, **16**, 409–27.

Soule, J.M. (1980). Engineering geologic mapping and potential geologic hazards in Colorado. *Bulletin of the International Association of Engineering Geology*, No. 21, 121–31.

Sridharan, A. and Allam, M.M. (1982). Volume change behaviour of desiccated soils. *Proceedings of the American Society of Civil Engineers, Journal of the Geotechnical Engineering Division*, **108** (GT8), 1057–71.

Stacey, T.R. and Page, C.H. (1986). *Practical Handbook for Underground Rock Mechanics*. Trans Tech Publications, Clausthal-Zellerfeld.

Steers, J.A. (1946). *The Coastline of England and Wales*. Cambridge University Press, Cambridge.

Sundberg, A. (1956). The river Klaralren, a study of fluvial processes. *Geografiska Annaler*, **38**, 127–316.

Sutcliffe, G. and Mostyn, G. (1983). Permeability testing for the OK Tedi Project (Papua New Guinea). *Bulletin International Association Engineering Geology*, Nos 26–27, 501–8.

Tan, T.K. (1988). Fundamental properties of loess from north western China. *Engineering Geology*, **25**, 103–22.

Taylor, C.L. and Conwell, F.R. (1981). BART — influence of geology on the construction conditions and costs. *Bulletin of the Association of Engineering Geologists*, **18**, 195–20.

Taylor, R.K. (1988). Coal Measures mudrocks: composition, classification and weathering processes. *Quarterly Journal of Engineering Geology*, **21**, 85–100.

Terzaghi, K. (1946). Introduction to tunnel geology. In *Rock Tunnelling with Steel Supports* (eds R. Proctor and T. White). Commercial Stamping and Shearing Company, Youngstown, Ohio, pp. 17–99.

Terzaghi, K. (1950a). Geological aspects of soft ground tunnelling. In *Applied Sedimentation* (ed. P.D. Trask). Wiley, New York, pp. 193–209.

Terzaghi, K. (1950b). The mechanisms of landslides. In *Applications of Geology to Engineering Practice* (ed. S. Paige). Berkey Volume, Memoir, Geological Society of America, pp. 83–124.

Terzaghi, K. (1962). Stability of steep slopes on hard unweathered rock. *Geotechnique*, **12**, 251–70.

Terzaghi, K. and Peck, R.B. (1968). *Soil Mechanics in Engineering Practice* (2nd edn). Wiley, New York.

Underwood, L.B. (1967). Classification and identification of shales. *Proceedings American Society Civil Engineers, Soil Mechanics and Foundations Division*, *93*, *SM6*, pp. 97–116.

Varley, P.M. (1990). Susceptibility of Coal Measures mudstone to slurrying during tunnelling. *Quarterly Journal of Engineering Geology*, **23**, 147–60.

Varnes, D.J. (1978). Slope movement, types and processes. In (eds Schuster, R.L. and Krizek, R.J.), *Landslides, Analysis and Control*, National Academy of Science, Report 176, Washington, D.C., pp. 11–35.

Vaughan, P.R. (1990). Characterizing the mechanical properties of *in situ* residual soils. *Proceedings of the 2nd International Conference on Geomechanics in Tropical Soils, Singapore*. Balkema, Rotterdam, **2**, 469–87.

Vaughan, P.R., Maccarini, M., and Mokhtar, S.M. (1988). Indexing the properties of residual soil. *Quarterly Journal of Engineering Geology*, **21**, 69–84.

Vick, S.G. and Bromwell, L.G. (1989). Risk analysis for dam design in karst. *Proceedings of the American Society of Civil Engineers, Journal of the Geotechnical Engineering Division*, **115** (GT6), 819–35.

Ward, W.H., Burland, J.B., and Gallois, R.W. (1968). Geotechnical assessment of a site at Mundford, Norfolk, for a large proton accelerator. *Geotechnique*, **18**,

399–431.

Weaver, J.M. (1975). Geological factors significant in the assessment of rippability. *The Civil Engineer in South Africa*, **17**, 313–16.

Winkler, E.M. (1973). *Stone Properties, Durability in Khan's Environment*. Springer-Verlag, New York.

Wittke, W. (1973). Percolation through fissured rock. *Bulletin International Association of Engineering Geology*, No. 7, 3–28.

Williams, A.A.B. and Donaldson, G. (1980). Building on expansive soils in South Africa. *Proceedings of the 4th International Conference on Expansive Soils, Denver*, **2**, 834–8.

Williams, A.A.B. and Pidgeon, J.T. (1983). Evapotranspiration and heaving clays in South Africa. *Geotechnique*, **33**, 141–50.

Williams, G.M. and Aitkenhead, N. (1991). Lessons from Loscoe: the uncontrolled migration of landfill gas. *Quarterly Journal of Engineering Geology*, **24**, 191–208.

Wolden, K. and Ericksen, E. (1990). Compilation of geological data for use in local planning and administration. *Engineering Geology*, **29**, 333–8.

Index